The Nanoscope

ISBN 978-0-578-22307-0

The Nanoscope

Stephen Winters-Hilt

Meta Logos Systems
Denver, Colorado

ISBN 978-0-578-22307-0

Golden Tao Publishing
USA

Preface

In this manuscript I describe Nanopore Detectors and "The Nanoscope."

This manuscript is one of three texts that are closely related and being published at about the same time. The Channel Current Cheminformatics material presented in Ch. 4 is much more extensively described in the parallel publication of the textbook "Data Analytics, Bioinformatics, and Machine Learning". The Machine Learning methods used for pattern recognition are also described in more detail in that textbook, as well as in a third textbook publication titled "Informatics and Machine Learning, from Martingales to Metaheuristics".

Acknowledgement

This book wouldn't have been possible without support of family and friends. So I'd like to thank my wife and sons foremost: Cindy, Nathaniel, and Zachary. I'd also like to thank my mother, sister, and brothers: Sybil, Teresa, Eric and Joshua. Josh also helped with some of the experiments and software. I'd also like to thank my grandparents and my aunts and uncles: Wilbur, Beulah, Bruce, Richard, Diana, Mark, Susan, and John. I'd also like to thank Rob and family.

This book wouldn't have been possible without having several 'genuises' as advisors or instructors over the years (including Nobel Prize winners), whose insights I was inspired to try to capture in some instructional form. During my time at Caltech I was fortunate to have as advisor or instructor: Kip Thorne (graduate advisor), Richard Feynman, Murry Gell-Mann, David Middlebrook, Amnon Yariv, Ron Drever, and Barry Barish (undergraduate advisor). During my time at Oxford I was fortunate to have as advisor or instructor: Roger Penrose and Nick Woodhouse. During my time at UWM I was fortunate to have as advisor or instructor: Nick Papastamatiou, Leonard Parker, John Friedman, and Jourma Louko. During my time at UCSC: David Haussler, David Deamer, and Mark Akeson.

I'd also like to thank the student researchers, postdocs, and lab technicians that I've advised and worked with: Eric Morales, Andrew Duda, Iftelhar Amin, Amanda Alba, Amanda Davis, Evenie Horton, Joshua Morrison, Anand Prabhakaran, Alex Ortiz, Raja Iqbal, Srikanth Sendamangalam, Charlie McChesney, Matthew Landry, Molly Oehmichem, Kenneth Armond Jr., Sepehr Merat, Daming Lu, Hang Zhang, Carl Baribault, Zuliang Jiang, Alexander Churbanov, and Alexander Stoyanov.

Contents

Chapter 1

Introduction

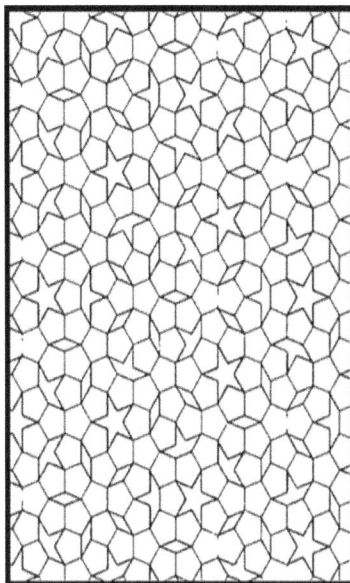

The Nanoscope functions as a device that can observe the states of a single molecule or molecular complex via linkage to a channel modulator. For the Nanoscope apparatus the observation is, thus, not in the optical realm, such as with the microscope, but in the molecular-state classification realm.

The Nanoscope description and operation is inherently interdisciplinary, and not coincidentally, the Nanoscope works at a scale where physics, chemistry, and biomedicine methodologies intersect. In some applications the Nanoscope can functions like a biosensor, but with detection at the single-molecule scale, e.g., a single transducer molecule is drawn from solution at a time (electrophoretically) and upon channel-capture it rattles around in the single channel, making transient bonds to its surroundings, and the binding kinetics of those transient bonds is directly imprinted on a surrounding, electrophoretically driven, flow of ions. The observed channel current blockade patterns are engineered or selected to have distinctive stationary signal statistics ("stationary statistics"), and changes in the channel blockade stationary statistics are designed to occur for a transducer upon introduction of its interaction target. The Nanoscope can see more subtle effects than changes upon binding, however, it can also track an individual molecule's conformational state (which, in turn, impacts binding).

With the Nanoscope we are, thus, reaching past the structure-function paradigm to the molecular state and molecular interaction tracking itself. The Nanoscope also offers 'native' information on the test molecule in that the Nanoscope can, typically, operate in the same buffer as the test molecule's native solution. The Nanoscope is inexpensive (compared to the other methods). The Nanoscope, thus, may offer the benefits of HPLC, MS, NMR, crystallography, and gels, all on one platform, and with more accessible protocols for standardization.

The relation between protein structure and function is well known and minor changes in protein folding or isoform variants, or surface modifications such as glycosylations, can impact that protein function. To complicate matters further, many proteins are inherently dynamic, so their structure-function relationship can give rise to dynamic functionality, with selection sometimes favoring very dynamic proteins. What is needed is a means to track protein conformation and its role in protein function, binding in particular, and this suggests we need a means to track the conformational state of a *single* protein. A method for using a nanopore transduction detector (NTD) is proposed for such an application. NTD transducers are molecules that serve to transduce the conformational or binding state of a molecule of interest into different channel current modulations, where the molecule of interest is tethered to a nanopore channel modulator. In previous work, using inexpensive biomolecular components, such as DNA hairpin channel-modulators, antibodies, and immuno-PCR linkages to antibodies, experiments were done to analyze individual antibodies and DNA molecules. Three complications were indicated before a general-use NTD platform could be established: (1) the convenient DNA-based modulators were often too short-lived for the binding study of interest; (2) the transducer's *bound* state often didn't modulate; and (3) the binding target often had a pI that didn't favor being drawn to the channel-tethered study molecule. In recent work NTD operation has been demonstrated for a wide range of pH, chaotrope concentration, and in the presence of interference agents, such that problem (3) is solved. While in the latest results shown here the other problems are solved as well: very long-lived channel modulators are demonstrated using locked nucleic acid (LNA) nucleosides; and induced modulations are demonstrated by engineering transducers to receive laser-tweezer impulses by means of a linked magnetic bead (another commoditized component). A general-use NTD platform is thereby possible using an alpha-hemolysin nanopore detector and performing the critical transducer engineering with readily available, inexpensive (commoditized), biomolecular components.

Analysis of individual molecular behavior
Proteins, such as enzymes, can have a high degree of variability. It has been demonstrated that enzyme turnover rate, for example, can differ at the single molecule level (details in later chapters), with a single enzyme observed with one constant turnover rate, while another enzyme, differing only in conformation, or possibly by a difference in glycation, has a different, but still constant, substrate processing rate. And this is a simple example where there is only one interaction region and it is (mostly) unchanging in its conformation for the individual protein examined. Some allosteric proteins, on the other hand, with multiple binding sites for a particular target, change their binding affinity according to how many ligands they have bound. Antibodies are known to change conformation during binding to one (or two) antigens in such a significant manner that this is the basis for activation of the complement cascade of the adaptive immune response. The nanopore transduction detector (NTD) is presented here as a general method for informing our understanding of protein structure-function relationships at the single molecular variant/isoform level.

Previously it has been observed that many antibodies directly exhibit modulatory channel blockades, and upon introduction of their antigen, their bound-state is directly transduced as a notably different channel modulation [1]. Determining the glycosylation profile of antibodies, and Fc glycosylation in particular, is critical to understanding antibody efficacy and blood circulation half-life, so the nanopore platform and the same signal processing methods for understanding NTD transducers can be directly applied to profiling antibody glycosylation blockade signals where the antibody is treated as an NTD transducer in and of itself.

Direct antibody profiling would likely only work for part of the glycosylation (or glycation) profile, however, since the Fab N-terminus neutral glycosylation and glycations would probably still need to be assayed by use of antibody intermediates (as with the standard HbA1c test).

Isomer characterization

With the Nanoscope we can study molecular characteristics that are obscured in ensemble-based measurements. Ensemble averages lose information about the true diversity of behavior of individual biomolecules. For complex biomolecules, in particular, there is likely to be a tremendous diversity in behavior, and in many cases this diversity may be the basis for their function.

There are dozens of 21 carbon isomers in the alpha acids family of molecules from hops, for example, pyrethrin also has 21 carbons (a natural insecticide), and naturally occurring 20 carbon diterpenes can be present as well. Hops plant samples will, thus, have mixtures of the 20-21C molecules, many of them isomer variants, that need to be differentiated to establish a useful profiling capability.

We can use the NTD to resolve the different 21C alpha acid isomers, and similar molecular weight variants, via the inherited specificity of the antibody or aptamer used to transduce molecular events. Multiplexing for simultaneous detection on multiple analytes, via the unique blunt-ended dsDNA termini on modulators giving rise to unique modulator signals, may be possible as well. Development of plant (isomer) profiling standards is in high demand.

The robust nature of the electrophoretic detection platform affords a lot of flexibility. If the reporter molecule used to bind to target (if present) is phosphorylated, then it can be selectively drawn out of the sample to the detector by the underlying electrophoretic mechanism used by the NTD apparatus. Just as nucleic acid can be drawn from a dirty blood sample off of the floor at a crime scene, the same electrophoresis separation advantages are inherited by the NTD as well – so simply using phosphorylated transducers can eliminate much of the preprocessing that might normally be needed for an HPLC setup.

The NTD platform has a lower event detection error rate than nanopore translocation-based methods and other methods (HPLC, MS), since it has a fundamental accuracy-vs-observation time engineering trade-off advantage over standard nanopore signals that are based on a fixed-level channel blockade of short duration to identify events. This is because the NTD signaling is based on channel modulators whose signal can be observed for as long as desired and thereby provide a trade-off between signal-to-noise needed and observation time on the individual signals.

Note: The band separation in the HPLC column could be further resolved with the NTD, thus the methods could be complementary in some sample processing applications. The separation of the molecular types can also be much more focused from the outset, on a specific molecule or collection of molecules of interest, thereby allowing simpler pre-processing. The nanopore detection identifies and quantifies individual molecular events as seen at the channel opening, and is inherently more quantitative and reproducible than HPLC.

The Nanoscope description that follows, and in later chapters, will not only describe the device physics and signal processing methodologies [1-100], but will also enhance learning and understanding of informatics, biochemistry, and single-molecule biophysics, including real-world applications in the health and medical sciences [1,54]. Biochemistry instruction relates to the set-up and analysis of protein channel current data, including a review of the underlying biophysics and related fundamental noise analysis. (We will see that the Nanoscope is operated near its noise limits, so highly optimized in this regard.) The computational instruction will include development and deployment of computer software for pattern recognition, feature extraction, statistical analysis and kinetics analysis. Applications in the medical sciences include the ability to track in real-time a patient's hemoglobin and TSH glycosylation levels, etc. Add in trace-level detection capability, on troponin for example, and you have a state-of-the-art diagnostic tool for trace biomolecule detection and characterization (real-time profiling) of transient protein complexes.

The nanopore detection method uses the stochastic carrier wave signal processing methods developed and described in prior work [1,54], and comprises machine learning methods for pattern recognition that can be implemented on a distributed network of computers for real-time device operation, feedback, and sampling control [20].

The Nanoscope, a.k.a., The Nanopore Transduction Detection (NTD) Platform
Nanopore molecular sensing, characterization, and sequencing, using nanopore transduction detection ("NTD"), in general uses electrophoresis to transport an unknown sample to, and possibly through, an orifice. A nanopore detection system includes an electrolytic solution so that when a constant electric field is applied, an electric current can be observed in the system. The magnitude of the electric current density across a nanopore surface depends on the nanopore's dimensions and the composition of any molecule, such as deoxyribonucleic acid ("DNA") or ribonucleic acid ("RNA"), that is occupying the nanopore. Sensing, characterization, and sequencing is made possible because, when close enough to nanopores, samples cause characteristic changes in electric current density across the nanopore.

Further, a nanopore 'filter', or channel detection device, can be used to detect one or more molecules of interest through unique signals imprinted on a nanopore blockage current. One example of such a system has been referred to as a "Coulter Counter", and has been used to count pulses to measure the bacterial cells passing through the aperture using hydrostatic pressure.

Often the molecule of interest in a channel detection device is attached to another molecule (referred to as a "reporter molecule") through a chemical bond. The reporter molecule and the molecule to which it is attached are sensed as they enter and blockade together as a single unit in a channel or nanopore system.

Some of the detection systems use a pore or channel which is large enough to allow the molecule of interest and a reporter molecule to pass completely through the pore and measure signals as a result of that passage, with the passage through the pore being referred to as a "translocation". Such translocations often occur very quickly, at an uncontrolled rate and at a random orientation, and therefore may not provide a signal with enough information to indicate the structure of the molecules translocating.

Some of the detection systems use a pore or channel which does not permit translocation, but partial intercalation into the channel instead (for "capture", or "end capture" if a lengthy polymer). In these cases the channel size can lead to multiple specific interactions with the intercalated analyte such that only a small set of intercalation, or "capture" states exist. Such configurations allow "transduction" of molecular capture state, or state histories, directly into different blockade current measurements.

Some of the detection systems use a pore or channel which permits limited translocation, via slow, high interaction, intercalation into the channel. In these cases the channel size lead to multiple specific interactions with the intercalated analyte such that only a controlled set of intercalation, or "sliding capture" states exist (possibly reduced to simply the number of n-mers in the detection zone, e.g. 4^n such states). Such configurations allow "transduction" of molecular sequence directly into different blockade current measurements.

The NTD platform [1] comprises a single nanometer scale channel that allows a single ionic current flow across a membrane and an engineered, or selected, channel blockading molecule. The channel blockading molecule is engineered or selected such that it provides a current modulating blockade in the detector channel when drawn into the channel. The channel is chosen such that it has inner diameter at the scale of that molecule or one of its molecular-complexes. For most biomolecular analysis implementations this leads to a choice of channel that has inner diameter in the range 0.1-10 *nanometers* (see Fig. 1.1, this material is revisited more extensively in Ch. 5). Given the channel's size it is referred to as a nanopore in what follows.

In order to have a *capture* state in the channel with a *single* molecule, a nanopore is needed. In order to establish a coherent capture-signal exhibiting non-trivial stationary signal statistics the nanopore's limiting inner diameter typically needs to be sized at approximately 1.5nm for duplex DNA channel modulators (precisely what is found for the alpha-hemolysin channel). The modulating-blockader is captured at the channel for the time-interval of interest by electrophoretic means.

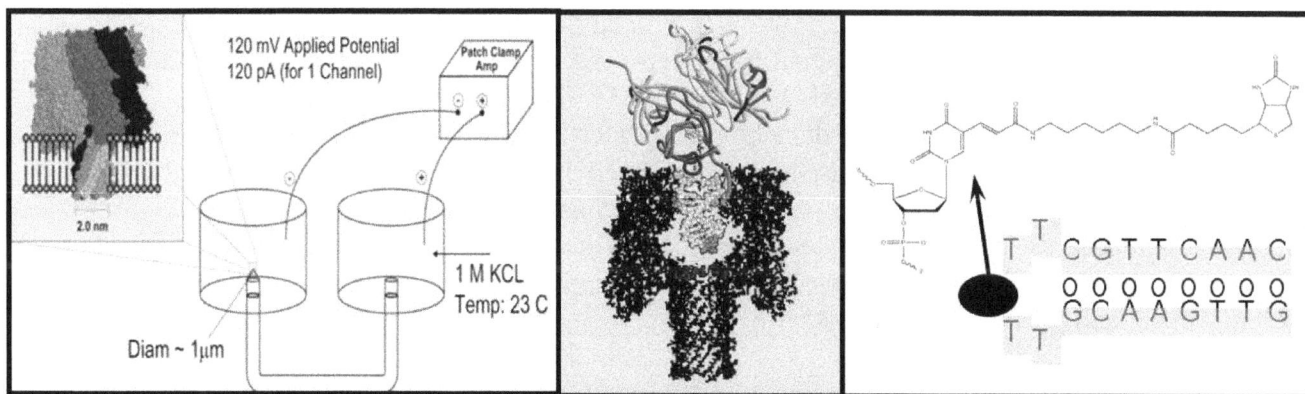

Figure 1.1. Schematic diagram of the Nanopore Transduction Detector.
Left: shows the nanopore detector consists of a single pore in a lipid bilayer which is created by the oligomerization of the staphylococcal alpha-hemolysin toxin in the left chamber, and a patch clamp amplifier capable of measuring pico Ampere channel currents located in the upper right-hand corner. **Center**: shows a biotinylated DNA hairpin molecule captured in the channel's cis-vestibule, with streptavidin bound to the biotin linkage that is attached to the

loop of the DNA hairpin. **Right**: shows the biotinylated DNA hairpin molecule (Bt-8gc) [17,27].

The NTD molecule providing the modulating blockade in what follows has a second functionality (Fig. 1.2 Right), to specifically bind to some target of interest such that its blockade modulation is discernibly different according to binding state. Thus, the NTD modulators are engineered to be bifunctional in that one end is meant to modulate the channel current, while the other end is engineered to have different states according to the event detection, or event-reporting, of interest. Examples include extra-channel ends linked to binding moieties such as antibodies or aptamers.

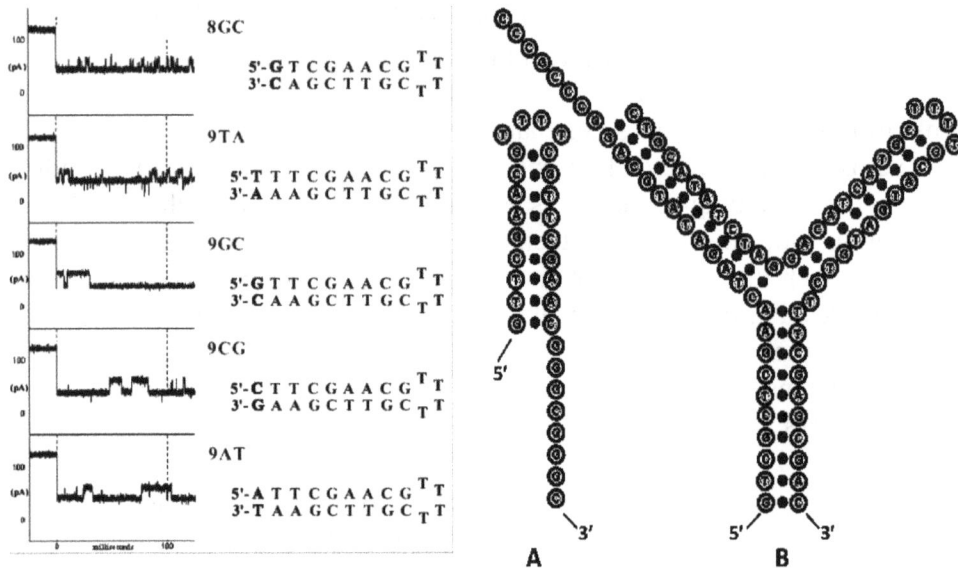

Fig. 1.2, Left Panel. DNA hairpin controls and their diagnostic signals [28,45,47]. The secondary structure of the DNA hairpins is shown on the right, with their highest scoring diagnostic signals shown on the left. Each signal trace starts at approximately 120 pA open channel current and all blockades are in a range 40-60 pA upon "capture" of the associated DNA hairpin. Even so, the signal traces have discernibly different blockade structure, which is extracted using an HMM. The signals are aligned at their blockade starts and the demarked time-trace is for 100 ms. **Right Panel.** A Y-shaped, NTD-aptamer is shown (labeled 'B'), where one arm is loop terminated such that it can't be captured in the channel, leaving one arm with an 8-base ssDNA extension for annealing to complement target (part of molecule 'A'), and where the base of the Y is designed to be captured in the channel and provide modulations.

In Fig. 1.3, the paired regions {1,9}, {2,4}, and {5,8} are meant to be complements of one another (with standard Watson-Crick base-pairing), and designed such that the annealed Y-transducer molecule is meant to be dominated by one folding conformation (as shown). The region 3 is a loop, typically 4 dT in size, that is designed to be too large for entry and capture in the alpha-hemolysin channel, such that the annealed Y-transducer only has one orientation of capture in the nanopore detector. The base region, comprising regions {1,9}, is designed to form a duplex nucleic acid that produces a toggling blockade when captured in a nanopore detector. The typical length of the base-paired regions is 8-10 base-pairs. Region 6 denotes the aptamer (and region 7 is its nucleic acid linker).

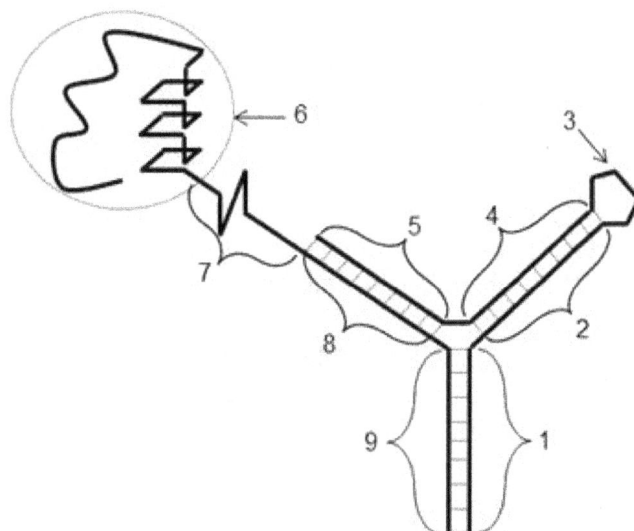

Fig. 1.3. Y-transducer for high-specificity aptamer binding detection. The Y-transducer is meant to have a high-specificity aptamer attached by a single stranded, possibly abasic (non-base-pairing), nucleic acid linker, region 7, to an aptamer in region 6. The sketch of the aptamer in region 6 is meant to suggest the 3D conformational aspect of the aptamer, where stacking of G-quadruplexes is a common, but not necessary, feature of aptamers.

In some embodiments an applied potential is used to establish current through the single, modulated ion-current, channel. In other embodiments periodic laser modulations are used to induce channel-state dependent stochastic modulations in the channel current. In other embodiments an applied potential and periodic laser modulations are used to induce channel-state dependent stochastic modulations in the channel current. In other embodiments multiple channels resident in the membrane are made possible, where only one or a few channels are modulated with a modulating channel blockade, and where any other channels present only offer a steady, unmodulated, noisy, current source that is easily filtered out by the use of the effective stochastic carrier wave heterodyning capabilities from hidden Markov model with binned duration model ("HMMBD") and meta-hidden Markov model ("HMM") methods.

In general, with embodiments, the NTD devices can be made "smarter" according to use of their "noise" information (in fact, this is true for devices in general, and this is given as a case study in that context in many descriptions). Consider, for example, a device that has two states, A and B, and associated noise profiles N(A) and N(B). A "noise profile" for a device state includes observations of device operational parameters while in that state. Using this information it is possible to track device operational state according to learning and classification of the device noise state. Part of the engineering task when enabling such a system is to establish a device with observational parameters that reveal system noise in a useful manner, such that noise "tracks on state". This noise is provided in NTD devices where a large variety of biomolecules are found to blockade the channel in a modulatory manner highly sensitive to their molecular state. In other complex systems, such as a car, this is possible too, such as what a good mechanic can do by simply listening to the car (noise) under different test conditions.

However, the system noise may not be significantly strong or distinctively tracking on states of interest. In this instance, injection of modulatory signal (periodic or stochastic) can be used to induce distinctive system noise as desired.

Assume it is desired to observe and classify (track) the state of a complex system. Assume there is a means to couple a signal generator to that complex system such that signal generation, or signal noise, is different according to complex system state. Then, state tracking is accomplished by pattern recognition and classification of the different signal types seen.

Often reporter/transducer molecules involve a dsDNA hairpin modulator. In these instances, the $4^4=256$ different duplex ends can be used in a 256-way multiplexing scheme (in general, 4^n labels for n-mer length tag). This is significant as the probe-set descriptions that follow often require 100 to 200 different analytes to be assayed simultaneously – these 256-end labels then provide a means for multiplex detection with mixtures undergoing sequential analyte measurement from a mixture solution (nanopore 'tasting' by electrophoretically drawing in molecules, identifying them, then ejecting via brief voltage reversal and repeating).

Fundamentally, the weaknesses of the known standard ensemble-based binding analysis methods are directly addressed with this single-molecule approach. The role of conformational change during binding, in particular, could potentially be directly explored in this setting. Embodiments also offer advantages over other translation-based nanopore detection approaches in that the transduction-based apparatus introduces two strong mechanisms for boosting sensitivity on single-molecule observation: (i) engineered sensitivity in the transduction molecule itself; and (ii) machine learning based signal stabilization and highly sensitive state resolution. NTD used in conjunction with novel pattern recognition informed sampling capabilities greatly extends the usage of the single-channel apparatus (including learning the avoidance of blockades associated with channel failure, when contaminants necessitate; and nanomanipulation, where we have a single-molecule under active control in a nanofluidics-controlled environment). For medicine and biology, NTD methods in accordance with embodiments can aid in understanding multi-component interactions (with cofactors or adjuvants), and aid in designing co-factors according to their ability to result in desired binding or modified state.

Nanopore transduction detection may also provide an inexpensive, quick, accurate, and versatile method for performing medical diagnostics. It is hypothesized that NTD biomarkers can be developed for early stage disease detection with femtomolar to attomolar sensitivity for doing the standard clinical tests of the future. The potentially incredible sensitivity of the NTD targeting on biomarkers also provides a significant new tool for public health.

The overall market impacted by this technology includes the 1,200-company pharmaceutical industry, the 1,000-company biotechnology industry, and the basic science research fields of biochemistry and molecular biology, which receive National Institute of Health (NIH) funding. Overall, this technology impacts a trillion dollar industry at the nexus of pharma, biotech, and academic research.

Chapter 2
Electrochemistry with Membranes and Channels

In this chapter are gathered some of the fundamental biochemistry and biophysics concepts relevant to channel-based detection methods.

2.1 Ionic Solutions
2.1.1 Thermodynamics in Biophysics
2.1.1.1 Equilibrium

The existence of equilibrium states, to which matter in bulk tends to approach spontaneously, is deduced from hundreds of years of experimentation in thermodynamics [101]. Conceptually, microstates describe the system (full atomic scale specification) between collisions, and the sequence of observed microstates tends towards those representing fairly uniform distribution of pressure and temperature (similarly for other state variables). This is presumably because the number of microstates with such uniform macroscopic properties far exceeds the number of microstates that retain some information about their prior history. The behavior that results is usually described as evolution towards microstates with maximum disorder, where entropy is a measure of that disorder. Entropy is the extensive variable that pairs with temperature, like volume pairs with pressure. Thermodynamic phenomenology works with the observed microstate values (intensive), and the separately measured extensive variables (such as volume), but direct measurement of entropy is rarely done (because it's rarely well-defined, changes in entropy, however, are typically measured) [102].

The three suppositions of Statistical Mechanics [101]:

(1)   Existence of equilibrium states (same as classical thermodynamics).

(2)   Equal probabilities postulate: all states equally likely (the asymptotic equipartition theorem). This postulate could result from the Ergodic hypothesis and either the principle of detailed balance (quantum mechanics) or Liouville's theorem (classical mechanics).

(3)   Entropy is defined by $\mathbf{S = k \log \Gamma}$, where $\mathbf{\Gamma}$ is an enumeration of the possible states of the system, and $\mathbf{k}$ is Boltzmann's constant. The latter definition of entropy ties into the information-theoretic definition presented by Shannon.

2.1.1.2 Non-equilibrium

Systems that are not in equilibrium are governed by the law of increase of entropy (second law of thermodynamics). Steady flows are an important sub-class of non-equilibrium (near equilibrium) systems. In steady flow situations it is often possible to describe phenomeno-logical equations of flow [103]. Examples of this include the Fourier Law (heat flow is linearly related to the gradient of temperature), Ohm's Law (electric current is proportional to the applied potential), and Fick's Law (diffusion rate is determined by the negative gradient of concentration). In developing the theory of steady-flow phenomenology, Onsager [104,105] and others [106-108], found that there could exist couplings between phenomeno-logical forces and flows. These couplings have been found to exist in all cases where the natural dimensionful constants involved lead to measurable forces. A coupling relevant to nanopore analysis is the electroosmotic flow in the channel, particularly when partially obstructed. Electroconstriction of the membrane is also observed (and may be important in modulation studies).

2.1.2   Fluid Flow

In fluid mechanics [109], conservation of matter, or continuity, is described by $\partial\rho/\partial t + \mathrm{div}(\mathbf{j}) = 0$, where ρ is the fluid density, $\mathbf{j}=\rho\mathbf{v}$ is the mass flux density, and $\mathbf{v}$ is the fluid velocity. The equation of motion for volume element in fluid, or Euler's Equation, is: $\partial\mathbf{v}/\partial t + (\mathbf{v \cdot grad})\mathbf{v} = -\mathbf{grad}(p)/\rho$ (valid for ideal fluids, where fluid viscosity and thermal conductivity negligible). For steady flow $\partial\rho/\partial t = 0$ and $\partial\mathbf{v}/\partial t = 0$, which yields Bernoulli's equation. For incompressible flows, ρ=constant, and continuity reduces to $\mathrm{div}(\mathbf{v})=0$.

Viscous fluids (with viscosity coefficients, scalars, that do not change noticeably in the fluid) are described by the Navier-Stokes equation: $\rho[\partial\mathbf{v}/\partial t + (\mathbf{v \cdot grad})\mathbf{v}] = -\mathbf{grad}(p) + \eta\,\Delta\mathbf{v} + (\zeta + \eta/3)\,\mathbf{grad}\,\mathrm{div}(\mathbf{v})$, where η is the coefficient of viscosity and ζ is the second viscosity. If the fluid is regarded as incompressible, the term with the second viscosity coefficient drops out.

Steady-state viscous flow, as described by the Navier-Stokes equation, is specified (for given flow geometry) by three parameters: kinematic viscosity, $v_k = \eta/p = [cm^2/sec]$, smallest linear dimension of flow environment, $l=[cm]$, and velocity of the main stream, $u=[cm/sec]$. Only one dimensionless quantity can be formed from the above, $R = ul/v_k$, known as the Reynolds number. The Reynold's number describes the ratio between inertial forces and viscous forces. Similarity of flow is possible with rescaling in flow geometry (changing l), if kinematic viscosity and flow velocity are changed such that the Reynolds number is unchanged. In nanometer-scale ionic channels, Reynolds numbers can easily be smaller than 10^{-10}, making classical string-like polymer motion on the DNA very unlikely, for example.

For non-steady viscous flows, assume a time constant, τ=[sec], characteristic of a changing flow. An added dimensionless quantity can now be constructed, sometimes called the Strouhal number, $S = u\ \tau\ /\ 1$ [109]. By adjusting the modulation of a current flow it may be possible to introduce a flow environment similar to very large u with small τ, to one with small u, i.e., one more amenable to examination in terms of slow flows (this is important to sequencing with slow-down on translocation via PLL modulation – where PLL's (phase locked loops) are described in [1] in the tradition EE sense).

2.1.3   Simple ions in solution
2.1.3.1 Water Clustering
Water has structure even when not in a crystalline phase (ice), as revealed in early X-ray analysis. Most water molecules reside in clusters: at the melting point the clusters are sized from 90 to 650 molecules, while at the boiling point the clusters are sized from 25 to 75 molecules. "The clusters are in constant movement. An H-bond oscillates with a frequency of about 0.5×10^{13} Hz. The mean lifetime of a single cluster lasts approximately 10^{-10}-10^{-11} s. Therefore, during the oscillation every H-bond will form 100-1000 times with the same oxygen atom, before it connects to another one. Because of these dynamics, the name flickering cluster is used [110]."

2.1.3.2 Hydration Radius
Due to the strong dipole moment of H_2O (1.85 D), the electric field around an ion will lead to a strong orientation of nearby water molecules. The orienting force from the electric field of the ion is in competition with the influences from other water molecules and the general destructive activity of thermic noise. There are two areas of interaction between ion and water molecule (similar to two layer model in electroosmosis): near the ion is the primary hydration zone, where the water dipoles are strongly oriented. The secondary hydration zone involves water molecules not oriented by field (they have primary zone shielding and greater distance), yet interacting sufficiently that a clear disturbance to the normal structure of water is observed (a fault zone). If an ion is taken as a macroscopic sphere with a hydrophilic surface (assuming outer hydration regions comprise sphere boundary), then Stokes' Law can be applied to derive an effective hydration radius from measurements of viscosity and mobility (under a driving force). For K and Cl the hydration radius is approximately 0.125 nm [110].

2.1.3.3 Debye Radius
In 1 M KCL solution the mean distance between ions is about 1 nm. A spherically symmetrical potential around a central ion forms the basis for the Debye-Huckel theory. Boltzmann's law of energy distribution then provides a relation between the local concentration and the local potential. The effective potential that results involves a Coulomb term (as if there were no shielding effects by the electrolyte) and an exponential decay term with radial distance. The length parameter (e-folding distance) for the exponential decay is known as the Debye-Huckel radius (another effective radius parameter), and is dependent on concentration. For 1 M KCL solution at 25 C (and other standard conditions), the Debye-Huckel radius is 0.304 nm [110].

2.2 DNA and polymer ions in solution
Polymers in solution are in a constant state of thermal agitation. Brownian motion has successfully described many observed phenomena, including viscoelasticity, diffusion, and birefringence [111]. Stochastic processes are usually employed to describe the phenomenol-

ogy, as originally done by Einstein. A limitation with this (non-computational) approach is that it assumes that the polymer solute moves on a much slower time-scale than the solute, as well as having a much longer length-scale than the scale of the solute diameter. The phenomenological equations that result can be expressed in two, different, formalisms: the Smoluchowski and Langevin equations. "The Smoluchowski equation is derived from a generalization of the diffusion equation and has a clear relevance to the thermodynamics of irreversible processes. The Langevin equation, on the other hand, has no direct relationship to thermodynamics, but is capable of describing wider classes of stochastic processes [111]."

Early research on polymer solutes and their motions examined bulk (macroscopic) properties via NMR spectroscopy, circular dichroism, and X-ray diffraction. In the early 1990's single molecule studies were begun with a variety of new approaches: atomic force microscopy, laser tweezers, and fluorescence microscopy. When applied to single DNA molecules this led to determination of the force required to break Watson-Crick base-pairs [112], the force required to stretch ssDNA and dsDNA through different conformational phases (B-form to S-form conversion, for example [113], and the force exerted by polymerases working on polynucleotides. Examination of single DNA molecules with these methods, and recent nanopore methods, indicates that almost every chemical bond can be significant in its dynamics and detector interactions: Covalent, Ionic, Polar Covalent, Dipole-Dipole, Hydrogen, Van Der Waals, Hydrophobic Effect, Coordinate, Watson-Crick Bonds, Stacking Effect, etc. Also of critical importance is an understanding the DNA structure on a collective level, in particular, the DNA helix, hairpin, protein-binding, and phase dynamics (B-form DNA ends may bend/twist to A-form, for example).

2.2.1 DNA Structural predictions based on crystallography and NMR spectroscopy

The first X-ray crystal structure of a DNA oligomer (the 'Dickerson dodecamer') was published in 1981 [114]. It established substantial deviation among base pairs in terms of propeller twist, rise per base pair, and sugar pucker. Numerous attempts have been made to understand the structural basis for these differences. As is true for models used to predict thermodynamic stability of duplexes, models based on dinucleotide steps have been reasonably successful. Although crystal structures have provided fundamental information that helps illuminate how DNA can bend and twist when bound to proteins, the approach has limitations. For instance, close packing of DNA in crystals is known to alter structure relative to solution phase, and the cryogenic temperatures used for high resolution may lead to under-representation of conformers that are common at physiological temperatures. NMR spectroscopy can overcome these limitations because the experiments are typically run at 1 mM concentration and ambient temperature. Whether structural averaging by NMR or X-ray approximation by a crystallized form, neither approach provides a clear picture of the conformational *history* of a free molecule in solution at physiological temperature.

2.2.2 Structure and Dynamics of Duplex Ends

The structure and dynamics of DNA duplex ends can influence numerous enzyme-dependent processes. Some of the most biologically important of these are integration of transposons and retroviral dsDNA into target chromosomes. Two well studied examples are transposition of the phage Mu genome, and integration of HIV dsDNA copies into target chromosomal DNA. In both cases, a consensus CA dinucleotide step at or near the duplex terminus is believed to confer flexibility on the viral DNA that is required for processing and strand transfer.

DNA duplex ends are significantly under-represented in NMR and crystal structure studies despite their critical importance in biology. For example, Hassan and Calladine's landmark study [115] was based on X-ray crystal structures for 60 oligomers. A•T pairs appeared only twice in the terminal dinucleotide step of the 120 duplex ends. This under-representation may be due to a historical bias since the Dickerson dodecamer contains only G•C pairs in the four base pair termini. But it may also be due to recognition of a built in bias in crystal structures because the helix ends are known to overlap, and interpretation of their structure is therefore ambiguous. NMR studies of DNA structure have also been biased toward the Dickerson dodecamer and its variants. Highly accurate examination of dsDNA duplex ends is done with a Nanopore Detector in what follows.

2.3 Membranes and channels

Analysis of the fossil record indicates that life probably began about 3.8 billion years ago (give or take a few hundred mill.). How a biological evolutionary process should arise from billions of years of cosmological and planetary (geologic) evolution is unclear. What seems clear, however, is that primitive replicating molecular systems at some point, if not right away, required good transport barriers to retain, yet replenish, their metabolic processes. Lipid bilayers seem a likely critical step in this process in that they self-assemble in water because of lipid's hydrophobic tail and hydrophilic head. Lipid bilayers can also be designed to be about 5 nm thick (similar to actual cell membrane), which reduces the evolutionary transport problem to selection of protein-size molecules that can stably reside in the membrane and aid in transport of nutrients (influx) and waste products (efflux). Physiological flux experiments reveal two types of transport classes: pores and carriers (with many of the latter simply involving very small pores, where kinetic features, such as a translocation-binding metastates, are discernible). It appears that nature has devised specialized channels/pumps for every major ion (seen in ocean water, at least, [116]). In further evolutionary refinements, channels became excitable, laying the foundation for intracellular communication and, eventually nervous systems. In most cells, channel excitation can also result from coupling to G-Protein coupled receptors (GPCRs), both directly and indirectly via secondary messenger. Together with GPCRs, ion channels form the core of cellular transport/communication. Notable from an RNA World perspective is that channels are protein-based, although construction of DNA-based ('origami') channels are known to be possible [117].

2.3.1   The prototype nanopore: a heptameric protein channel

Alpha-hemolysin is a toxin secreted by *Staphylococcus aureus*. Alpha-hemolysin monomers have the property that they coalesce into heptameric protein channels in cell walls (or lipid bilayers, which have a comparable width of 5 nm). X-ray diffraction analysis [118] of crystallized (heptameric) α-hemolysin revealed a protein channel with a *cis*-side aperture diameter of approximately 2.6 nm. The length of the channel is approximately 10 nm: 5nm of wider channel (the *cis*-side vestibule) plus 5 nm of narrow (trans-membrane) channel. A ring of threonine residues is located at the *cis*-entry. From the *cis*-entry the vestibule diameter opens to approximately 3.6 nm, then shrinks back to the limiting aperture diameter of 1.5 nm. The residues defining the limiting aperture consist of an alternating ring of lysine and glutamate residues. Beyond the limiting aperture is the trans-membrane part of the channel, which consists of neutral residues, except for a ring of leucine residues (hydrophobic). The trans-membrane channel diameter ranges between 1.8 nm and 2.0 nm, and opens at the *trans*-end to a 2.2 nm diameter aperture. The *trans*-side aperture consists of a ring of alternating lysine and aspartate residues.

2.3.1.1 Nanopores in Lipid Bilayers: Biomolecular Analysis in Solution

A nanopore is a trans-membrane channel with dimensions ranging from ~1 nm to 100 nm. Important spherical and cylindrical/helical diameters in that size range include the B-form DNA helical diameter (2 nm), and the water molecule diameter (0.3nm). Viruses at 100 nm diameters are too large to interact well with the nanopore (almost always too large to translocate, for example), while bacteria at micron size never translocate a nanopore. A nanopore detector, as discussed here, comprises a single nanopore in a membrane that separates two chambers of electrolytic solution. The detector is based on ionic current observations when a potential difference is applied across the membrane. Objects drawn into the nanopore cause ionic current blockades that form the basis of the molecular observations (i.e., observations derive from the ionic current imprints of captured or translocating molecules). With channel current detection, particle analysis can be done on solutions to obtain particle concentrations, solution mixture composition, and even molecular dynamics (at nanometer scale).

Early channel current detectors had millimeter diameters (0.1 mm) and were used to count cell concentrations and mixture compositions [119]. Information obtained about the excluded cell volume was used in classifying blood cells as red or white, the ratio of which provided important data for medical diagnostics. The 100 μm-scale pores of Coulter were devised in the early 1950's. It wasn't until the early 1970's that nanometer-scale pores were examined [120-122]. At that time, Bean made a nanometer-scale channel from crystalline structures (mica) that had defect tracks (from fission events). When etched with HF the normally impervious mica is removed along the defect-track in its crystalline structure. Depending on how this process is controlled, pores have been obtained with diameters ranging down to 6 nm (50 nm diam. pores commercially available). Although this technology has been used for observations on uncharged particles (polystyrene spheres with 90 nm diameter, [121]), it doesn't work as well with charged molecules (like DNA). Another complication is that the etching method for pore construction inevitably leads to long tunnel-like channels, which doesn't provide the best configuration for detector uses. Detection of biomolecules with biologically-based nanometer-scale pores also showed promise at about this time with the work by Hladky and Haydon [123]. They showed that a biological channel, the bacterial antibiotic gramicidin, could self-assemble in a lipid bilayer to form a functional channel (with currents of order 1 pA). This potentially solved two of the mica-channel problems: the lipid bilayers are very thin, about 5 nm, and the protein-based, biologically functional, nanometer-scale pore seemed better suited to passing charged biomolecules. Gramicidin was too small to detect most biomolecules, however, since it could barely pass molecules the size of the water molecule. It wasn't until 1994 [124] that a sufficiently large pore was studied, α-hemolysin. In the 1994 paper, Bezrukov *et al.* studied the blockades resulting from a charge-neutral polymer: polyethylene glycol (PEG). Later modifications to the gramicidin pore permitted its use as an antibody-modulated (on-off) biosensor, while modifications to the α-hemolysin pore enabled its use as a metal biosensor [125], among other things [126,127].

In 1995 and 1999, alpha-hemolysin was successfully used for DNA homopolymer translocation studies and classification [128]. In certain situations, intramolecular, Angstrom-level, features are beginning to be resolved by channel based detection [42,45,47]. For nanoscopic channels, interactions between channel wall and translocating biomolecules cannot, usually, be ignored. On the one hand this complicates analysis of channel blockade signals, on the other hand, tell-tale on-off kinetics are revealed for binding between analyte and channel, and this is what has allowed the probing of intramolecular structure on single DNA molecules [42,45,47]. Nanometer-scale pores are being developed in solid-state media [129,130] in hybrid media and with refinements to the biologically-based nanopore device [42,45,47]. This provides rich opportunities for the future because at nanometer scale a wealth of new prospects arise, from assaying solutions, to recognizing indivdual molecular motions.

2.3.1.2 The α-hemolysin channel

The α-hemolysin Nanopore Detector is based on the α-hemolysin transmembrane channel, formed by seven identical 33 kD protein molecules secreted by *Staphylococcus aureus*. As mentioned previously, the total channel length is 10 nm and is comprised of a 5 nm *trans*-membrane domain and a 5 nm vestibule that protrudes into the aqueous *cis* compartment [131]. From the *cis*-entry the vestibule diameter opens to approximately 3.6 nm, then shrinks back to the limiting aperture diameter of 1.5 nm. The trans-membrane channel diameter ranges between 1.8 nm and 2.0 nm, and opens at the *trans*-end to a 2.2 nm diameter aperture. The *trans*-side aperture consists of a ring of alternating lysine and aspartate residues (see Fig. 2.1).

A single strand of DNA is about 1.3 nm in diameter (and an eight nucleotide segment is about 5.4 nm long when fully stretched). Given that water molecules are 0.15 nm in diameter, this means that one hydration layer separates ssDNA from the amino acids in the limiting aperture. This places the charged phosphodiester backbone, hydrogen bond donors and acceptors, and apolar rings of the DNA bases within one Debye length (0.3 nm in 1 M KCl) of the pore wall. Not surprisingly, DNA and RNA strongly interact with the α-hemolysin channel. For applied potentials greater than about 60 mV, however, ssDNA is able to overcome those interactions and is found to translocate. Conclusive evidence describing ssDNA translocation was given by [128], where it was shown that the number of blockade events correlated with the number of translocated ssDNA molecules. Double-stranded DNA, on the other hand, is not found to translocate for any potential. Although dsDNA is too large to translocate, about ten base-pairs at one end can still be drawn into the large cis-side vestibule (see Fig. 2.1). This actually permits the very sensitive experiments, as the ends of "captured" dsDNA molecules can be observed for extensive periods of time to resolve features [22,42,45,47].

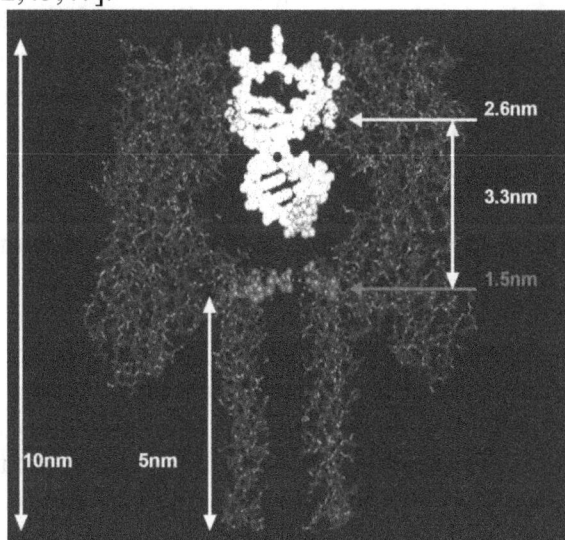

Fig. 2.1. Alpha Hemolysin Channel, with a nine base-pair DNA hairpin shown captured.

2.3.1.3 Membrane Environment and Channels for use in Biosensing and Screening

α-hemolysin is a membrane protein that forms an aqueous transmembrane pore. In order for such a pore to form in a lipid bilayer it is necessary for lipid molecules to be laterally displaced. For microbial pore-forming toxins, the energy used to drive the pore-formation process is thought to be solely provided by conformational changes in the toxin molecules themselves (i.e., the process is ATP-independent). For alpha-hemolysin, the energy needed for the pore-formation is shown, in [132], to be due to the oligomerization of toxin monomers to form the channel heptamer complex. Although pore-forming toxins, and membrane-permeabilizing molecules in general, have incredibly diverse sequence and structure, they all share in the same mechanism. They either directly intercalate into target membranes, or bind to particular target molecules in the membranes, and do so from a solution soluble form. They then assemble into multimeric, membrane-spanning pores. Attributes of the membrane, other than specific binding molecules, are often critical to this process, such as cholesterol-rich microdomains or lipid rafts. The role of cholesterol as a specific binding agent for individual formation events is well documented, and is required for pore-formation by many toxins (in natural setting, otherwise, channel formation can be activated by introduction of solvent, e.g., n-decane), including the Anthrax pore-forming toxin [133-135]. But cholesterol-rich microdomains also play a role in channel formation in a non-specific manner. This is due to the microdomains acting as concentration platforms that can aid in the assembly of proto-channel multimers as is described for the aerolysin heptamer in [136]. Once a proto-channel multimer has formed, such microdomains can also aid in the last step of transmembrane channel formation. This is due to the junctions between cholesterol-sphingolipid-rich domains and fluid-phase phosphoglyceride domains having locally favorable (weakened) bilayer characteristics that favor membrane penetration [137]. The opposite effect is also known to be medically relevant: unknown membrane constituents, or the lack thereof, can block an alpha-hemolysin heptamer complex from inserting a transmembrane functional domain. This is found to be the case for human granulocytes [138], where the agent preventing channel formation is unknown (although it might simply be a matter of membrane thickness being too great for the channel to bridge).

Cholesterol not only factors into protein-channel based detection as a specific and general pore-formation co-agent, but also, paradoxically, as a membrane strengthening agent in the nanopore detector setting in that it allows for greater vibrational shock resistance in the bilayer. For this reason, even though cholesterol is not required for α-hemolysin channel formation in bilayers, it is sometimes introduced in nanopore experiments purely as a bilayer-strengthening agent, which also serves to reduce the (bandwidth limiting) membrane noise contribution in the nanopore detector by approximately 35%.

2.3.2   Membrane Receptors: the other signal transduction mechanism

G-protein coupled receptors (GPCRs) are membrane-bound protein receptors formed by amino acid sequences that self-assemble in membranes with a characteristic seven membrane crossing. Functional behavior depends on residues in the (often alpha-helical) transmembrane sections, as well as residues in the amino acid sections that loop out of the membrane, including the ends extending into the cells interior and exterior. The other key characteristic of the GPCRs is the G-protein to which they couple (on the membrane's inner wall), of which there are many varieties, as there are many varieties of receptors. GPCRs do group into families, however, as there are a number of common traits among certain subgroups. Some GPCRs bind ligand (such as neurotransmitters and hormones), and some are excited by photons, and all work by initiating cascades of (intracellular) second-messenger signaling. Not surprisingly, GPCRs are thought to exhibit pleiotropic (having multiple action) effects. As Hille [116] notes: "a wide variety of these receptors activate GTP-binding proteins that serve as timer-switches in the cascade, remaining active until their intrinsic GTPase activity cleaves the bound GTP. Activated G proteins selectively turn on at least three groups of enzymes, adenylyl cyclase, phospholipases, and a phosphodiesterase; adenylyl cyclase can also be inhibited." Another key property of GPCRs is that they can act directly on some ion-selective channels (as well as indirectly by second-messenger, such as processes that lead to phosphorylation of the channel protein itself).

2.4 Temperature dependence of the rate of a chemical reaction

The temperature dependence of the rate of a chemical reaction was described by S. Arrhenius by the following equation: $k_R = A\exp(-E_A/RT)$, where k_R is the rate constant, E_A is the activation energy, T is the temperature, and R is the gas constant (i.e., working with mole quantities of matter). A is an empirical factor. This equation describes not only the temperature dependence of chemical reactions, it can also be used to describe reaction rates of other physiochemical processes, such as diffusion, kinetics of phase transitions, etc. At certain temperature intervals, complicated biological processes, like growth rate, can also be fitted in this way.

In some reactions the energy of activation (E_A) itself is a function of temperature. In these cases the straight line in the Arrhenius plot gets kinks. The reasons for this can vary. In complex processes it is possible that at different temperatures, different reactions with different energies of activation will become rate limiting. In other cases such transitions may occur as the result of conformational changes of one of the components of the reaction, such as conformational change in an enzyme. In reactions of membrane-bound enzymes, sometimes phase transitions of the enzyme near lipids may cause such effects.

Statistical mechanics can derive the Arrhenius equation. This is based on the theory of absolute reaction rate as developed by H. Eyring. Chemical reactions presuppose splitting of chemical bonds. To analyze such processes it is necessary to know the interaction energy of a given atom in the molecule as a function of distance to each of its neighbors. This is an intricate function that can be expressed mathematically as an n-dimensional space corresponding to the n possible ways of geometrical approach. In this space a line may be found which shows the way from state A to state B over the smallest peaks of the activation energies. This line is called the reaction coordinate. It is possible to represent the energy level along this line in a two-dimensional graph [110].

2.5 Electrochemistry Electronics and Noise
2.5.1 Voltage Clamp

Patch clamps formed on biological membranes [139] have been studied for over 40 years. Together with artificial planar bilayers formed across micron-scale apertures they can be used to examine the structure and dynamics of protein channels. To apply a "voltage clamp" across the membrane is to control the potential difference across that membrane. This is not as straightforward as it sounds as a "battery" can't simply be connected up to provide that potential difference given the rapid voltage fluctuations in the electrodes, solutions, and from membrane capacitive fluctuations (from which a number of noise sources coupled in, including acoustic). Since nanopores typically pass currents in the hundreds of picoamperes, however, the voltage clamp circuit can usually drive the electrodes themselves. The equivalent circuit for the electrochemical part of the current pathway is, thus, dominated by the membrane capacitance and the channel conductance, i.e., a capacitor in parallel with a resistor. The access resistance between amplifier headstage and electrodes (all in series) is usually about $10k\Omega$ (including electrodes), together with the 2pF membrane capacitance for a small bilayer this indicates an RC time-constant of 20 ns, or a signal bandwidth of approximately 50MHz. With more specialized hardware, the access resistance can probably be reduced two orders of magnitude, likewise for smaller membrane capacitance from development of smaller bilayers. Thus, the useful bandwidth for detector use will generally not be dictated by charge transfer (capacitive) limitations in the amplifier/nanopore circuit. Rather, the useful bandwidth will be dictated by limitations due to fundamental noise sources.

2.5.2 Electronic Noise Sources

The accessible frequencies for nanopore detection are closely tied to how fundamental noise sources are managed. The noises result from unavoidable thermal noise properties of the circuit elements as well as low-current shot noise effects. In the end, the bandwidth accessible to detection determines how long observations must be made in order to discern molecular events (minimally, for Coulter counter operation) or perform classifications. Improving the device physics behind the nanopore detector has led to efforts to shrink the bilayer size and cool the aqueous chamber in which the experiment takes place. Since it is important to understand the fundamental noise sources that essentially delimit the capabilities of the nanopore detector, key relationships for Johnson noise and shot noise, among other noise sources, will be briefly derived and explained.

Nanopore channel current measurements exhibit the full range of noise types seen in electronic systems: (i) noise from random thermal motion on the charge carriers (Johnson noise), (ii) noise from the discrete nature of the charge carriers themselves (shot noise), and (iii) a low-frequency noise, 1/f noise (also known as flicker noise), for which no universal explanation is known. In addition to the usual electronic noise sources, noise also results from changes in the protein/bilayer geometry and from the motions of captured analytes (if present). The non-electronic noise sources often appear as 1/f noises (or Lorentzians if two-state Markov processes, or random decay processes are resolved singly).

A single α-hemolysin channel conducts 120pA under 120mV applied potential (with *cis*-side at negative potential, in 1M KCL, at room temperature), effectively a $1G\Omega$ resistor. The specific capacitance of lipid bilayers is approximately $0.8 \ \mu F/cm^2$ (very large due to molecular dimensions [116]), and the specific conductance is approximately $10^{-6} \ \Omega^{-1}cm^{-2}$. In order for bilayer conductance to produce less RMS noise current than fundamental noise sources, such as Johnson noise and shot noise (described below), the leakage current must be a fraction of a pA (under the conditions above). This is the first key noise problem to be resolved and it is done by reducing to a $200\mu m^2$ bilayer area [140], for which a 0.24 pA leakage current results and for which total bilayer capacitance is 2pF. Ignoring rectification properties (by keeping the *cis*-side negatively biased), the equivalent circuit for charge flow in the α-hemolysin nanopore/bilayer geometry is simply a resistor in parallel with a capacitor, for which the RC time constant is 2 ms. Such a long time constant would normally cause patch clamp protocols that involve stepping through various voltage levels to suffer from slow (2ms) relaxation times at each step. Hardware packages are already standard, however, that use small ($10k\Omega$) shunt resistances to greatly shorten those relaxation times (Axopatch 200B).

2.5.2.1 Johnson noise

The Johnson noise relation for resistors, $<V^2>=4kTR\Delta f$ [141,142], was first explained by Nyquist [143,144], where $<V^2>$ is the mean square voltage of the thermal noise across the resistor, k is Boltzmann's constant, T is temperature, R is the resistance, and Δf is the bandwidth of the detector (10kHz on nanopore detector described above). A derivation of Johnson noise along the lines of Nyquist begins with two resistors in a loop, each with the same resistance, R. The argument is then made that detailed balance implies that the noise power spectral density flowing from one resistor must match that flowing from the other. If the connecting lines between the resistors are stretched out to form transmission lines, with line impedance matching that of the (terminating) resistors, there will be no reflections between resistor and line. The same detailed balance argument (with no reflectance) then implies that noise on the transmission line must balance. Noise voltage waves on the transmission line can be written in terms of the traveling modes, $\exp[i(kx-\omega t)]$, that have phase velocity $v=\omega/k$. For a line of length L with periodic boundary conditions ($V(0)=V(L)$), the mode relations are: $k=2\pi n/L$, for n any integer. The 1-d density of modes is then $\sigma(\omega)=(dn/d\omega)/L=1/(2\pi v)$. The mean energy in the modes is obtained from the Planck formula, $<e(f)>=hf/(\exp(hf/kT)-1)$, for the classical limit $hf \ll k_BT$, i.e., $<e(f)> \approx k_BT$. The energy density per unit frequency, E(f), then satisfies $E(f)=k_BT/v$, and the power spectral density on absorption at a resistor at one end is $S_V(f)=vE(f)= k_BT$. For the circuit described, $S_V(f)=<(V(f)/2R)^2>R$, thus $<V^2(f)>=4RkT$. Integrating over available bandwidth then gives the familiar Johnson noise equation $<V^2>=4RkT\Delta f$.

The description of thermal noise for a parallel RC circuit (encountered with nanopore/bilayer systems) is actually simpler [145-147]. Recall that a capacitor stores energy $E_c=CV^2/2$, thus, at equilibrium, we expect the probability, dP, of voltage between V and (V+dV) to be: $dP=N_0\exp(E_c/kT)dV$, where N_0 is a constant that normalizes such that total probability is unity: $N_0=(C/2\pi kT)^{1/2}$. The mean-square value of thermal voltage fluctuations across the capacitor then follows by evaluation of the second moment (i.e., integration of $d<V^2>=N_0V^2\exp(E_c/kT)dV$), which leads to the relation $<V^2>=kT/C$. Noise fluctuations thus introduce thermal energy $C<V^2>/2= kT/2$ at the capacitor. The Thevenin equivalent for a noisy resistor is an ideal (noise-free) resistor in series with a noisy emf source [148]. Assuming a white noise spectrum (instead of detailed balance), the noisy emf has $<(V_E)^2>=S_V(0)\Delta f$, where $S_V(0)$ is the voltage spectral density of the resistor noise. A simple voltage divider relationship exists between the resistor noise voltage and the capacitor voltage: $V=V_E(1/i2\pi fC)/(R+(1/i2\pi fC))$. Thus, $d<V^2>=S_V(0)(1/2\pi fC)^2/(R^2+(1/2\pi fC)^2)df$, which integrates to yield $<V^2>=S_V(0)/4RC$. Combined with $<V^2>=kT/C$, the spectral density is found to satisfy $S_V(0)=4kTR$, or, in terms of the mean square voltage at the resistor observed in a bandwidth Δf, $<V^2>=4kTR\Delta f$.

Thermal noise due to the 1 GΩ channel resistance (and Δf=10kHz) has an RMS noise voltage of 0.4 mV (or RMS noise current of 0.4 pA). Likewise, the RMS thermal noise voltage from the capacitor is 0.5 μV, or 0.5 fA RMS noise current (note that less capacitance means greater mean square noise current, i.e., a smaller charge reservoir has greater charge variance). Although negligible for the channel geometry described, the thermal noise from the capacitor may become important in future nanopore detectors. If the bilayer area is reduced from 200 to 1μm^2, the thermal noise from the capacitor is 0.1 mV (0.1 pA), which would be comparable to the resistance noise (recall that C is proportional to A/d, where A is the bilayer area and d is the bilayer thickness). This indicates that bilayers with area less than the 1μm^2 reach a critical design trade-off between capacitive noise power tolerance and larger bandwidth. Overall, working at lower temperature can reduce Johnson noise, but that's little help in experiments involving nanopores/bilayers, where temperatures are generally restricted to be at least a few degrees above the freezing point of water.

2.5.2.2 Shot noise

Shot noise is the result of current flow based on discrete charge transport [149]. If current flow is not biased in one direction (by an applied voltage), the shot noise is typically negligible compared to thermal noise. For applied voltage resulting in current I, the mean square current due to shot noise is $<I^2> = 2q<I>\Delta f$. A simple derivation of shot noise [145-147] begins by defining $N=\int^{\tau} n(t)dt$, where N is the total number of anions crossing a flow cross-section in time τ (cation species treated similarly, if they exist), and n(t) is the anion crossing rate at time t. By the Ergodic theorem it then follows that $<N>=<n>\tau$, where $<...>$ is the ensemble average. (Ergodic theory indicates that time averages equal ensemble averages, at equilibrium, see [150] and [151] for more detail.) Shot noise arises from fluctuations in N and can thus be expressed in terms of $\Delta N=N-<N>$, i.e., via the random process $N_{\tau}= \Delta N/\tau$. As ensembles, we expect to see Poisson statistics on charge flow fluctuations, i.e., the law of small numbers is applicable [152]. For Poisson processes the mean and variance are equal, which leads to $<n> = <N>/\tau = var(N)/\tau = <(\Delta N)^2>/\tau = \tau<(N_{\tau})^2>$. By the Wiener-Khintchine theorem (autocorrelation function and power spectral density are Fourier transform pairs, it then follows that $S_n(0) = \lim_{\tau\to\infty}(2\tau<(N_{\tau})^2>) = 2<n>$. Since I(t)=qn(t), it follows that $<I>=q<n>$. The current power spectral density is then $S_I(0) = q^2S_n(0) = 2q<I>$, and application of the Wiener-Khintchine theorem again (with bandwidth Δf) yields $<I^2> = 2q<I>\Delta f$. For the total channel current, $<I_T>=120pA$, and at 1M KCL we can attribute about 60 pA to K^+ flow and 60pA to Cl^- flow. Treated as separate shot noise sources, and adding current noise in quadrature, this leads to $<(I_T)^2> = 2q<I_T>\Delta f$, the same expression with all current simply treated as single source. During 120pA nanopore operation with 10KHz bandwidth there is thus a 0.6 pA noise due to the discreteness of the charge flow.

2.5.2.3 1/f and telegraph

For open channel current flow in the α-hemolysin channel example there is typically no 1/f noise and the RMS noise current ranges between 1.0 and 1.5 pA. (Most protein channels exhibit 1/f noise, which is usually associated with channel gating.) Adding the shot noise (0.6 pA) and Johnson noise (0.4 pA) contributions (in quadrature) then accounts for 0.7 pA of that RMS noise. The nanopore operation described above is thus operated at approximately twice its fundamental noise limit.

Noise is 1/f-type if it has power spectrum S(f) proportional to $1/f^{\alpha}$ at low frequencies, with $0.5<\alpha<1.5$ (1/f noise is generally overtaken by Johnson noise in the 100Hz to 1MHz range [153]). By the Weiner-Khintchine theorem, for the $\alpha<1$ type noise sources, the associated autocorrelation function, C(τ), is proportional to $|\tau|^{\alpha-1}$. Signals with long time-scale correlations, such as those with $|\tau|^{\alpha-1}$ fall-off at large τ (α near 1), are found to be ubiquitous if not universal. It has been suggested from the inception of the subject that 1/f noise might be understood in terms of a *superposition* of simple exponential-relaxation or switching (Markov) processes [145,154-156]. A single exponential relaxation process satisfies: $F\{e^{-\lambda\tau}u(t)\} = (\lambda+i2\pi f)^{-1}$, and $S(f) = (\lambda^2+(2\pi f)^2)^{-1}$, so is actually of type $1/f^2$ for much of its range (a Lorentzian spectrum). Although not as obvious, two-state switching (random telegraph noise) also has a Lorentzian power spectrum, $S(f) = 4p_1p_2(x_1-x_2)^2\tau(1+(2\pi f\tau)^2)^{-1}$ [153,157], where p_1 is the occupation probability for discrete value x_1 (similarly for p_2 and x_2), and τ^{-1} is the sum of transition rates back and forth (i.e., in terms of time constants for the transitions, $(\tau)^{-1} =$

$(\tau_1)^{-1}+(\tau_2)^{-1}$. If the states are not known beforehand, observation of occupation probabilities, and extracting τ^{-1} from the break frequency in the signal's power spectral density, leads to the following expressions for time constants for the states: $\tau_1 = \tau/p_2$, and $\tau_2 = \tau/p_1$.

2.6 Channels and Polymers
2.6.1 Partitioning and Translocation in Channels -- The Free Energy Barrier
In order for a polymer (or large molecule) to enter a channel opening that polymer must first overcome a free energy barrier. If the channel is much larger than the molecule (a micron-scale channel), enthalpic contributions can be ignored, leaving only an entropic barrier to entry. At nanometer scale, however, enthalpic contributions can't be ignored. An example of this is the maltoporin channel, where partitioning of sugars reveals a blockade signal with a Lorentzian power spectrum [158]. They describe how average residence time of the sugar in the maltoporin pore and the average time between capture events can be inferred. The reciprocal of the time constants then gives the rate constants k_{off} and k_{on} respectively. In turn, the ratio of these rate constants gives the equilibrium binding constant K= k_{on}/k_{off}. In their study, Kullman, et al., found that the binding constant to the maltoporin channel increased about six-fold whether the polymer was added from the cis or trans sides. k_{on} differed substantially for polysaccharide addition to the two sides, while k_{off} was identical. This was taken to indicate that the bound sugar occupies the same domain within the pore regardless of side of entry. Capture of polymer, PEG (polyethylene glycol), in α-hemolysin has also been studied [159]. This is one of several studies that focused on capture of DNA and RNA in single α-hemolysin channels. Foremost in those early studies were questions about the collision rate of molecules with the pore entrance.

Estimates of ssDNA collision with the α-hemolysin pore begin with the assumption that the molecules are point charges diffusing toward a perfectly adsorbing disk. Thus, the collision rate is taken to be equal to flux to the pore, J, and J = 4DdC, where D is the diffusion coefficient of ssDNA in aqueous buffer, d is the diameter of the pore mouth, and C is concentration. Calculated values for this flux with no applied potential ranged from 50 $\mu M^{-1}s^{-1}$ [160] to 200 $\mu M^{-1}s^{-1}$ [161]. Although the mass distribution of a DNA molecule is certainly not concentrated at one point, a 30 bp DNA duplex is more that 10 nm long, for example, the charge distribution is generally much less uniform. In fact, a conducting object will tend to have more charge concentrated at its "points." This is particularly noticeable at the ends of wires, where the equipotential surfaces that peel away from the conductor are very bunched at the ends, while the orthogonal field lines are correspondingly divergent. In the case of the non-conducting polymer in solution, it is not clear how much this will tend to concentrate charge, but for polymers captured in strong ionic current it is quite likely that the polymer charge will redistribute (as if the polymer were conducting) giving rise to charge concentration at the end (this assumption will be important for later force calculation on that captured tip).

The measured capture rate in dilute solutions under standard conditions (120 mV applied potential, 23 degrees C, 1 M KCl), is found to be between 1.5 $\mu M^{-1}s^{-1}$ [162] and 5.8 $\mu M^{-1}s^{-1}$ [161], which is a fraction of the estimated collision rate, as expected given the free energy barrier. In those studies, the capture rate was weakly dependent upon composition of short synthetic ssDNA polymers [161] but strongly dependent upon voltage [160-162]. An interesting complication is that two voltage-dependent capture regimes are observed [161,162], with a sharp transition between them at about 140 mV. One hypothesis for the transition [161] is that the field at the pore mouth extends a critical distance into the bulk phase and that this field influences ssDNA diffusion toward the pore. An alternative that has not yet been studied is that the heptamer channel conformation undergoes a small change at voltages above 140 mV.

Henrickson et al. [160] measured the rate of ssDNA entry into the α-hemolysin pore as a function of applied voltage and ssDNA concentration. A capture event was defined as reduction of current by at least four standard deviations of the open channel noise. At 120 mV applied potential and 1 μM ssDNA, the measured capture rate from the cis-side was 5 μM^{-1} sec^{-1} at ambient temperature. Nakane and co-workers [162], on the other hand, examined ssDNA capture by the α-hemolysin channel with attention to the possible coupling to microcapillary devices. In their case, a capture event was defined as a drop in current below 80% of the open channel current. Under standard conditions (120 mV, 23 degrees C, 1 μM ssDNA) this criterion yielded 1.5 capture events s^{-1}.

2.6.2 ssDNA partitioning/translocation in α-hemolysin
2.6.2.1 Primitive channel blockade mechanism -- ignoring bond-formation events
Those molecules that surmount the free energy barrier remain captured until they either translocate or escape by diffusion (back into the cis-chamber). Polymers too large to translocate can still be captured at one end (dsDNA), but the focus in this section is on the blockade mechanism for translocating molecules, ssDNA in particular. In that instance, blockades have three main attributes: (1) blockade amplitude, usually expressed as a fraction of the open channel current, I/I_o, (2) blockade duration, and (3) blockade RMS noise. Blockade amplitudes for homopolymers range from 84% for poly(A) RNA to 95% for poly(C) RNA [140], while blockade durations range from 1 μs per nucleotide monomer for poly(dC) to 20 μs per nucleotide monomer for poly(A) RNA. The RMS noise attribute is a much weaker feature than the first two, and won't be discussed further here. The main explanation for the different blockade amplitudes is the different volumes (per base) occupied by the different homopolymers. Blockades then correspond to reduction in ionic flow according to the excluded volume of the homopolymeric strands. The difference in excluded volume corresponding to polypurine and polypyrimidine deoxyoligonucleotides is approximately 0.3 nm^3, or a 6% difference in total ionic flow volume (after correcting for waters of hydration [40]). This difference is in approximate agreement with the observed blockade amplitudes. The mechanism behind the blockade durations are described in the next section, where correlations are made to the polynucleotide composition.

2.6.2.2 Blockade duration and polynucleotide composition
The simplest model for the different polynucleotide blockade durations is that they should be proportional to polynucleotide length (assuming that the polymer length is much greater than the nm pore dimensions). This has been verified for the following RNA molecules: poly(U): 6 μs/base; poly(C): 2 μs/base; and poly(A): 16 μs/base. Together with the blockade amplitude information this provided the means for the first nanopore-based polymer sequencing, with the molecule $A_{(30)}C_{(70)}Gp$ [140]. That particular polymer was chosen because it would have given a balanced blockade step if the homopolymer characteristics above carried over in a proportionate manner. The bilevel blockade isn't balanced, however, simply because the overall oligo length is short (violating assumptions), so much so that transitions in secondary structure along the oligo are probably important (as well as end effects). Secondary structure is likely to be important since such effects are acknowledged to exist in the explanation by Akeson et al. [140] for the polyC-polyA blockade paradox. In

that study it was noted that polyC, a pyrimidine, is smaller than polyA, a purine, and therefore should have smaller blockade amplitude, but the opposite is true (95% vs. 84%). This can be understood if the single-stranded helical secondary structure must be unraveled in the case of polyA (with diam. 2.1 nm, [163]) but not in the case of polyC (helical diam. 1.3 nm [164]). The result is a longer blockade for polyA, due to the energy required to extend the 2.1 nm helix, and a reduced blockade amplitude, due to the elongated, non-helical structure. Note: biological polymers, such as protein, DNA, and RNA, are not ideal random coils in solution (polyU RNA is most random). Bases stack to their neighbors, form intra- and inter-strand hydrogen bonds, and produce helical windings and duplex regions such as hairpins.

2.6.2.3 Temperature effects

Studies on ssDNA/RNA translocation over temperatures ranging from 15°C to 40°C strongly suggest contributions from interactions with the pore [161]. Viscosity changes alone can't explain the translocational temperature dependence: $\tau \sim a/T^2 + b$, where **a** is a constant that depends on the polymer type, **T** is the temperature in °C and **b** is an additive constant. For example: poly(dC)$_{100}$ was 3.2 times faster than poly(dA)$_{100}$ at 15°C, but only 2.1 times faster at 40°C.

2.6.3   Forces acting on polymers captured or translocating in a nanopore

Forces acting on polymers in nanometer-scale pores include direct electrical interactions, coupled electrical interactions, and transport processes having to do with the nonequilibrium steady-state flow of the electrolyte. The direct electrical interactions are mainly influenced by electrophoresis and chemical bond formation. The coupled electrical interactions include electroosmosis (described by phenomenological equations of non-equilibrium thermodynamics [103]), and also electroconstriction via constrictions on the bilayer that couple into polymer-pore interactions mechanically. The main transport forces derive from diffusion, conductivity, and viscosity, and all have temperature dependences according to classic non-equilibrium stationary thermodynamics [103,106-108].

At the voltages used to obtain channel translocations (70-300mV), electrophoretic forces dominate polymer motion in the solution-phase. At the onset of translocation new forces come into play, with initiation of translocation described using reaction kinetics phenomenology, i.e., for the polymer to initiate translocation of the pore it must first overcome an energy barrier. The activation energy that leads to capture/translocation has both entropic and enthalpic contributions. Entropic contributions to the activation energy dominate for larger channels, where interactions are negligible (in which case the terminology "entropic barrier" is often used). Nanometer scale biological pores, on the other hand, usually have substantial interactions, i.e., non-negligible enthalpic contributions (such as binding sites and electrostatic attraction or repulsion at the mouth of the channel). From observations of polymer capture rates over a range of voltages it is possible to ascertain a number of interesting properties of that polymer, including an estimate of the difference between the electrostatic potential energy of the polymer in solution state and in its translocation activation state.

Polymers impinging on the pore will meet the translocation activation-energy requirement at a rate proportional to the Gibbs factor, $\exp[-\Delta G^*/kT]$, where ΔG^* is the barrier height in units of energy, and k and T are the Boltzmann constant and the temperature (in Kelvin) [110]. If a range of applied potentials is examined, activation energy is expected to decrease with increase in electrostatic potential energy, ΔU. The phenomenology with ΔU variation has Van't Hoff-Arrhenius factors, $\exp[-(\Delta G^*-\Delta U)/kT]$. If an applied voltage results in a voltage drop, ΔV, between solution state and activation state, and the unshielded charge is z, then the change in electrostatic potential energy is $\Delta U = ze\Delta V$, where e is the is the fundamental unit of charge. If a change in voltage does not result in coupled changes in activation energy (such as from electroosmotic or electroconstrictive contributions), or direct changes due to binding, then capture rates on the polymer should fit a curve of the form: $R = \kappa v \bullet exp[-(\Delta G^* - ze\Delta V)/kT]$, known as the Van't Hoff-Arrhenius equation (R is the blockade rate in events min^{-1}, κ is a probability factor, v is a frequency factor). In terms of the parts dependent on the voltage this is then $R = R_0 exp[ze\Delta V/kT]$, from which the $ze\Delta V$ value can be determined by a log-linear fit. Such an analysis is done in [160], where it is assumed that $\Delta V = V$, where V is the voltage applied to the electrodes. The fit in such a circumstance has a z value of 1.9 on the *cis*-side (vestibule side), and z=1.4 on the *trans*-side.

2.6.4 Polymer-Pore Interactions

If interactions with the pore walls are ignored the free energy barrier for polymer translocation reduces to an entropic barrier (enthalpic contributions ignored). In such a circumstance, a closed form solution exists [165]. When simplified for translocation along a strong potential gradient (compared to the partition entropy) it was found that translocation time is proportional to the length of the polymer (for sufficiently long polymer). In further work, [166], it is shown that diblock copolymers can discerned in terms of their translocation times in such a circumstance. Actual sequencing prospects with such a scenario seem remote, however, as positional permutations of bases in the body of a polymer (away from its ends) should, according to the theory, have the same translocation times. Regardless, interactions between polymer and channel appear to be common in nanometer scale pores. Even a charge-neutral polymer, such as PEG, is seen to interact with the channel protein α-hemolysin [159]. In the PEG study, enthalpic contributions are thought to arise solely from depletion of waters and charges in the channel [167].

Given that enthalpic contributions can arise simply from water and charge depletion (particularly where the polymer is threading the narrowest regions of the channel), interactions between polymer and channel seem unavoidable in nanopores with dimensions like α-hemolysin. Kasianowicz *et al.* [128] and Akeson *et al.* [140] showed that translocation times for charged polymers (RNA/ssDNA) are much greater than that indicated by free solution diffusion. Meller *et al.* [161] investigated ssDNA translocations and found that polymers shorter than a certain critical length (12 bases, or ~5nm) display different behavior than their longer counterparts. This was significant because the most restrictive part of the α-hemolysin channel was thought to extend for approximately 5nm. Meller *et al.* then showed that translocation time for ssDNA bases<12 drops noticeably faster than the linear extrapolation for ssDNA oligos with large N values. This adds further support to the notion that strong polymer-pore interactions are taking place. Shorter polymers experience only a portion of those interactions (i.e., lower effective channel drag) leading to a reduced translocation time. A phenomenological treatment of polymer translocation with strong channel interactions has been done by [168], where the prospects for sequencing seem doubly remote due to degeneration to stick-slipping between deep minima in the polymer-pore interaction (such minima are usually more than one base apart) [169], and there is still no clear explanation for the abnormally broad distribution of translocation times.

2.6.5 dsDNA and DNA hairpin partitioning

Rapid ssDNA/RNA translocations, ranging between 1 and 20 μs/base, will probably need to be slowed more than three magnitudes to have reasonable single-base pattern recognition (at least initially). Resolving the sequencing stick-slip problem mentioned above would benefit from a translocation braking mechanism also. One idea is to use a processive enzyme coupling as a brake. For dsDNA the problem is much simpler, translocation simply won't occur, providing the opportunity for extensive dsDNA-terminus identification (i.e., terminal base-pair identification, and possibly identification of a few adjacent base-pairs). For DNA hairpins with 3-9 base-pair stems (and similarly short dsDNA strands), it is possible for simultaneous dissociation of bonds to result in ssDNA conformation(s) that can translocate. For the DNA hairpins there is thus richer information than with (long) dsDNA terminus reads, with time to dissociation (and translocation) related to the energy of that dissociation, which is related to the base composition of the hairpin. Hairpins are sequences of single-stranded nucleic acids that fold back on themselves, bonding in the fold, or stem, with the usual Watson-Crick base-pairs as they would in a regular section of dsDNA [170,171]. At the head, or bend, of the hairpin there is no base pairing. The particular choice of hairpin to study was guided by the extensive characterization of a DNA hairpin with a six base-pair stem and a four-deoxythymidine loop [172]. When the six base-pair hairpin is captured by the α-hemolysin nanopore a current blockade occurs that lasts hundreds of milliseconds until the molecule dissociates and translocates (seen as a spike).

2.6.6 Sequence-specific DNA oligonucleotide detection using engineered nanopores

Another approach to detection using α-hemolysin involves oligonucleotide attachments to α-hemolysin monomers [239], with the eventual formation of a heptamer with one of the mutant monomers (the oligo in the DNA-protein complex is designed to reside in the vestibule). The resulting biosensor can then recognize an oligo that exactly pairs with the "DNA-nanopore." Howorka *et al.* claim, on the basis of DNA duplex lifetimes, to be able to discriminate between DNA strands up to 30 nucleotides in length. Actual sequencing is not likely on this developmental path, however, since it requires specially engineered proteins to identify each oligo. It also doesn't fit the tape-player paradigm, where the translocational blockades of the different bases are read like magnetic signals at the head of a tape player. Sequence-specific detection also derives no benefit from the powerful pattern recognition possibilities that result from the inherent sensitivity of the device and modern machine learning methods (instead, detection is largely accomplished without computation, using purely chemical pattern recognition).

Chapter 3

Channel current blockade detection (*translocation* detection)

A nanopore detector is based on a membrane that separates two chambers of electrolytic solution with a single nanopore providing a channel across that membrane. The detector is based on ionic current observations when a potential difference is applied across the membrane. Objects drawn into the nanopore cause ionic current blockades that form the basis of the molecular observations (i.e., observations derive from the ionic current imprints, or blockades, due to captured or translocating molecules). With channel current detection, particle analysis can be done on solutions to obtain particle concentrations, solution mixture composition, and even molecular dynamics. Early channel current detectors had millimeter diameters (0.1 mm) and were used to count cell concentrations and mixture compositions [119]. Information obtained about the excluded cell volume was used in classifying blood cells as red or white, for example, the ratio of which provided important data for medical diagnostics. The 100 μm-scale pores of Coulter were devised in the early 1950's.

It wasn't until the early 1970's that nanometer-scale pores were examined [120-122]. Bean made a nanometer-scale channel from crystalline structures (mica) that had defect tracks (from fission events). When etched with HF the normally impervious mica is removed along the defect-track in its crystalline structure. Depending on how this process is controlled, pores have been obtained with diameters ranging down to 6 nm (50 nm diam. pores commercially available). Although this technology has been used for observations on uncharged particles (polystyrene spheres with 90 nm diameter, [121]), it doesn't work as well with charged molecules (like DNA). Another complication is that the etching method for pore construction inevitably leads to long tunnel-like channels, which doesn't provide the best configuration for detector uses. Detection of biomolecules with biologically-based nanometer-scale pores also showed promise at about this time with the work by Hladky and Haydon [123]. They showed that a biological channel, the bacterial antibiotic gramicidin, could self-assemble in a lipid bilayer to form a functional channel (with currents of order 1 pA). This potentially solved two of the mica-channel problems: the lipid bilayers are very thin, about 5 nm, and the protein-based, biologically functional, nanometer-scale pore seemed better suited to passing charged biomolecules. Gramicidin was too small to detect most biomolecules, however, since it could barely pass molecules the size of the water molecule. It wasn't until 1994 [124] that a sufficiently large pore was studied, α-hemolysin. In the 1994 paper, Bezrukov et al. studied the blockades resulting from a charge-neutral polymer: polyethylene glycol (PEG). Later modifications to the gramicidin pore permitted its use as an antibody-modulated (on-off) biosensor, while modifications to the α-hemolysin pore enabled its use as a metal biosensor [125], among other things [126,127]. In 1995 and 1999, α-hemolysin was successfully used for DNA homopolymer translocation studies and classification [128,140]. Nanometer-scale pores then began to be developed in solid-state media [129,130]. Nanopores provide rich opportunities for the future because at nanometer scale a wealth of new prospects arise, from characterizing just about anything that can form a colloidal suspension in electrolytic solutions, to polymers like DNA, to the molecular motions that indicate molecular identity [45,47,48].

One of the key strengths of nanopore detectors is that they analyze populations of molecules individually. With signal processing and pattern recognition, this information enables a new type of cheminformatics based on channel current measurements. Single molecule observations are also of interest in biophysics; binding/conformational changes on captured dsDNA end regions, for example, might be tracked and understood using the nanopore blockade signal. DNA regions away from the ends may eventually be studied in a similar manner, using pore-translocation confinement to reveal distinctive conductance/binding properties on those bases threading the pore's limiting aperture constriction. Single molecule classifications permit a number of technical innovations. For sequencing, the single molecule basis of measurement may permit Sanger-type sequencing [163] on DNA molecules separated by capillary electrophoresis. If DNA can be translocated slowly enough, through a limiting aperture with dominant contributions to resistance spanning only two or three nucleotides length (about 20 Angstroms for ssDNA, 10 Angstroms for dsDNA), then DNA sequencing of a single molecule may eventually be possible. For single nucleotide polymorphism (SNP) identification, small sample volumes can be used, such that PCR amplification may not be needed. Expression analysis and disease identification (for individualized therapeutics) are just a few of the possibilities. Non-PCR expression analysis may even offer a new level of experimentation on live cells using patch-clamp methods.

3.1 Standard Apparatus
Nanopore detection is based on a nanometer-scale ion channel that can report on the channel-interactions of individual, nanometer-scale, biomolecules. The reporting is via measurements of ion flow through the channel when there is only a single channel, i.e., there is only one conductance path [20,45,47,140,124,128,167,174].

Each nanopore experiment described in what follows was conducted using one alpha-hemolysin channel inserted into a diphytanoyl-phosphatidylcholine/hexadecane bilayer, where the bilayer was formed across a 20-micron diameter horizontal Teflon aperture [47] (see Fig. 3.1). The bilayer separates two seventy-microliter chambers containing 1.0 M KCl buffered at pH 8.0 (10 mM HEPES/KOH). A completed bilayer between the chambers was indicated by the lack of ionic current flow when a voltage was applied across the bilayer (using Ag-AgCl electrodes). Once the bilayer was in place, a dilute solution of α–hemolysin (monomer) was added to the *cis* chamber. Self-assembly of the α–hemolysin heptamer and insertion into the bilayer results in a stable, highly reproducible, nanometer-scale channel with a steady current of 120 pA under an applied potential of 120 mV at 23°C (± 0.1 °C using a Peltier device). Once one channel formed, further pores were prevented from forming by thoroughly perfusing the *cis* chamber with buffer. Molecular blockade signals were then observed by mixing analytes into the *cis* chamber.

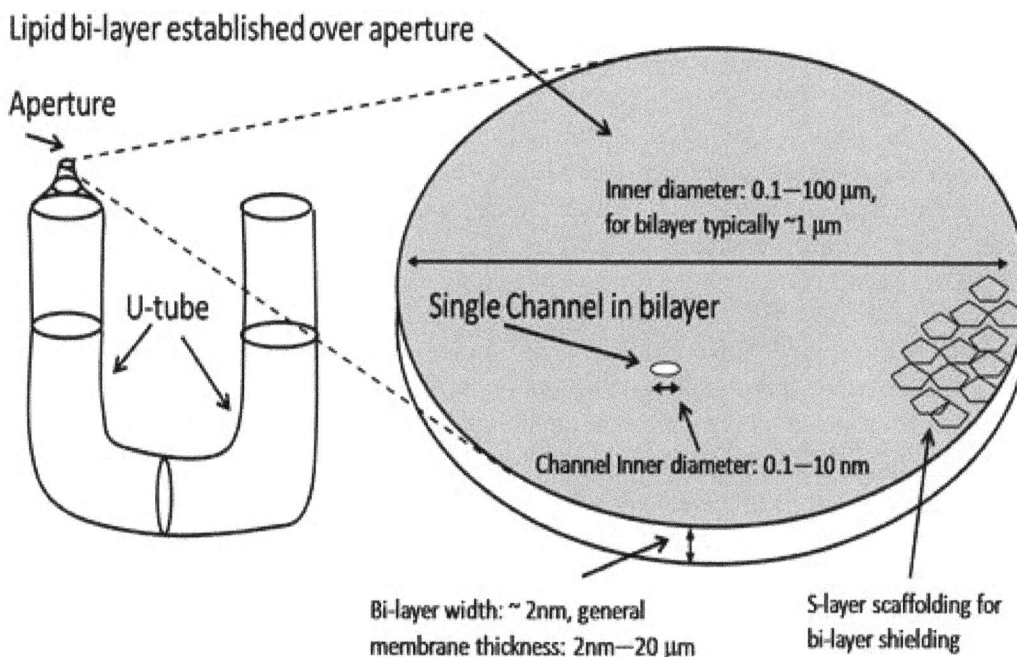

Fig. 3.1. A schematic for the U-tube, aperture, bilayer, and single channel, with possible S-layer modifications to the bi-layer.

3.1.1 Standard operational and physiological buffer conditions

The standard buffer condition for the nanopore detector is 1.0 M KCl with a pH of 8.0. This buffer was found to be most conducive to channel formation and to channels that do not gate. At significantly lower pH the channel is known to gate, if it even forms in the first place, which complicates use of the nanopore detector at physiological conditions (pH 7.0, 100mM NaCl). Since the pH of blood is usually in the range 7.35 to 7.45, and channel formation has been observed at 250 mM KCl, nanopore operation at the high pH and high salt end of the physiological range, relevant for antibody function in the bloodstream, may be possible with minimal alteration to the experimental parameters. Evaluation of antibody/antigen binding efficacy in a physiological buffer environment is particularly important if the nanopore/antibody detector is to be used for clinically relevant screening on the efficacy of antibodies to a given antigen. Biochemically relevant screening on enzyme activity, for example, requires working with physiological buffer testing.

3.1.2 α-Hemolysin channel stability – introduction of chaotropes

The α-Hemolysin channel is stable up to high salt concentrations ($MgCl_2$ above 2M and KCl up to 4M) and presence of some other additives (urea up to 7M in some experiments, glycerol 5%) at pH around 8.0. Typical pattern of current rise with increase in background electrolyte, KCl is observed. Specifically, the current versus KCL concentration is obtained in running buffer with composition 1M KCl, 20 mM HEPES (pH 8.0), with HEPES concentration maintained constant as content of KCl is increased.

A limitation in the utility of the nanopore/antibody antigen-binding tester (similarly for a nanopore/aptamer tester) is that once antigen is bound by a channel-captured antibody it is very difficult to effect the release of that antigen. This is a complicating issue in acquiring a large sample of antibody-antigen binding observations. A buffer-based solution to this problem is already known from purifying antibodies through a column containing antigen, where the release of antibodies bound in the column is effected by perfusion with 1.0 M $MgCl_2$. This presents the possibility of weakening the antibody-antigen binding by some choice of buffer in order to obtain large sample sets of binding events. The limitation of this is that the parameters will have likely deviated substantially from the physiological norm. Alternatively, a balanced stoichiometric ratio of antibody to antigen could be rapidly sampled, with lengthy sampling acquisitions only on antibody captures that occur without bound antibody and that then wait to observe antigen binding.

3.2 Controlling Nanopore Noise Sources and Choice of Aperture

The accessible detector bandwidth is limited by noise resulting from 1/f (flicker) noise, Johnson noise, Shot noise, and membrane capacitance noise. In Fig. 3.2, upper left, the current spectral density is shown for the typical bilayer, an open α-hemolysin channel, and a channel with DNA hairpin blockade. For 1.0 M KCl at 23C, the α-hemolysin channel conducts 120 pA under an applied potential of 120 mV. The thermal noise contribution at the 1 GΩ channel resistance has an RMS noise current of 0.4 pA, consistent with Fig. 3.2. Shot noise is the result of current flow based on discrete charge transport. During nanopore operation with 120pA current (with 10KHz bandwidth) there is, similarly, about 0.6 pA noise due to the discreteness of the charge flow. As with Johnson noise, the Shot noise spectrum is white, consistent with Fig. 3.2. The specific capacitance of lipid bilayers is approximately 0.8 $\mu F/cm^2$ (very large due to molecular dimensions), and the specific conductance is approximately 10^{-6} $\Omega^{-1}cm^{-2}$. In order for bilayer conductance to produce less RMS noise current than fundamental noise sources (under the conditions above), the leakage current must be a fraction of a pA. This problem is solved by reducing to less than a 500μm^2 bilayer area, for which less than 0.6 pA leakage current results and for which total bilayer capacitance is at most 4pF. This indicates that a decrease in bilayer area by another magnitude is about as far as this type of noise reduction can go. Preliminary attempts to do this, however, lead to a very unpredictable toxin intercalation rate (possibly due to surface tension factors), among other difficulties.

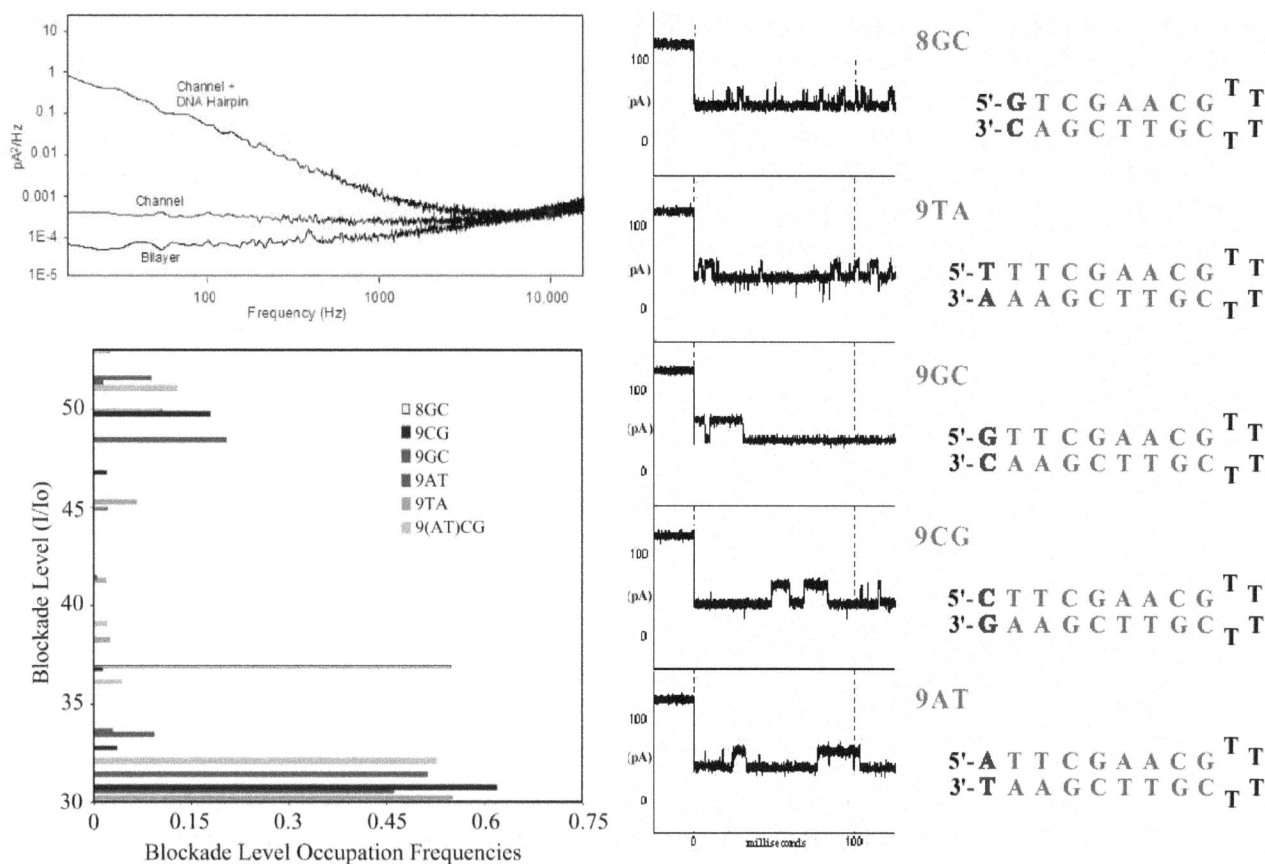

Fig. 3.2. Left. The top panel shows the power spectral density for signals obtained. The bottom panel shows the dominant blockades, and their frequencies, for the different hairpin molecules. **Right. DNA hairpin controls and their diagnostic signals.** The secondary structure of the DNA hairpins is shown on the right, with their highest scoring diagnostic signals shown on the left [45]. Each signal trace starts at approximately 120 pA open channel current and all blockades are in a range 40-60 pA upon "capture" of the associated DNA hairpin. Even so, the signal traces have discernibly different blockade structure, which is extracted using an HMM. The signals are aligned at their blockade starts and the demarked time-trace is for 100 ms.

The five DNA hairpins shown in Fig. 3.2 have been studied in [45,47], where they have been carefully characterized, and are used in other experiments as highly sensitive controls. Use of the controls entails testing a channel, especially an oddly behaving channel, with a known nine base-pair DNA hairpin control. If the familiar, visibly discernible, control blockade signals do not occur, the channels viability is then looked into further. The nine base-pair hairpin molecules examined in the prototype experiment share an eight base-pair hairpin core sequence, with addition of one of the four permutations of Watson-Crick base-pairs that may exist at the blunt end terminus, i.e., 5'-G•C-3', 5'-C•G-3', 5'-T•A-3', and 5'-A•T-3'. Denoted 9GC, 9CG, 9TA, and 9AT, respectively. The full sequence for the 9CG hairpin is 5' CTTCGAACGTTTT CGTTCGAAG 3', where the base-pairing region is underlined. The eight base-pair DNA hairpin is identical to the core nine base-pair subsequence, except the terminal base-pair is 5'-G•C-3'. The prediction that each hairpin would adopt one base-paired structure was tested and confirmed using the DNA mfold server.

31

3.3 Length resolution of individual DNA hairpins

The α-hemolysin geometry was probed using a series of hairpins, with stem lengths ranging from three to nine base-pairs. The six-base-pair hairpin is described in what follows, while those hairpins longer than six base pairs shared the six-base-pair stem/head at their core. Those hairpins with stems less than six base pairs were constructed by removing base pairs from the six base-pair core sequence. Starting from the three-base-pair hairpin, each base pair addition resulted in a measurable increase in median blockade shoulder lifetime that correlated with the calculated ΔΔG° of hairpin formation (Fig. 3.3 Left).

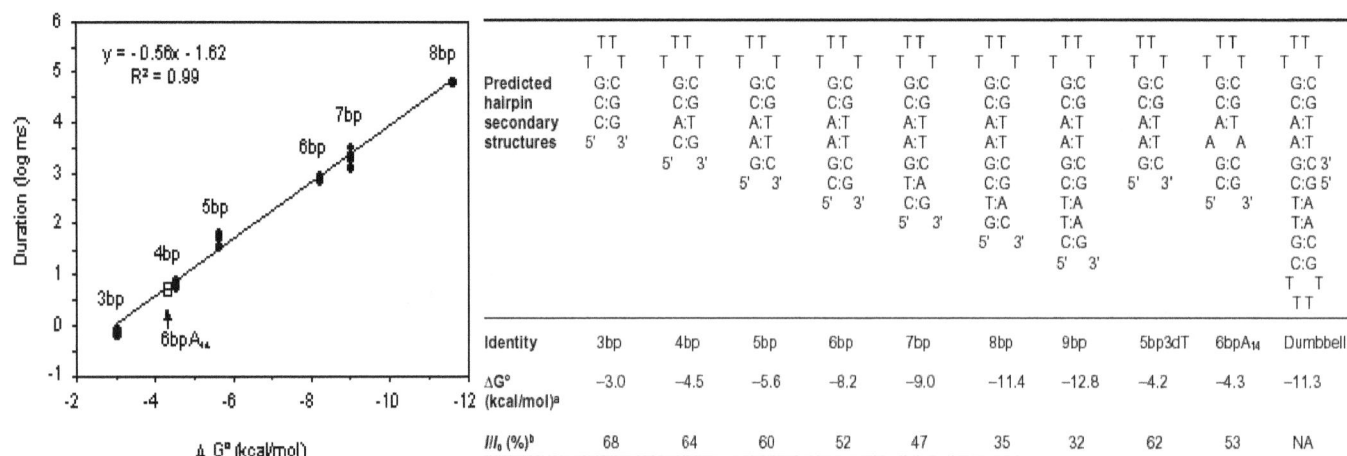

Predicted hairpin secondary structures (from left to right: 3bp, 4bp, 5bp, 6bp, 7bp, 8bp, 9bp, 5bp3dT, 6bpA14, Dumbbell):

```
      TT      TT      TT      TT      TT      TT      TT      TT      TT      TT
     T  T    T  T    T  T    T  T    T  T    T  T    T  T    T  T    T  T    T  T
     G:C     G:C     G:C     G:C     G:C     G:C     G:C     G:C     G:C     G:C
     C:G     C:G     C:G     C:G     C:G     C:G     C:G     C:G     C:G     C:G
     C:G     A:T     A:T     A:T     A:T     A:T     A:T     A:T     A:T     A:T
    5'  3'   C:G     A:T     A:T     A:T     A:T     A:T     A A     A A     G:C 3'
            5'  3'   G:C     G:C     G:C     G:C     G:C     G:C     G:C     C:G 5'
                    5'  3'   C:G     C:A     C:G     C:G    5'  3'   C:G    5'  3'
                            5'  3'   C:G     T:A     T:A                    T:A
                                    5'  3'   G:C     T:A                    T:A
                                            5'  3'   C:G                    G:C
                                                     C:G                    C:G
                                                    5'  3'                  T  T
                                                                            TT
```

Identity	3bp	4bp	5bp	6bp	7bp	8bp	9bp	5bp3dT	6bpA14	Dumbbell
ΔG° (kcal/mol)[a]	−3.0	−4.5	−5.6	−8.2	−9.0	−11.4	−12.8	−4.2	−4.3	−11.3
I/I_0 (%)[b]	68	64	60	52	47	35	32	62	53	NA

Figure 3.3. Left. Standard free energy of hairpin formation vs. shoulder blockade duration. **Right.** DNA Hairpins studied on the prototype nanopore detector.

Standard free energy of hairpin formation was calculated using the mfold DNA server (see Fig. 3.3 Right), and correlated with median duration of hairpin shoulder blockades (solid circles). Each point represents the median blockade duration for a given hairpin length acquired using a separate α-hemolysin pore on a separate day. Median blockade durations and ΔG° for the equivalent of the 6 bp hairpin with a single mismatch (6bpA14, Fig. 3.3 Right) are represented by open squares. All experiments were conducted in 1.0 M KCl at 22 ±1 °C with a 120 mV applied potential. (Increasing stem length also resulted in a 10 μs increase in median duration of the terminal spike, consistent with longer, but still microsecond time-scale, ssDNA translocation on the dissociated hairpin.) A downward trend in shoulder current amplitude was also observed from I/I_0 equal to 68% for a 3 bp stem to I/I_0 equal to 32% for a 9 bp stem (Fig. 3.3 Right).

Figure 3.4 shows an example where blockade events caused by six-base-pair hairpins were classified against blockades caused by 3,4,5,7 and 8 base-pair hairpins. The time-domain FSA passed 529 of the six-base-pair hairpin events to the SVM and 3185 of all other events. Because selectivity was relaxed at the FSA, there were many ambiguous signals with SVM scores near zero. Using an additional set of independent data, the SVM can be trained to exclude these by introducing a rejection region (the region between dashed lines in Fig. 3.4). The events that were rejected were primarily fast blockades similar to those caused by dumbbell hairpin (which can't translocate, see [48] for further details) or acquisition errors caused by the low selectivity threshold of the FSA. When 20% of the events were rejected in this manner, the SVM scores for the six-base-pair hairpin discrimination achieved a sensitivity of 98.8% and a specificity of 98.8%. Sensitivity is defined as true positives/(true positives + false negatives), and specificity is defined as true positives/(true positives + false positives). (A true positive is an event in the test data that comes from the positive class and is assigned a positive value; a false positive occurs when the SVM assigns a positive score to an event in the test data when that event actually comes from the negative class. A false negative is an event that is assigned a negative value, but actually comes from the positive class.) Similar results were obtained for each class of hairpins depicted in Fig. 3.4. Overall the SVM achieved an average sensitivity of 98% and average specificity of 99%. Thus, the stem length of an individual DNA hairpin was determined at single base-pair resolution.

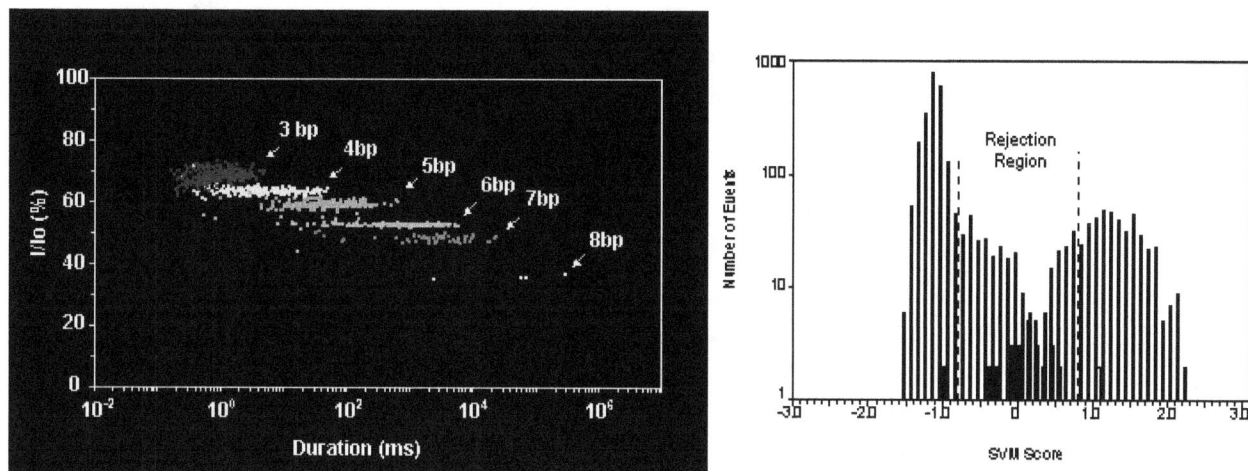

Figure 3.4. Left. Event diagram for DNA hairpins with 3 to 8 base-pair stems. Events were selected for adherence to the shoulder-spike signature. Each point represents the duration and amplitude of a shoulder blockade caused by one DNA hairpin captured in the pore vestibule. The data for each hairpin are from at least two different experiments run on different days. Median I/I_o values for each type of hairpin varied by at most 2%. The duration of the 9 bp hairpin blockade shoulders were too long for us to record a statistically significant number of events. Control oligonucleotides with the same base compositions as the DNA hairpins, but scrambled, caused blockade events that were on average much shorter than the hairpin events and that did not conform to the shoulder-spike pattern. **Right.** Classification of the 6bp hairpin (solid bars) versus all other hairpins (open bars) by SVM. Note the log scale on the Y axis. The dashed lines mark the limits of the rejection region. The boundaries of the rejection region were determined by independent data, not *post hoc*, on the data shown. The events that were rejected were primarily fast blockades similar to those caused by loops on the dumbbell hairpin (Fig. 3.3 Right) or acquisition errors caused by the low selectivity threshold of the FSA.

3.4 Detection of single nucleotide differences (large changes in secondary structure)

DNA hairpins with single nucleotide differences were examined in a context where those differences were expected to lead to significant differences in the hairpin molecule's secondary structure. The first hairpin considered involved alterations to the loop of the standard five-base-pair hairpin with a 4-deoxythymidine loop (**5bp4dT** in Fig. 3.3 Right) to one with a 3-deoxythymidine loop (**5bp3dT** in Fig. 3.3 Right). The **5bp3dT** hairpin caused pore blockades in which the shoulder amplitude was increased ~2 pA and the median shoulder duration (21 ms) was reduced 3-fold relative to the same hairpin stem with a 4-deoxythymidine loop (**5bp** in Fig. 3.3 Right). Typical events are illustrated in Fig. 3.5. The FSA acquired 3500 possible five-base-pair hairpin signals from ten minutes of recorded data. The SVM classification for this data set (Fig. 3.5) gave sensitivity and specificity values of 99.9% when 788 events were rejected as the unknown class. The second example involved the hairpin stem. Introduction of a single base-pair mismatch into the stem of a six-base-pair hairpin ($T_{14} \rightarrow A_{14}$, **6bpA$_{14}$** in Fig. 3.3 Right) caused an approximately 100-fold decrease in the median blockade shoulder duration relative to a hairpin with a perfectly matched stem (**6bp** in Fig. 3.3 Right). Typical events are shown in Fig. 3.5. This difference in duration is consistent with the effect of a mismatch on ΔG° of hairpin formation (Fig. 3.3 Left), and it permitted a 90% separation of the two populations using the manually applied shoulder-spike diagnostic. When analysis was automated, the FSA acquired 1031 possible events from ten minutes of recorded data (Fig. 3.5). With the aid of wavelet features that characterized the low frequency noise within the shoulder current, the SVM was able to discriminate the standard six-base-pair hairpin from the mismatched six-base-pair hairpin with sensitivity 97.6% and specificity 99.9% while rejecting only 42 events.

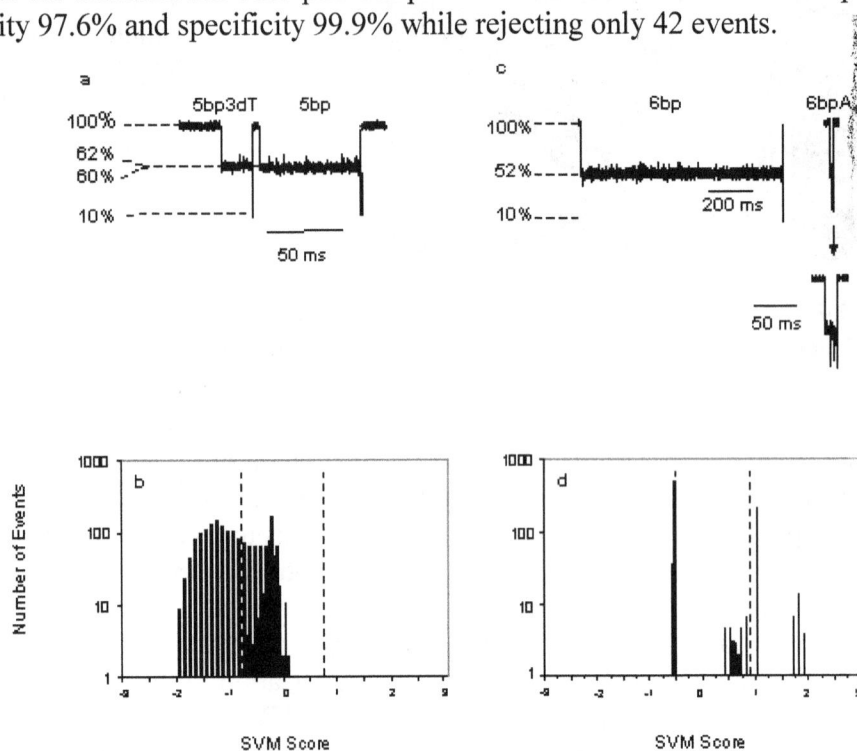

Figure 3.5 Detection of single nucleotide differences between DNA hairpins. **A,** Comparison of typical current blockade signatures for a 5bp hairpin and a 5bp hairpin with a three-dT loop. The standard 5bp hairpin event has a two percent deeper blockade than the 5bp3dT hairpin. **B,** Histogram of SVM scores for 5bp hairpins (filled bars) versus 5bp hairpins with three-dT loops (clear bars). **C,** Comparison of typical current blockade signatures for a standard 6bp hairpin and a 6bp hairpin with a single dA_3-dA_{14} mismatch in the stem. The 6bpA$_{14}$ event is expanded to show the fast downward spikes. These rapid, near-full blockades and the much shorter shoulder durations are the main characteristics identified and used by SVM to distinguish 6bpA$_{14}$ hairpin events from 6bp hairpin events. **D,** Histogram of SVM scores for 6bp hairpins (filled bars) versus 6bpA$_{14}$ hairpins (clear bars).

3.5 Blockade Mechanism for 9bphp

More involved than the classification or sequencing of a molecule is the actual understanding of that molecule's kinetic behavior as revealed by the ionic current blockade information measured by the nanopore detector. In the discussion of blockade mechanism that follows, for the nine-base-pair hairpins, the remarkable sensitivity of the nanopore device becomes apparent. This indicates that the α-hemolysin nanopore detector is likely to be an important tool for single-molecule observation and manipulation.

Ionic flow through the α-hemolysin channel was strongly modulated by the terminal base-pair on DNA hairpins with stem-length nine or more base-pairs. This modulation was most apparent on the nine base-pair hairpins, where the blockade states were discerned with the shortest time-constants (lifetimes). The lower level blockade states for the DNA hairpins are found to have lifetimes that correlate with the energy of dissociation on the terminal bond [47]. An anti-correlation with terminal bond energy is found for the density of lower level blockade spikes.

A working model has been developed to explain the mechanisms underlying the current transitions for the observed 9bp hairpin blockades (Fig. 3.6). The model requires that the 9bp duplex stem is long enough so that the terminal base pair can interact with amino acids in the vestibule wall and that a frayed end can reach the limiting aperture (at lysine-147). This is a reasonable assumption because circular dichroism assays indicate that the 9bp hairpin stem is a B form duplex in bulk phase. The length per base-pair of B form DNA is 3.38 angstroms, therefore the total stem length is approximately 30.4 angstroms. The distance between the narrowest part of the vestibule mouth at threonine-9 and the pore limiting aperture at lysine-147 is 33 angstroms. Therefore, if the hairpin loop is perched at the ring formed by threonine-9, the 9bp stem would reach within 3 angstroms of the limiting aperture. Given the uncertainty about the exact position of the hairpin loop and the 1.9Å precision of the α-hemolysin X-ray crystal structure [118], not to mention effects from waters of hydration and ion fixed-layers, this distance is probably accurate within ±1 bp. Upon capture of a 9bp blunt hairpin, the initial conductance state ($I_{IL}/Io = 35\%$, where IL stands for intermediate level) is caused by orientation and immobilization (on the millisecond time scale) of the hairpin due to an electrostatic bond formed between the terminal base pair of the hairpin stem and residues in the vestibule wall. The predominant interaction is binding between the nucleotide in the 3' position and the protein. This state initiates virtually all events because it is entropically favored. The dwell time, τ_{IL}, for the intermediate conductance state is largely independent of base pair identity or orientation because the bases are hydrogen bonded to one another and the interaction with the surface is due to the terminal 3' phosphodiester anion. If, however, the 3' nucleotide is unpaired (i.e. a dangling nucleotide) its identity does matter in terms of the duration of τ_{IL}. Preliminary results show that single nucleotide 3' overhangs have dwell times in the IL state with order dA>dC. This suggests that the unpaired bases hydrogen bond or stack against residues in the pore vestibule. The IL state invariably transitions to the upper conductance state, UL. This state ($I_{UL}/Io = 48\%$) corresponds to desorption of the terminal base pair from the protein wall and orientation of the hairpin stem along the axis of the electric field and the axis of ionic flow. Current is higher in this state because the low resistance path along the major groove leads relatively unimpeded from the pore mouth to the limiting aperture. From the UL conductance state, the hairpin may return to the IL state or it may transition into a third conductance state, LL, where the residual current is equal to 32% of the open channel current. Residence time in this state is dependent upon terminal base pair identity and orientation. In this state, it is hypothesized that the nucleotide at the 5' end of the duplex stem is adsorbed to the pore wall so that the 3' nucleotide is positioned directly over the pore-limiting aperture. Thus, when the duplex end frays, the 3' strand may extend and penetrate the limiting aperture resulting in the transient spikes.

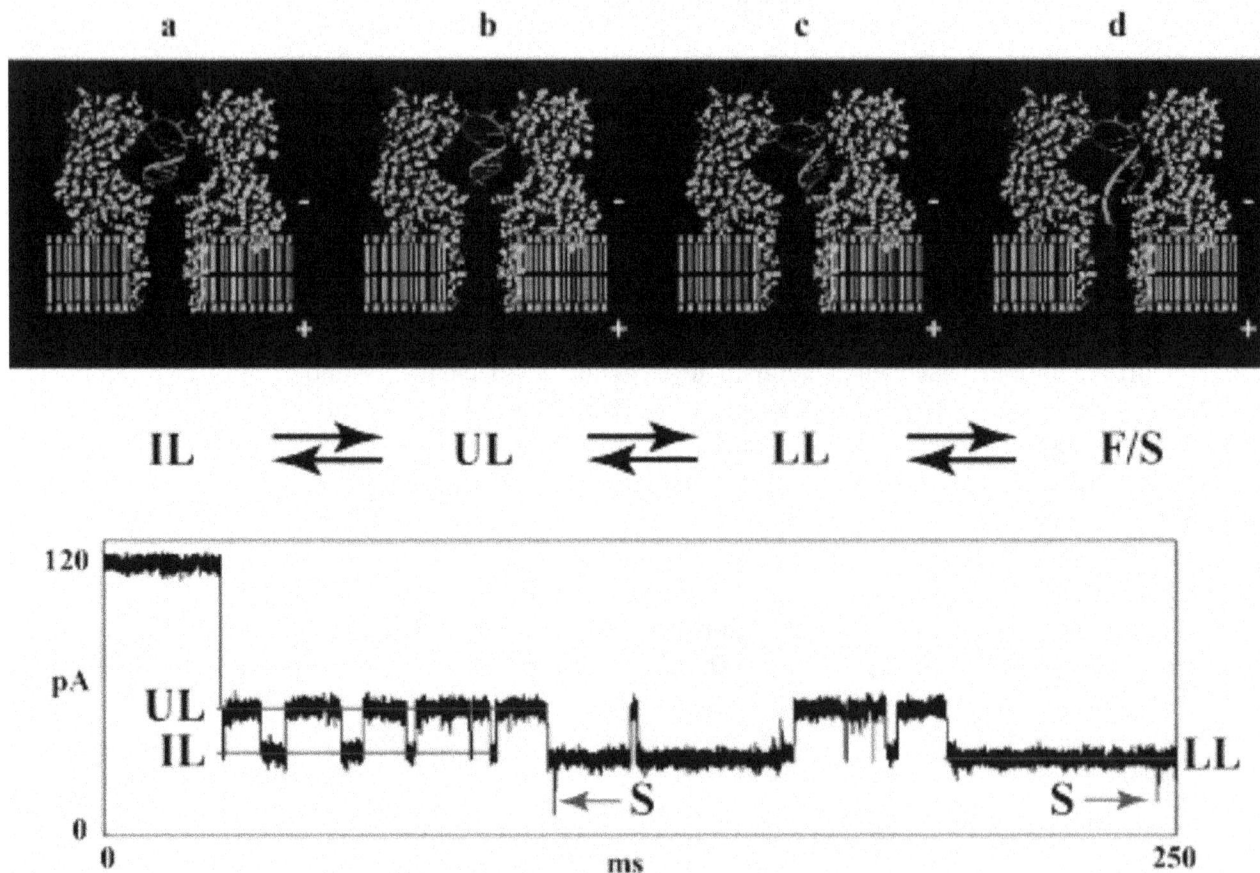

Figure 3.6 Blockade Mechanism. The intermediate level (IL) conductance state initiates most blockades and always transitions to the upper level conductance state (UL). This is explained by binding of the hairpin terminus to the vestibule interior (IL) followed by desorption of the DNA from the protein wall and orientation of the stem along the axis of the electric field (UL). Transitions from the UL state were either back to the IL state or to the lower level conductance state (LL). From the LL state there were brief transitions to nearly full blockade, denoted by F/S for fray/spike conductance state. The LL and F/S states are both thought to involve binding between the hairpin's terminal 5' base and the pore's limiting aperture. The brief F/S state behavior is explained by a terminus-fraying event that is accompanied by extension by the terminal 3' base into the limiting aperture. Part of the evidence for this is a strong spike (fraying) frequency correlation with the different terminus binding energies. Asymmetric base addition or phosphorylation (at the terminal 3' and 5' positions) is part of the evidence for the asymmetric roles for 5' binding (LL and F/S) and 3' fraying/extension (F/S).

Beyond hydrogen bonding and steric considerations, some of the terminus dynamics was influenced by the nearest-neighbor base-pair (i.e. stacking energies), which indicated that the penultimate base-pair might be readable as well. Next-to-nearest-neighbor influence on the terminal base-pair dynamics was thought to be much less, and this was consistent with the minor changes in blockade signatures on hairpins whose ends were the same but that have different base-pairings further up their stems. This drop-off in sensitivity bodes well for terminus classification on *generic* duplex DNA. At the same time, gross changes in stem base-pairs, such as a change from a Watson-Crick base pair to a Hoogsteen (or wobble) base-pair, resulted in distinctly different blockade signatures [47], which bodes well for SNP assaying schemes.

Residual channel current decreases as blockading DNA hairpins increase their stem length from 3 to 8 base-pairs. For DNA hairpins with stems shorter than 8 base-pairs, multiple states were not clearly discernible, presumably because the hairpins were too short to bind to the channel favorably or interact with the current/force constriction near the limiting aperture. For 9 base-pair hairpins, and longer, a clear 1/f noise (flicker noise) is discernible (Fig. 3.2) – a preliminary indication of the single-molecule binding kinetics described in Fig. 3.6 (and in detail in [47]). HMM/EM characterization on the five classes of hairpin signatures revealed the existence of two major conductance blockade levels, one minor level intermediate between them, and one to three other statistically relevant levels depending on the hairpin (a pre-processed form, found by HMM/EM level identification, and use of EVA-projection, is shown in Ch. 4). By examining the transition probabilities between the various levels it was found that blockades typically began in the less common intermediate level and from there almost always transitioned to the UL blockade level. The mechanism described in Fig. 3.6 hypothesizes that the upper level (UL) blockade state is unbound. A result that strengthens this hypothesis, is that the UL blockade levels are approximately the same for 8, 9, 10, 11, and 12 base-pair DNA hairpins. This plateau occurs well before that of the other blockade levels – the lower level (LL) blockades, for example, continue to become greater as the hairpin stem length is increased from 8 base-pairs to 10, beyond which it plateaus as well. Beyond 10 base-pairs the hairpin is simply longer than the depth of the channel's *cis*-vestibule, so further base addition causes it to "stick-out further", but cause negligibly greater occlusion of flow than that caused by the fully blockaded *cis*-vestibule. With base-pair addition, however, there arrives greater residual charge, thus greater force drawing the molecule into the vestibule (and slightly deeper channel blockades are seen consistent with this) and dominance of the LL state in the blockade ("toggling") signal. The explanation for the early UL plateau centers on the tight flow geometry between channel and captured hairpin. In such a geometry, much of the ionic flow is confined to be in or near the grooves of the captured DNA molecule. For the unbound molecule, this groove flow can be directed towards the limiting aperture by appropriate orientation of the hairpin molecule. The unbound molecule, thus, appears to cause a gap junction "short circuit" effect, where the contribution to the ionic current is not significantly altered as the hairpin is extended across a 3 base-pair (approx. 1nm) gap separating the hairpin terminus from the vestibule's limiting aperture.

A critical understanding derived from the 9-base-pair DNA hairpin analysis is that if the UL blockade state is unbound at its terminus there is the possibility that conformational kinetics might be observable at the pore-captured polymer end. This motivated examination of a set of dsDNA termini that had already been examined using NMR. Results (in [47]) show agreement with NMR via number of low energy conformational states observed [47].

3.6 Conformational Kinetics on Model Biomolecules

Two conformational kinetic studies have been done, one on DNA hairpins with HIV-like termini, the other on antibodies. The objective of the DNA HIV-hairpin conformational study was to systematically test how DNA dinucleotide flexibility (and reactivity) could be discerned using channel current blockade information (see [35], for a complete description of the results pertaining to this study). The structural and physical properties of DNA depend upon nucleotide sequence, as is manifest in differences in three dimensional structure and anisotropic flexibility. Despite the multitude of crystallographic studies conducted on DNA, however, it is still difficult to translate the sequence-directed curvature information obtained through these tools to actual systems found in solution. Information on the DNA molecules' variation in structure and flexibility is important to understanding the dynamically enhanced DNA complex formations that are found with strong affinities to other, specific, DNA and protein molecules. An important example of this is the HIV attack on cells: one of the most critical stages in HIV's attack is the enzyme mediated insertion of viral into human DNA, which is influenced by the dynamic-coupling induced high flexibility of a CA dinucleotide step positioned precisely two base-pairs from the blunt terminus of the duplex viral DNA. This flexibility appears to be critical to allowing the HIV integrase to perform its DNA modifications. The CA dinucleotide presence is also a universal characteristic of retroviral genomes [265,266]. The behavior of the DNA hairpins containing the CA dinucleotide at different positions relative to their blunt-end termini, is studied in [35] using a nanopore detector. The nanopore detector feature extraction makes use of HMM-based feature extraction and SVM-based classification/clustering of "like" molecular kinetics. We hypothesized that the DNA hairpin with CA dinucleotide, positioned two base-pairs from the blunt terminus, would have "outlier" channel current statistics qualitatively differentiable from the other DNA hairpin variants. This is found to be the case, where the UL state, corresponding to the unbound terminus state, has shortest life for hairpin labeled CA_3 (with the CA dinucleotide step two basepairs from the blunt terminus). Since the UL state is hypothesized to be unbound, the fact that it has the shortest lifetime on average is an indication of the associated molecule's propensity to be bound to the channel (binding site is unknown at this time, although the work in [47] suggests some likely binding sites on the channel). In other words, CA_3 has strongest interaction with channel (and surroundings), as hypothesized, and neighboring variants (CA_2, CA_4), that have GC pairs shifted one base-pair shifted closer and further from the terminus, share this property to a lesser extent. Note: the "CA" notation refers to a dinucleotide step along the backbone of the self-annealed ssDNA strand in the hairpin molecule, while the GC base-pair described is part of that strand annealing, with the 'G' base-paired to the 'C' referred to in the CA-step. The molecules with GC pairs that are more than 1 base-pair distant behave similarly to eachother; the DNA hairpin with no GC pair also separates with its own characteristic curve.

3.7   Biomolecular classification

Protein channels represent a wealth of prospects for future detector implementations. At the moment focus is on the larger pores, like α-hemolysin, and the analytes that they might be useful in examining, particularly for rapid sequencing on DNA. As this field develops, however, the smaller, more ion-selective channels will probably be put to use in examination of smaller, non-polymer, solutes. Such efforts will probably range from study of sugars, and other solutes native to the cell-pore environment, to synthetic molecules, such as ligands that might strongly bind to GPCRs (such ligands are of interest because 50% of modern medicines are based on them). A key feature of biologically based pores, for detector implementation, is that they self-assemble and do so with very high structural reproducibility. For non-gating pores, like α-hemolysin, there is also long-term retention of channel geometry. Again, due to their tried-and-tested biological heritage, the same pores can often interact (examine) molecules without interacting too strongly (i.e., sticking, or worse, forming strong bonds that are effectively permanent on the timescale of the experiment). Solid-state channels will probably overtake biologically-based channels in detection utility at some point, but the time-horizon for that occurrence has become unclear due to difficulties with solid-state channel stability/reproducibility and problems with DNA sticking during translocation through solid-state channels. That said, carefully engineered solid-state nanopores offer the best long-term prospects for rapid DNA sequencing and other such large-scale enterprises. For all nanopore implementations a key innovation simply derives from the, typically, single molecule observation process. Conformational changes on captured dsDNA end regions, for example, can be tracked and understood using the nanopore blockade signal [47]. For sequencing, the single molecule basis of measurement, together with a rapid sampling of molecules, can be used to classify DNA molecules separated by capillary electrophoresis (using the present, biological-channel based, technology). As the field develops further it is of particular interest as to whether single molecule translocation measurements may permit single molecule sequencing. As the technology develops, and incorporates more optical/florescent sensing technology, it seems likely that nanopores will still be very useful in a single-molecule manipulation role. Other innovative applications of interest include single nucleotide polymorphism (SNP) identification where small sample volumes can be used (such that PCR amplification may not be needed). With SNP identification, and more general expression analysis, disease identification for individualized therapeutics is just one of many possibilities. Eventually, non-PCR expression analysis may even provide a new level of experimentation on live cells using patch-clamp methods.

3.8 DNA sequencing (Translocation-based)
DNA sequencing with a channel is currently being explored using two paradigms. The first is an outgrowth of the classic Sanger-style sequencing approach [163], where copies of the DNA molecule to be sequenced are cut to different lengths, terminated with nucleotide-dependent fluorophores, and separated by length using capillary electrophoresis. The sequencing task then reduces to reading the fluorophores tags on the electrophoresis-separated oligonucleotide populations. The modification with use of the nanopore is in performing those terminus reads without fluorophores, using channel-captured terminus identification instead. The prospects for substantially accelerated DNA sequencing with this approach are limited, however, as the main measurement bottleneck is in the common electrophoretic separation. The Sanger-nanopore approach, however, does offer the prospect for reduction to "lab-on-a-chip" technology and, consequently, extensive parallelization. If length information on an end-captured oligonucleotide can be obtained by some means inherent to the single molecule end-capture event, rather than electrophoresis, then truly rapid DNA sequencing (not merely parallelization) may eventually prove possible in this paradigm. It is in pursuit of such hopes that length-dependent reptation critical phenomena, and other possible length-dependent effects, are being considered. There are subtleties resulting from the low Reynold's number fluid environment, however, that make the prospects slim. In particular, it may prove necessary to impart substantial tension to the end-captured polymer, such as via a magnetic bead attached to the free end, etc., in order to elicit, length-dependent, critical phenomena, if it's even found to exist.

In the second sequencing paradigm, DNA sequencing is accomplished during translocation through the nanopore. Enzymes, proteins designed by nature to act as nano-manipulators, may play a critical role in such channel-translocation based approaches to DNA sequencing. This is because there is a problem, at present, with controlling the DNA translocation rate in nanometer scale channels. Enzymes such as lambda-exonuclease, for example, may provide a critical braking role for DNA as it is belayed though a nanometer-scale channel. (Another braking mechanism, based on phased-locked-loop methods that employ modulation of the applied potential, may offer an alternate, non-enzyme-based, approach at some point.) It is also hoped that working with enzymes in the single-molecule nanopore setting will prove informative about enzymatic properties as well. Similarly, DNA-binding proteins (non-enzymatic), and/or ssDNA to synthetic-dsDNA pairing (with appropriately engineered Watson-Crick variants), may be useful in slowing translocation and/or improving identification of nucleotides. This is the paradigm that most of the solid-state efforts are beginning to explore. In work using α-hemolysin, DNA-hairpins with ssDNA overhangs have been used to probe the channel's limiting aperture. One of the goals of the latter effort is to explore another complication with the translocation-based paradigm: the base-deconvolution problem. Suppose a stretched strand of ssDNA crosses the 5nm neck of the pore channel with 7 bases, this leads to a difficult deconvolution task in order to determine the sequence: 4^7=16,384 patterns to discriminate (in very noisy environment with decoys). It may be that the α-hemolysin nanopore only has short "pinches" upon entering and exiting its transmembrane region, however, in effect requiring deconvolution on only 3 bases (64 patterns), which is probably manageable with the current device sensitivity. Likewise, for synthetic pores the goal is to engineer the constriction zone at the neck to be only a few bases across (picture an hourglass shaped pinch). In the end, an outgrowth of the translocation-based paradigm may be successful, but only because the difficult base-calling task is decoupled from the single-molecule handling and left to some external, possibly florescence-based, method. An example of this would be FRET detection between a fluorophore fixed at the pore's trans-side and a nucleotide-attached fluorophore in the process of translocating out that side.

3.9 Improving the sensitivity of the nanopore detector

Nanopores in bilayers are usually studied by controlling the voltage on the bilayer. Since nanopores typically pass currents in the hundreds of picoamperes, or less, the voltage clamp circuit usually drives the electrodes themselves. The equivalent circuit for the electrochemical part of the current pathway is dominated by the membrane capacitance and the channel conductance, i.e., a capacitor in parallel with a resistor. The access resistance between amplifier headstage and electrodes (all in series) is usually about 10kΩ (including electrodes), which, together with the 2pF membrane capacitance for a small bilayer, indicates an RC time-constant of 20 ns, or a signal bandwidth of approximately 50MHz. With more specialized hardware, the access resistance can probably be reduced two orders of magnitude, likewise for smaller membrane capacitance from development of smaller bilayers. (Improvements to the device physics have already resulted from efforts to shrink the bilayer size and cool the aqueous chamber in which the bilayer resides.) Thus, the useful bandwidth for detector use will generally not be dictated by charge transfer (capacitive) limitations in the amplifier/nanopore circuit. Instead, as mentioned in Ch. 2, the accessible frequencies for nanopore detection are closely tied to how fundamental noise sources are managed. The noises result from thermal noise contributions from the circuit elements as well as low-current shot noise effects. In the end, the bandwidth accessible to detection determines how long observations must be made in order to discern molecular events (minimally, for Coulter counter operation) or perform classifications. One objective in further refinements to the nanopore detector is to directly introduce high frequency excitations to the region surrounding the nanopore. Such as laser-excitations introduced via a pulsed evanescent wave off a light fiber. It should then be possible to observe the excited molecular "ring-down" as it imprints on the surrounding ionic flow. In such an experiment the noise sources might be kept in abeyance while at the same time exciting the captured analyte over a wider range of frequencies. In general, modulation over a range of frequencies, such a low-to-high frequency "chirp," is sought to provide extensive side-information to aid in identification.

3.10 Single molecule biophysics: estimation of forces acting on DNA hairpin probes
The dsDNA blockade for the vestibule traversing 9 base-pair hairpin is 32% residual current. If a similar threading of ssDNA corresponds to the activation state for *cis*-side entry, with half of the dsDNA blockage used to estimate that ssDNA blockade, then residual current would be approximately 66%. If conditions are such that baseline current is 120pA at 120mV, then such a residual current implies a resistance of 1.52 $G\Omega$, or a change in resistance of $\Delta R=0.52$. The voltage drop for the activation state, ΔV, is approximated as that across ΔR, i.e., $\Delta V=41mV$. For the *cis*-side at 120mV applied voltage the ΔU value is 3.6×10^{-20} Joules [160], which indicates a z value of 5.6. Similar arguments for the *trans*-side, with 86% blockade taken for activation state, leads to z=1.6, relatively unchanged from [160] due to the similar voltage drop assumptions. If the distance between impinging solution state and the molecules activation state is due to translocation to the limiting aperture, then the electrostatic potential energy spans a distance of approximately 5nm, resulting in a force of 7pN for *cis*-side capture. If the distance between impinging solution state and activation state on the *trans*-side is on the order of one base-pair entry into the pore (about 4Å), then a rough approximation to the, transient, *trans*-side initial capture force is 67pN. A variety of other assumptions are equally valid, with noticeably different results, so force calculations remain imprecise at this time.

If the entire asymmetry between the two forces is attributed to electroosmosis (however unlikely), then as much as 30pN of force can be attributed to electroosmosis (flowing from *trans* to *cis*), with the 7pN and 67pN values taken to pertain to the net force due to the applied potential. (Note: with a buffer concentration of 10mM, the diffuse double layer critical to electroosmosis phenomenology is approximately 1nm thick [175], which envelops almost the entire α-hemolysin channel interior). If the actual electroosmotic force is anything close to 30 pN it will be non-negligible. In which case, as the polymer begins to thread the pore the electroosmotic force is not expected to change greatly from the *trans*-side, while from the *cis*-side there is probably increased force when the limiting aperture is reached (so the *cis*-side force is brought on par with the *trans*-side eventually). The profile on the potential voltage, in contrast, changes greatly throughout the translocation process, in order to track regions of reduced ionic flow over which the voltage drop must redistribute. Once translocation reaches the point where the channel is traversed, the voltage drop is dominated by the approximately 12 base segment of ssDNA that threads the narrow 5nm transmembrane section of the α-hemolysin channel. If the ssDNA charge is not greatly changed upon *cis* entry, and the possible electroosmotic force is ignored, then the force drawing the ssDNA through the entire pore reduces to only 7pN.

3.11 Sticking problem and use of 'Un-sticking' Excitations
A practical limitation of the nanopore device is that molecules to be detected, or specially designed blockade gauges (for indirect attachment) can blockade the channel and get "stuck" in one blockade state. One way of seeing how this might occur is to consider DNA hairpins with stems having 10, and more, base-pairs. As the hairpins are extended from 9 base-pairs, the clearly discernible toggle signal slows, and resides longer in the lower level.

When 14, or more, base-pair stems are attempted the resulting molecular blockades appear stuck in their lower level (at the time-scales of the experiment). The trend on blockade signals going from 14-20 base-pairs was not elaborated on in the analysis of the 20 base-pair dsDNA molecule that was the studied in [41], however, since that paper focused on how to recover highly structured blockade information from a minimally informative "stuck-state" situation. What was accomplished in [41] was to observe channel blockades due to a 20 base-pair DNA hairpin designed to have the same 9 base-pair terminus as the "sensitive" DNA gauge used in [42,45,47]. As expected, the 20 base-pair dsDNA molecule no longer modulates the channel flow, being stuck in just one blockade state. It is shown in [41], however, that the molecule can be re-excited to its telegraph signaling via a bead attachment which is periodically tugged by a laser beam. In this way, the larger DNA hairpin has its blockade signal "re-awakened" to reveal binding kinetic information at its channel-captured end. This offers the prospect that simple excitations to the channel environment may enable having an analyte in a highly sensitive interaction (and detection) regime vis-à-vis the channel detector.

3.12 Comparison with Established Methods for Single Molecule Characterization

Angstrom precision structures for numerous DNA, RNA, and protein molecules have been revealed by X-ray diffraction analysis and NMR spectroscopy. These approaches rely upon average properties of very large numbers of molecules and are often biased towards crystallization and NMR conformer structures different from those present in solution under any conditions, physiological conditions in particular. With the introduction of atomic force microscopy and laser tweezers in the early 1990's three direct measures of the force have been performed at the single molecule level: (1) the force required to break A•T or G•C base pairs [112,176,177], (2) the force required to extend single or double stranded DNA through distinct structural conformations, e.g., B form to S form DNA [113,178], etc., and (3) the forces exerted by polymerases working on polynucleotides [179]. Single molecule analytical techniques, however, have yet to offer a means to directly observe *single molecule binding histories* for molecules in solution. Brief snapshots of binding events, often with significant time-averaging of events, are possible with molecular beacon approaches, but these approaches come nowhere near matching the potential of a nanopore detector to observe single molecules for extensive periods, unmodified by chromophore attachment, etc. The nanopore detector also presents the possibility of observing conformational change *within* a molecule.

Chapter 4

Channel Current Cheminformatics

This chapter describes the signal processing pipeline for managing the datastream off of a nanopore detector. Some review of signal analysis concepts is done in Sec. 4.1-4.4. In Sec. 4.5 the signal acquisition is describe; in Sec. 4.6-4.8 the feature extraction; in Sec. 4.9 the main method for classification and clustering on the data is described (the SVM). In Sec. 4.10-4.11 the signal processing is described at a higher, system, level, to describe how a scalable, real-time, signal prcessing pipeline is possible.

4.1 Power Spectra and Standard EE Signal Analysis
Typical power spectra for captured nine-base-pair DNA hairpins are shown in Fig. 4.1, along with a spectrum for the open channel. Below 10 kHz, the current fluctuation caused by the captured DNA molecule (i.e. the blockade noise) is greater than all other noise sources. Such blockade noise typically arises from changes in DNA conformation (molecular structure), changes in DNA configuration (molecular orientation, including waters of hydration), and changes in chemical bonds (internally or with surrounding channel). The

power spectra for all the signals examined in [45] had approximately Lorentzian profiles, indicative of a predominately two-state switching process (seen as random telegraph noise). Discriminating between the DNA hairpins on the basis of their power spectral (or other Fourier transform properties, or wavelet properties) is possible for small sets of hairpins. For larger sets of hairpins, or for very similar hairpins like here, the HMM-based feature extraction proved critical, due to their strengths at extracting features from aperiodic (stochastic) sequential data. HMMs can be used for classification as well as feature extraction. Here, HMMs are used for feature extraction in conjunction with a fast, highly accurate, pattern recognition method, known as a Support Vector Machine (SVM). The resulting signal processing and pattern recognition architecture enabled real-time single molecule classification on blockade samplings of only 100 milliseconds [45]. With modern computational methods and hardware it is possible to extract the resolving power of the nanopore instrument for real-time classification and handling.

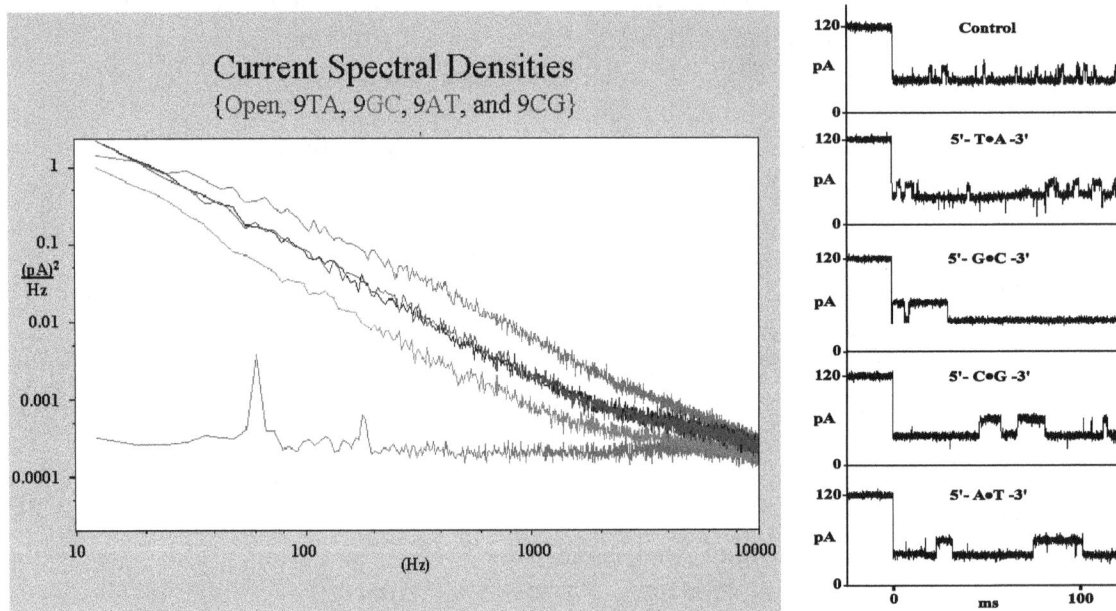

Figure 4.1 Left. Typical power spectra for captured nine-base-pair DNA hairpins and the open channel. **Right.** Typical blockade signatures for each of the five classes of DNA hairpins. The nine base-pair hairpins differ in only their terminal base pairs.

4.2 Channel current cheminformatics for single-biomolecule/mixture identifications
Using the testing protocol (described in what follows), we were able to determine which of five species of DNA hairpin had been added to the *cis* chamber of the nanopore device. This was achieved in less than six seconds with 99.6% accuracy. The five species of DNA hairpins consisted of a control hairpin and four hairpins that differed only in their terminal base-pairs (Fig. 4.1 Right). The variants were chosen to include the two possible Watson-Crick base pairs and the two possible orientations of those base pairs at the duplex ends. The core 8bp stem and 4 dT loop were identical with the primary sequence 5'-TTCGAACGTTTTCGTTCGAA-3', where the base-paired compliments are underlined. The eight base-pair hairpin that was used as a control had the primary sequence 5'-GTCGAACGTTTTCGTTCGAC-3'. *These results were for test data drawn from nanopores established on days other than those used to generate the training data (shown on next page).*

44

Figure 4.2 shows the scoring for multiple observation days, with the number of single molecule sampling/classifications ranging from 1 to 30. At 75% weak signal rejection, approximately 15 classification attempts were needed to classify the type of single-species solution being sampled; final solution classification was obtained in six seconds on average. If training and testing were done on data drawn from the same set of days of nanopore operation, albeit different samples, 99.9% calling was obtained with 15% rejection, and throughput was about one call every half second.

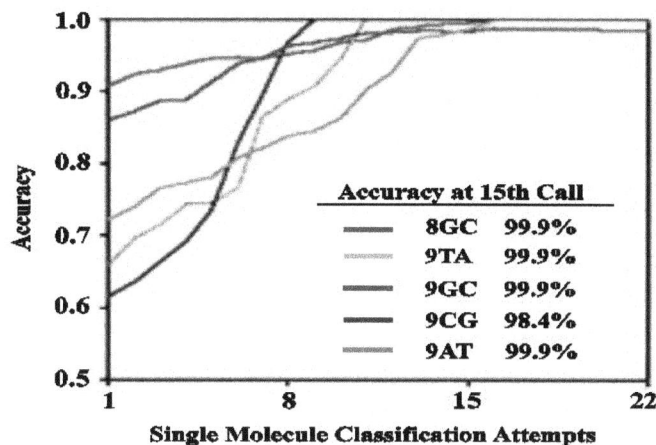

Figure 4.2 Accuracy for classification of single-species solutions of 9TA, 9GC, 9CG, 9AT, and 8GC. By the 15th classification attempt single-species solutions can be identified with high accuracy (inset).

Identification of two hairpins in mixtures was also attempted. Figure 4.3 shows the percentage of 9TA classification in a 3:1 mixture of 9TA to 9GC. (Although the mixture preparations are estimated to be ±10% of their stated mixture ratios, calibration and testing of aliquots from the same mixture compensates for such common error.) The assay on 9TA concentration asymptotes to 75% ± 1%, consistent with the 3:1 ratio, and the assay error drops to 1% after approximately 100 individual molecule classification attempts (completed in 40 seconds).

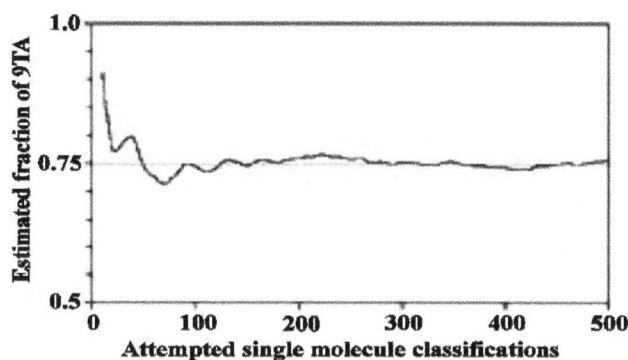

Figure 4.3 Classification on a 3:1 mixture of 9TA and 9GC hairpin molecules as a function of single molecule acquisitions. The 3:1 mole ratio is accurately identified within 1% error after 100 observations (about 40 seconds).

HMM/EM characterization on the five classes of hairpin signatures revealed the existence of two major conductance blockade levels, one minor level intermediate between them, and one to three other statistically relevant levels depending on the hairpin. By examining the transition probabilities between the various levels it was found that blockades typically began in the less common intermediate level and from there almost always transitioned to the greater conductance blockade level. (See Ch. 7 for further details.)

With completion of FSA preprocessing, an HMM is used to remove noise from the acquired signals, and to extract features from them. The HMM configuration used for control probe validation is implemented with fifty states that correspond to current blockades in 1% increments ranging from 20% residual current to 69% residual current [1]. In this HMM application the HMM states, numbered 0 to 49, corresponded to the 50 different current blockade levels in the sequences that are processed. The standard "grayscale" HMM, or 'generic HMM', feature extraction setup is then done: the state emission parameters of the HMM are initially set so that the state j, $0 <= j <= 49$ corresponding to level $L = j+20$, can emit all possible levels, with the probability distribution over emitted levels set to a discretized Gaussian with mean L and unit variance. All transitions between states are possible, and initially are equally likely. Each blockade signature is de-noised by 5 rounds of Expectation- Maximization (EM) training on the parameters of the HMM. After the EM iterations, 150 parameters are extracted from the HMM. The 150 feature vectors obtained from the 50-state HMM-EM/Viterbi implementation are: the 50 dwell percentage in the different blockade levels (from the Viterbi trace-back states), the 50 variances of the emission probability distributions associated with the different states, and the 50 merged transition probabilities from the primary and secondary blockade occupation levels (fits to two-state dominant modulatory blockade signals). Variations on the HMM 50 state implementation are made as necessary to encompass the signal classes under study.

The 150-component feature vector extracted for each blockade signal is then classified using a trained Support Vector Machine (SVM). The SVM training is done off-line using data acquired with only one type of molecule present for the training data (bag learning). Further details on the SVM and overall channel current cheminformatics signal processing are detailed in [1].

4.3 Channel Current Cheminformatics: feature extraction by HMM
The HMM-based profiling used for feature extraction provided better discrimination than the wavelet-based profiling used in previous efforts [48]. The improved signal resolution on channel blockades with HMMs is not new [180]. (The wavelet-domain FSA that generates the blockade-level profiling does have the advantage, however, of being hundreds of times faster than the HMM processing in this instance.) The better performance with HMM processing indicated that signal analysis benefited from parsing structural information in the stochastic sequence of blockade-states. Parsing structures in stochastic data is a familiar problem in gene prediction, where Hidden Markov Models (HMMs) have been used to great advantage [181,182]. Typically with gene prediction, however, HMMs are operated at a high level that parses coding starts and stops, etc., with feature scoring on starts and stops performed at a lower level by neural net or related statistical methods. For channel current analysis, the HMM extracts structural features without identifying them, effectively operating at the lower level, and used with EM [183], accomplishing de-noising on the blockade-state structure [180] prior to extracting those features.

A single HMM/EM process was used to perform the feature extraction in the experiments that follow. If separate HMMs were used to model each species, the HMM/EM processing could also be operated in a discriminative mode. This requires multiple HMM/EM evaluations (one for each species) on each unknown signal as it is observed. Increased computational burden would thus be added at the worst place: the expensive feature extraction stage. Ssemi-scalable, species-specific processing could be considered for the HMM/EM in an indirect manner, by using prior HMM/EM characterization of the species to identify a reduced set of features relevant to each species. The reduced feature set relates to physical characterizations of the captured molecule, such as level states, their time-constants, and allowed level transitions.

Samples using blockade signatures of longer duration (prior to truncation) require fewer rejections to achieve the same signal classification accuracy. A situation that would probably favor longer signal samples than the 100 ms used here was seen in attempts to read more of the DNA hairpin end-sequence than the terminal base pair. Preliminary indications are that the penultimate base pairs can probably also be identified using longer signal samples (17 species with control). Scaling the classification task from 5 to 17 species may also require refinements to the feature extraction, such as the species specific HMM feature extractions mentioned above.

Tests with mixtures of hairpins required an added calibration due to the nanopore's different acceptance rates for different hairpins (i.e., there are different free energy barriers to capture). This finding was consistent with a model for hairpin capture (see below) in which hairpins are captured by an entropically accessible binding site. It is also in agreement with the brief intermediate level state typically observed at the start of the signal blockades.

4.4 Bandwidth limitations
Nanopore-based detection is limited by the kinetic time-scale of the molecular blockade states, where the molecular blockade states typically correspond to binding and dissociation (analyte-channel binding, or antibody-antigen binding, for example), or due to internal conformational flexing. It is hypothesized that it is possible to probe higher frequency realms than those directly accessible at the operational bandwidth of the channel current based device, or due to the time-scale of the particular analyte interaction kinetics, by modulated excitations. This can be accomplished by chemically linking the analyte or channel to an "excitable object", such as a magnetic bead, excited by laser pulsations, for example. In one configuration, the excitable object can be chemically linked to the analyte molecule to modulate its blockade current by modulating the molecule during its blockade. In another configuration, the excitable object is chemically linked to the channel, to provide a means to modulate the passage of ions through that channel. Studies involving the first, analyte modulated, configuration, indicate that this approach can be successfully employed to keep the end of a long strand of duplex DNA from permanently residing in a single blockade state (see next Sec.). Similar study of magnetic beads linked to antigen may be used in the nanopore/antibody experiments if similar single blockade level, "stuck", states occur with the captured antibody (at physiological conditions, for example).

Examples of excitable objects include microscopic beads (magnetic and non-magnetic), fluorescent dyes, charged molecules, etc. Bead attachments can couple in excitations passively from background thermal (Brownian) motions, or actively by laser pulsing and laser-tweezer manipulation. Dye attachments can couple excitations via laser or light (UV) excitations to the targeted dye molecule. Large, classical, objects, such as microscopic beads, provide a method to couple periodic modulations into the single-molecule system. The direct coupling of such modulations, at the channel itself, avoids the low Reynolds number limitations of the nanometer-scale flow environment. For rigid coupling on short biopolymers, the overall rigidity of the system also circumvents limitations due to the low Reynolds number flow environment. Similar consideration also come into play for the dye attachments, except now the excitable object is typically small, in the sense that it is usually the size of a single (dye) molecule attachment. Excitable objects such as dyes must contend with quantum statistical effects (at the single-molecule level), so their application may require time averaging or ensemble averaging (where the ensemble case might involve multiple channels observed simultaneously). In both of the experimental configurations, a multi-channel platform may be used to obtain rapid ensemble information. In all cases the modulatory injection of excitations may be in the form of a stochastic source (such as thermal background noise), a directed periodic source (laser pulsing, piezoelectric vibrational modulation, etc.), or a chirp (single laser pulse or sound impulse, etc.). If the modulatory injection coincides with a high frequency resonant state of the system, informative low frequency excitations may result, i.e., excitations that can be monitored in the usable bandwidth of the channel detector. Increasing the effective bandwidth of the nanopore device greatly enhances its utility in almost every application, particularly those, such as DNA sequencing, where the speed with which blockade classifications can be made is directly limited by bandwidth restrictions.

4.5 Signal acquisition: the time-domain Finite State Automaton (tFSA)
In [184], a time-domain finite state automaton with eight states is used for signal identification and acquisition (shown below, Fig. 4.4, based on the first 100 ms of channel current blockade signal in Fig. 4.1 Right). Two states, sequentially connected, were used for reset and initialization on the FSA. Transition between the two states, from reset-start to reset-ready, was accomplished upon measuring a short section of acceptable baseline current (200 μs). An abrupt drop in current to 70% residual current (determined by the holistic tuning that is described in what follows), or less, then triggered transition from the reset-ready state to the signal-active state. From the signal-active state, processing advanced to one of two states (good- and bad-end-level states) according to an end-of-signal profile. The profile rule simply required that the last end-level-range observations had to have current above minimum-end-level-value. Satisfying the rule led to the good-end-level state, otherwise the bad-end-level state was reached. If there was a normal return to baseline (good-end-level state), or a signal-blockade scan exited due to truncation (bad-end-level state), the signal complete state was reached, otherwise further scanning was performed. Further scanning involved transition through the internal active state, where local signal properties, observation less than maximum-cutoff and observation greater than minimum-cutoff, were used to decide whether to exit (to the reset-end state) or continue the blockade scan (return to the signal-active state). Similar to the local blockade signal properties that determined how to transition from the internal-active state, transition to the acquire-signal state from the signal-complete state was based on several global properties of the signal trace: maximum blockade sample less than maximum-cutoff and greater than min-max-internal, minimum blockade sample greater than minimum-cutoff and less than max-min-internal, and signal duration greater than or equal to minimum-duration.

48

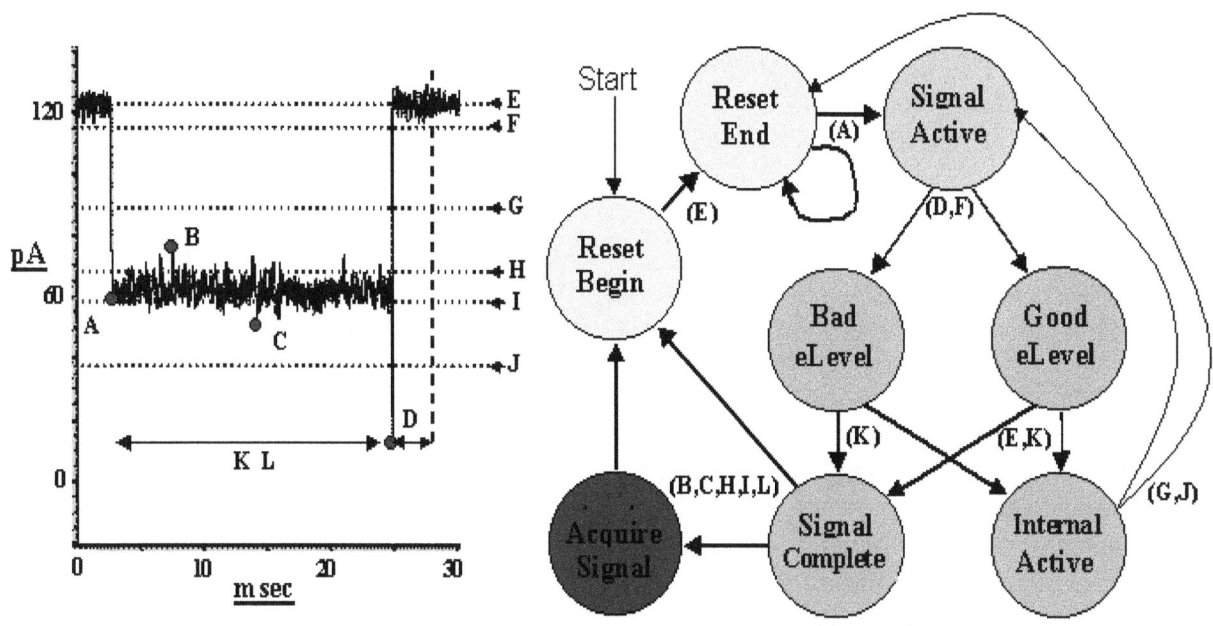

Fig. 4.4. Schematic for the finite state automaton used for acquisition of 6 base-pair DNA hairpin blockade signals observed in [184], where a sample signal is shown to the left. The letters label various types of feature extraction parameters and their placement in the FSA diagram indicate where the decision-making or thresholding is dependent on those parameters.

The FSA shown in Fig. 4.4 [1] was eventually tuned to operate such that it would rarely miss signal acquisitions (low false negatives) by allowing for large numbers of mistaken signal acquisitions (i.e., large false positives), followed by filtering to achieve high specificity (as described in Fig. 4.5). The acquisition bias was accomplished by imposing constraints on valid starts that were weak while maintaining constraints on valid interior and ends that were strong. The bias towards high sensitivity for *initiating* acquisition permitted tuning on FSA parameters with a simplified objective (part of the benefit of a multi-pass bootstrap tuning process). For the blockade signatures studied, the FSA parameters for maximal signal acquisition shared a broad, common range, allowing one set of FSA parameters (a single generic FSA) to acquire all signals.

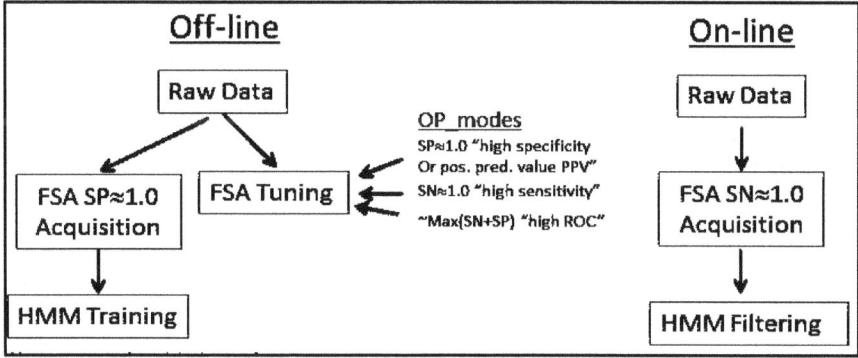

Fig. 4.5. FSA with alternating SP:SN optimized tuning. Step 1: Acquire signals with high specifity (SP=1, SN = whatever), obtain a 'gold standard' reference set (if you have an expert, have them provide as much of this as they

can manually). Step 2: Extract feature information from the gold standard set, to know what "it" looks like. (HMMs will often be used for this in what follows.) Step 3: Do acquisition with high sensitivity, followed by a specificity filter learned at Step 2. In other words, have (SN=1, SP= whatever) → (filter boosts to SP=1 with minimal drop in SN).

The FSA described in Fig. 4.4 enables acquisition of localizable channel current signals using 'holistic' tuning and 'emergent grammar' tuning. (Emergent grammar tuning, and use of wavelets, is described in [184].) When attempting to tune the FSA it can be viewed as a "holistic engine" of a multiply connected (not independent) set of variables, states, and their interactions. For acquisition we seek minimal feature identification comprising identification of signal beginnings and ends (and thus durations as well). *Holistic tuning is mainly done by testing global features for anomalous changes, or 'phase transitions'*. One of the main global features of the acquisition process is the number of acquisitions itself, made under a particular set of tuning parameters. In Fig. 4.6 is shown the result of a holistic tuning process on the start_drop_value parameter. A critical requirement for holistic tuning is having a viable initial tuning state to initialize the process, e.g., multiple parameters must be within their live 'lock range' on tuning parameters analogous to the PLL lock-range constraint [1]. The code description with the core tuning parameters highlighted is in Sec. 4.4.2.

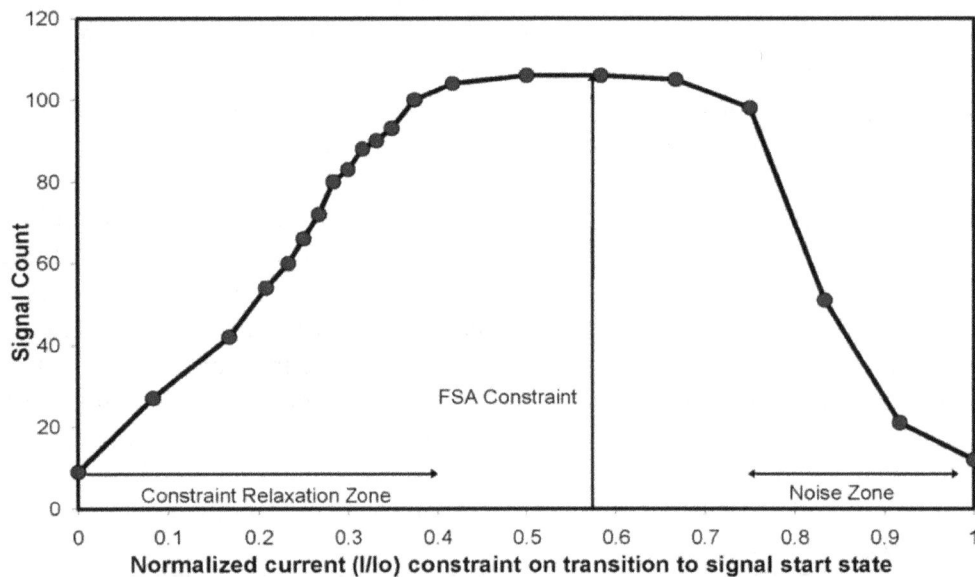

Fig. 4.6. Tuning on 'start_drop_value for a collection of blockade signals resulting from channel captures of DNA hairpins with 6 base-pair stem length. For baseline-normalized current constrained to drop to 0 channel current to trigger possible acquisition we see that very few acquisitions succeed (approximately 10 signal acquisitions shown). As we relax this start of acquisition constraint on possible signal acquisitions, we steadily see more signal counts until it plateaus starting at a baseline-normalized current of 0.4 to a baseline-normalized current of 0.7. The paradoxical seeming drop in signal acquisitions for the more hair-trigger acquisitions for baseline-normalized current drop, to only 0.8 or greater, is due to the FSA often triggering on noise, and eventually rejecting the indicated signal as invalid, but in doing so sometimes missing a valid signal start, resulting in fewer overall signal acquisitions. The holistic tuning process seeks the plateau region (that is not directly responsive to

change in cutoff over a broad range) as an indication of a robust acquisition setting, with the 0.57 value chosen in the example shown.

The O(L) time-complexity feature identification "scan" process can also be employed for simultaneous feature extraction on various statistical moments. Identification of sharply localizable 'spike' behavior can also be done in the scan process (still with only O(L) time complexity) based on a nonparametric method that is described next.

4.5.1 tFSA spike detector

A channel current spike detector algorithm can be used to characterize the brief, very strong, blockade "spike" behavior observed for duplex DNA molecular termini that occasionally fray in the region exposed to the limiting aperture's strong electrophoretic force region. (See [1] for details, where nine base-pair hairpins were studied, the spike events were attributed to a fray/extension event on the terminal base-pair.) A complication with the spike feature extraction is the blockade level from which the spike event occurs is not known, or too variable to use to identify the spike blockade event. To have a robust feature extraction a test-level-crossing heuristic was used, where for a fixed blockade level the number of signal crossings at that level are counted (such as from spikes). The test level used in the crossing analysis is then shifted to higher levels, with increasing crossing counts as the level passes thru the signal region. What results is linear increase in crossing count for actual spike features as the test level used in the crossing analysis is increased, until the main signal region is reached. In the case of the channel current analysis the various levels of blockade seen for a particular molecular blockade typically have Gaussian noise about the average of each level. Thus, as the line-crossing sweeps thru the signal blockade level and probes the tail of the Gaussian noise distribution about that signal blockade level an exponential increase in level crossings is seen (see Fig. 4.7). Focusing on the linearly increasing count region, and extrapolating to the counts up to the average of the signal blockade level from which the spike deflections are seen, a count on spike events (or a frequency on spike events) can then be robustly ascertained.

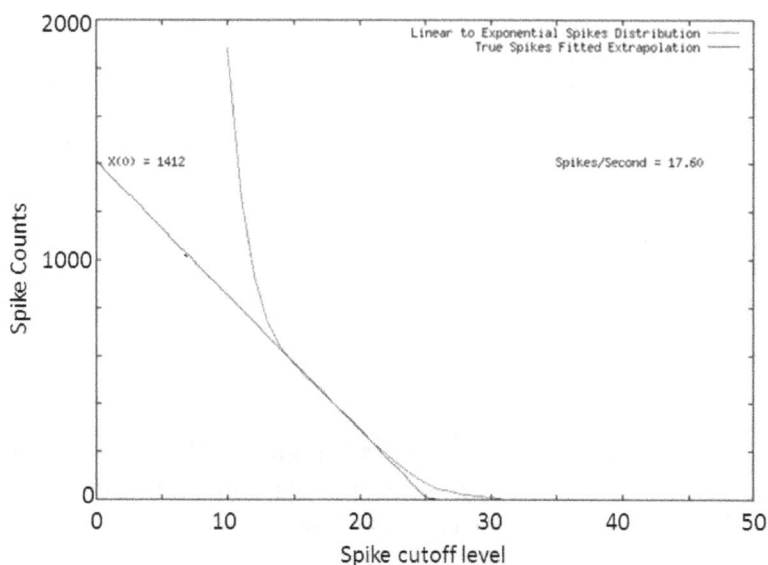

Fig. 4.7. Robust Spike feature extraction: radiated DNA. A time-domain FSA is used to extract fast time-domain features, such as "spike" blockade events. Automatically generated "spike" profiles are created in this process. One such plot is shown here for a radiated 9 base-pair hairpin, with a fraying

rate indicated by the spike events per second (from the lower level sub-blockade). Results: the radiated molecule has more "spikes" which are associated with more frequent "fraying" of the hairpin terminus--the radiated molecules were observed with 17.6 spike events per second resident in the lower sub-level blockade.

The spike detector software is designed to count "anomalous" spikes, i.e., spike noise not attributable to the Gaussian fluctuations about the mean of the dominant blockade-level. The extrapolations provide an estimate of "true" anomalous spike counts. Together, the formulation of HMM-EM, FSAs and Spike Detector provide a robust method for analysis of channel current data [1]. In Fig. 4.7 the plot is automatically generated for spike characteristics for blockade data for DNA hairpins examined: one with cross-linking radiation damage and one without damage. The plots are also automatically fit with extrapolations of their linear phases. By this method, the non-radiated DNA exhibited a full-blockade "spike" from its lower-level blockade with a frequency of 3.58 spikes per second (indicating a fraying of the blunt ended terminus of the molecule at that rate). For the radiated molecule the frequency of spikes was 17.6 spikes per second, indicating a much greater fraying rate (and associated dissociation of the terminal base-pair), consistent with that molecule being weakened by radiation such that its terminal base-pair frays more frequently.

The additional "spike" frequency feature is found to improve classification accuracy between two species of DNA hairpins by approximately 5% in the hairpin discrimination SVM tuning that is scored for various kernel parameters in Fig. 4.8. This is an example of how non band limited signal features can be extracted without the limitations of a HMM state quantization pre-processing (or Fourier transform method feature extraction from electrical engineering signal processing) to arrive at a more informed process than seems possible given the usual constraint of the Gabor limit, as mentioned previously.

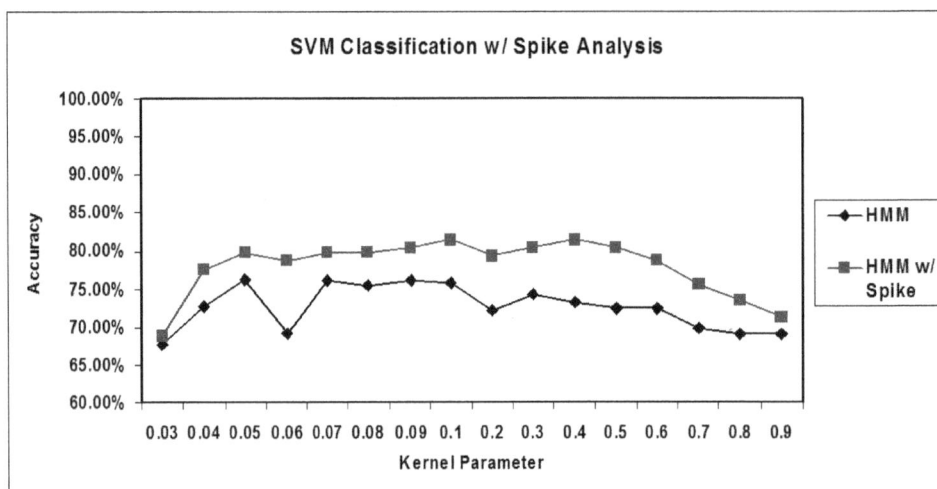

Figure 4.8. SVM classification results with and without spike analysis. Adding a spike feature significantly improves classification accuracy, by approximately 5%, over a wide range of kernel parameters.

Once the lifetimes of the various levels are obtained, information about a variety of other kinetic properties is accessible. If the experiment is repeated over a range of temperatures, a full set of kinetic data is obtained (including the aforementioned "spike" feature frequency analysis). This data may be used to calculate k_{on} and k_{off} rates for binding events, as well as indirectly calculate forces by means of the van't Hoff Arrhenius equation.

4.5.2 tFSA-based channel signal acquisition methods with stable baseline

The tFSA program (in C) begins with State="Reset Begin" (see Fig. 4.4 and comments in code below, Fig. 4.9) with a loop to self a minimum of 10 times on the sample data being scanned, where the data in [1,184] was sampled at 20μs, thus minimum time to advance from the "Reset Begin" state is 0.2 ms in the application shown (via baseline_to_reset=10 in the code shown below). In order to only do loop 10 times, the observed blockade value must exceed the open_channel_avg value on each sample of the 10 observations. Until 10 such observations exceeding the open_channel_avg value are tallied, whether consecutively or not, the loop will not advance from "Reset Begin" to "Reset End". The baseline_to_reset parameter is reset to 10 after each possible signal acquisition is resolved (with acquisition or rejection). The value of 10 is itself chosen by a 'holistic' tuning process.

Once at the State "Reset End", the blockade sample values are checked (shown as self-loop in Fig. 1) to see if they've dropped significantly from the reset condition (e.g., dropped below baseline). This will be the first of a series of instances where weak conditions are used on initiating signal acquisitions, while much stricter conditions must be satisfied later (when better informed about the signal) in order to complete, and fully acquire, the signal. A blockade sample observation is deemed to have "dropped significantly" from its reset condition if it drops below a cutoff named the "start_drop_value" in Fig. 1 and the code below, to arrive at the next state, "signal active", for initiating signal acquisition. Sometimes a blockade sample drops right through the floor, however, to large negative values, etc., due to noise or a shock, etc. These falsely triggered signals are excluded by excluding start drops that go below the "start_drop_limit" value. (Again, all parameters are tuned.) Once at the State "Signal Active", each subsequent blockade sample is read into an array, for possible signal acquisition and recording, and for use by O(1) data analysis algorithms (keeping the overall FSA operation O(L) on L observation samples). Such algorithms are used to calculate simple statistical properties of the blockade region (in an O(1) process), such as the maximum, minimum, duration, and (running) average of the blockade signal, and the (running) standard deviation of the blockade signal. The notion of a 'running' statistical evaluation is that the initialization of the statistical parameter may be O(N), for N length observation in the signal scan window, but that as the windowing on data used in the scan operation is slid along the data observation sequence, further updates on that sliding-window statistical parameter is only O(1). This sliding-window, or 'running', evaluation, then allows higher order statistical moments to be computed at O(L) on the full observation sequence under study (code implementations for this will be shown in detail for the first few statistical moments in what follows).

```
while (index<length) {
    // data is read from (binary) datafile, or taken from streaming (buffered)
    // data from a live experiment, and placed in the variable 'rescale'
    /* Now at State="Reset Begin" */
    if (baseline_to_reset>0) {
        if (rescale>open_channel_avg) { baseline_to_reset--; }
        index++; continue;
    }
    /* Now at State="Reset End".*/
    if (start_active<1 && rescale<start_drop_value && rescale>start_drop_limit) {
        signal_start[j] = index;
        signal_max[j] = rescale;
        signal_min[j] = rescale;
        sigindex=0;
        sigdata[sigindex] = rescale;
        start_active = 1;
        get_base_lead = 0;
        index++; continue;
    }
    if (start_active<1) { index++; continue; }
    else {
        /* Now at State="Signal Active". */
        sigindex++;
        sigdata[sigindex] = rescale;
        bad_end_level=0;
        for (i=0;i<end_level_range;i++) {
            if (data[i]<end_level_value) { bad_end_level=1; i=end_level_range; }
        }
        signal_end[j] = index-1-end_level_range;
        signal_length = signal_end[j] - signal_start[j] + 1;
        if (signal_length>max_length) {
            signal_end[j] += 1+end_level_range;
            signal_length += 1+end_level_range;
        }
        if ((bad_end_level<1&&rescale>open_channel_avg)||(signal_length>max_length)){
            /* Exit condition is obtained. */
            if (signal_length>min_length && signal_min[j]<max_min_internal) {
                /* Now at State="Acquire Signal". */
                t = sigindex-end_level_range;
                do_simple_profile(sigfile,signal_start,signal_end,signal_max,signal_min,t,sigdata,j);
                printf("signal %d processing complete\n",j);
                sigindex=0; //resets signal info
                j++;
            }
            baseline_to_reset=10;
            start_active=0; //resets
            /* Now Reset to State="Reset Begin". */
            get_base_lead = 1;
            index++; continue;
        }
        /* Now at State="Bad eLevel". */
        else if (((index-signal_start[j]>end_level_range) &&
                (data[end_level_range-1]>max_internal)) || rescale<min_internal) {
            start_active=0; //resets
            // Now at State="Reset End". Note, not a full reset to
            // State="Reset Start" since baseline_to_reset not reset to 10.
            get_base_lead = 1;
            index++; continue;
        }
        // Now fall-through to State="Signal Active", for another blockade
        // sample iteration, after some min and max evaluations and sweep
        // boundary avoidance.
        else if (mod_index>0 && ((index%mod_index<mod_index_range) ||
                        (index%mod_index>mod_index-mod_index_range))) {
            start_active=0; //resets
            get_base_lead = 1;
            index++; continue;
        }
        else {
            if ((index-signal_start[j]>end_level_range) && (data[end_level_range-1]>signal_max[j])) {
                signal_max[j]=data[end_level_range-1];
            }
            if (rescale<signal_min[j]) { signal_min[j]=rescale; }
            index++; continue;
        }
    }
    index++;
}
```

Fig. 4.9. The main while loop for signal scanning for the FSA diagram shown in Fig. 4.4.

As a preliminary step for each new sample acquisition, to minimize bad-acquisition blocking on good signal that might immediately follow, exit conditions are tested for channel blockade completion (i.e., a return to the baseline, open channel, current readings). In Fig. 4.4, the exit condition is obtained either by a return to baseline (bad_end_level=0, or State="Good eLevel"), or due to acquisition truncation (case with State="Bad eLevel", due to truncation, even though good for acquisition). Once an exit condition is reached without rejection, a proper signal acquisition has occurred (with data already loaded into the acquisition array), and we now arrive at the State "Signal Complete". A collection of signal conditions are then tested on the total signal data for the final acceptance or rejection decision.

In Fig. 4.4 we had onset of possible signal acquisition triggered by a significant drop in the blockade level average (evaluated O(windowsize) on data, so 'real-time' with minimal memory buffering needs). We could also trigger on change of blockade level standard deviation in that same window evaluation (still just O(windowsize) evaluation). A diagram showing the latter, and related methods, for acquisition of unstable signals, is given in Fig. 4.10:

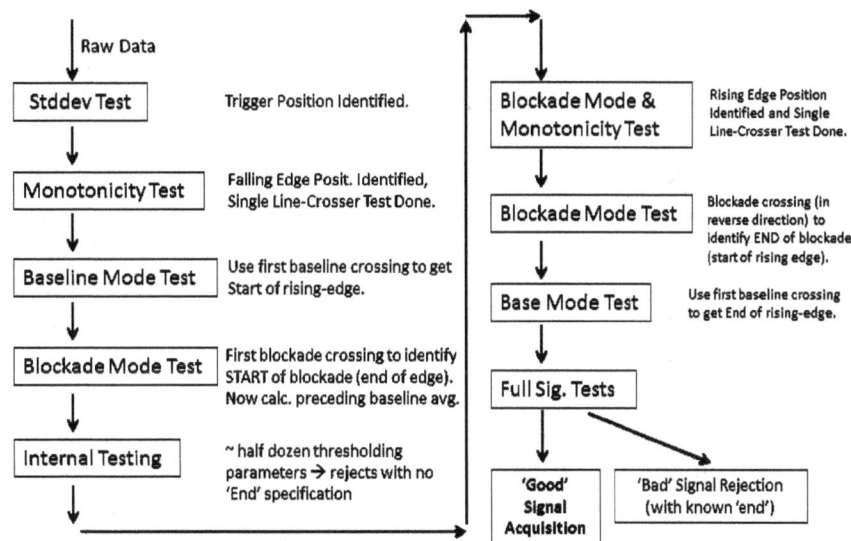

Fig. 4.10. FSA acquisition flowchart.

4.5.3 tFSA-based channel signal acquisition methods without stable baseline
In channel current blockade analysis, and electrical signal analysis in general, the tFSA signal acquisition is much more difficult if the reference baseline is not stable. For electrical signals one type of unstable baseline that typically results is from capacitive discharge/charge giving rise to an exponential rise (or fall) in the baseline current when event sampling is necessary before system relaxation can occur (to steady baseline). If the exponential rise/fall in the channel current signal was the same this could simply be factored out, but typically the charge/discharge reset is incomplete, and the capacitive properties themselves variable under load, giving rise to (effectively) different exponential rise/fall baseline references with every device reset. Even this extreme case can be handled with a properly designed tFSA (one making use of filters analogous to boxcar filters from electrical engineering, for example). A tFSA signal acquisition for stable baseline scenarios, but still with very challenging noisy data, is described first, followed by the enhancements needed for handling unstable baseline.

Using the start_drop_value parameter introduced already one seeks to identify onset of blockade events by their deviation, typically expressed by some multiple of standard deviations, from the baseline mean (see Fig. 4.10 above). A 'three-sigma' rule is often used, i.e., event acquisition onset is triggered when a channel current reduction by more than three standard deviations of baseline noise, from the baseline mean, is observed. The falling edge of the blockade onset can then be precisely fixed (down to a specific sample observation in many cases) by performing a monotonicity test on the falling edge, which can be done with an O(L) line-crosser analysis like that used in the spike analysis [1]. Fixing the start of blockade then depends on the data processing conventions adopted (often one chooses whatever convention yields the best classification/clustering at later SVM processing stages, if in use). Whatever the convention, the core, 'stable', information that guides the blockade level identification is a modal analysis on the different blockade levels seen. What is revealed by this is a mode identifying the baseline level (at least in the region just prior to the triggered signal acquisition) and the blockade level. If there is stable baseline, or even stable capacitive discharge baseline (i.e., a single exponential profile occurring with each sampling reset), then the modal analysis will directly reveal that baseline, or identification of the asymptotic, 'relaxed', exponential baseline level. Once a suspected blockade signal onset has occurred, passing the standard deviation, monotonicity, and mode tests, signal acquisition commences on further streaming signal, with ongoing 'running' O(L) measurement of statistical moments and max, min sample values, with rejection if the 'internal' signal statistics does not fall within the desired range of possibilities. Identification of end-of-blockade, i.e., identification of the rising edge back to baseline current with no blockade, is then much the same as the falling edge identification, with use of standard deviation, monotonicity and modal tests. Typically identification of return to baseline requires additional measurement of a minimum number of baseline samples after return to baseline to avoid premature truncation on signal acquisitions that are sufficiently noisy that their internal blockade fluctuations occasionally 'spike' back to baseline level.

Sample code for a modal scan is given in Fig. 4.11, where use is also made of integer-based variables for speedup on multiplications with integer variables instead of floating point variables. Working with integer-valued data is not as lossy as it might seem at first since the data encoding schemes used by third-party DAQs and amplifier developers, and by their efficient binary datafile encodings, are themselves integer-based with shift and float multiplication operations to bring the data back to the floating value and precision originally observed, with the resolution (number of significant figures) used in the recording or quantization process. Note: the hardest part of this process is often 'cracking' the binary file to learn the data encoding scheme and the shift and float-multiplication conversion values needed to recover the original data values direct from the binary data. This is typically not intended to be proprietary information by the third part equipment developers (strictly not propriety in some cases) but accessing the company technicians to learn the encoding specifications is often extremely difficult and time-consuming. So just figuring it out the binary encoding scheme using a 'known ciphertext' attack is often the quickest way to proceed.

```
sub Scan_Data_Modes { # used to do binary data read and identify stat
modes
    my ($self,$input_file,$type,$mode) = @_;
    if (!$input_file) { $input_file = "data_file"; }
    $self->{input_file} = $input_file;
    # generate conversion lookup array for speedup over float mult
    my $maxrawint; # must be set, depends on file-formatting
    my $rawint;
    my @conversion_array;
    my $factor; # must be set, depends on file-formatting
    $self->{factor}=$factor;
    for $rawint (-$maxrawint..$maxrawint) {
        $conversion_array[$maxrawint+$rawint]=$rawint*$factor;
    }
    # begin binary read and mode count
    # details and some initializations deleted
    my $Data_fh = new FileHandle "$input_file";
    binmode($Data_fh);
    while (read($Data_fh,$buffer,4)) {
        $index++;
        my ($value,$vvalue) = unpack 'ss', $buffer;
        $data_index = $index-$header_skip;
        $data[$data_index]=$value;
        my $binned_sample_value = $data[$data_index];
        $mode_count{$binned_sample_value}++;
    }
    close($Data_fh);
    # 'housekeeping' code that is omitted
}
```

Fig. 4.11. Code to scan channel current data and do a blockade modal analysis.

If the channel current blockade signals have an unstable baseline, then the window-based FSA shown in Fig. 4.12, and elaborations on it, can be critical to locking onto the signal, but may not be optimal at calling the edges, thus multiple passes may be needed (bootstrap acquisition), with early passes involving the window-based tFSA to get a preliminary lock so that the unstable base-line moving average can be subtracted, shifting to a simpler acquisition where a sample-based tFSA can then take over for the final signal acquisition with the most precise edge recognition possible.

```perl
while ($sample_index<$total_sample_count) {
    $sample_index++;
    my $sample_value = $data[$sample_index];
    if (!$reset_start) {
        if ($sample_value>$reset_value) { next; }
        else { $reset_start=1; next; }
    }
    elsif (!$reset_end) {
        if ($sample_value<$reset_value) { next; }
        else {
            $reset_end=1;
            $active{start}[$sweep_count]=$sample_index;
            print "active signal region starting at $sample_index... ";
            next;
        }
    }
    # do look-ahead for end of active signal region
    my $lookahead_sample_index = $sample_index+10;
    my $lookahead_sample_value = $data[$lookahead_sample_index];
    if ($lookahead_sample_index>$total_sample_count-1) {
        $lookahead_sample_value = $signal_end_cutoff+1;
    }
    if ($lookahead_sample_value>$signal_end_cutoff) {
        $active{end}[$sweep_count]=$sample_index;
        print "and ending at $sample_index\n";
        $reset_start=0;
        $reset_end=0;
        $sweep_count++;
        $lowpass_initialized=0;
        next;
    }
    # only here if ready to process active signal region
    ###
    # identify rising edge from baseline
    if ($lowpassdiff_value>$rising_diff_cutoff) {
        if ($sample_index-$prior_cutoff_index>20) {
            $prior_cutoff_index=$sample_index;
            my $dref = $self->{data_ref};
            my $std_dev = $Get_Std_Dev->($dref,$sample_index,-30,$window_size,$factor);
            if ($std_dev>$std_dev_cutoff) { $fail_count++; }
            else {
                my $edge_start=$sample_index;
                $signal{start}[$j]=$sample_index-5;
                $signal{start_baseline}[$j]=$lowpass_baseref;
                $start_flag=1;
                $running_signal_length=0;
                $running_signal_max=100;
                $running_signal_min=400;
                $running_signal_sum=0;
            }
        }
        else { $prior_cutoff_index=$sample_index; }
    }
    # identify falling edge to baseline
    if ($lowpassdiff_value<-$falling_diff_cutoff) {
        my $lowpass_baseref =
        $Get_Lowpass->($self->{data_ref},$sample_index,$pref,$window_size,$factor);
        if ($lowpass_baseref<($factor*$self->{highest_count_asympt_mode}+100) ) {
            my $dref = $self->{data_ref};
            my $std_dev2 = $Get_Std_Dev->($dref,$sample_index,50,$window_size,$factor);
        }
    }
    # perform running signal evaluation
    if ($start_flag==1) {
        $running_signal_length = $sample_index-$signal{start}[$j]-5;
        if ($running_signal_length>0) {
            my $rescaled_data = $sample_value*$factor-$signal{start_baseline}[$j];
            if ($rescaled_data<$running_signal_min) { $running_signal_min=$rescaled_data; }
            if ($rescaled_data>$running_signal_max) { $running_signal_max=$rescaled_data; }
            $running_signal_sum+=$rescaled_data;
            $running_signal_avg=$running_signal_sum/$running_signal_length;
        }
    }
}
```

Fig. 4.12. While loop for window-based tFSA signal acquisition.

4.6 Signal statistics (Fast): Mean, Variance, and Boxcar filter

Sometimes the more sophisticated window-based tFSA methods can be avoided entirely by use of a boxcar filter (a form of lowpass notch filter) as a preprocessing stage, which is shown in Fig. 4.13. In the worst case scenario, all of these methods need to be used, with the boxcar filter used in a post-processing validation method (as well as throwing the kitchen sink at the problem) in order to get the signal acquisition to work. The process that might be undertaken on a challenging signal acquisition, thus, might go as follows:

(1) scan for asymptotic baseline statistics

(2) do a preliminary window-based FSA scan to get a handle on the baseline

(3) estimate the baseline signal

(4) perform a sample-based tFSA scan on the baseline-subtracted signal

(5) perform repeated tFSA scans (since fast) with different biases to lock onto all signal regions

(6) perform boxcar filter on raw signal in indicated signal regions with identified baseline attributes used to determine the optimal boxcar filter.

(7) perform merge on signal acquisitions indicated at steps (5) and (6)

```perl
sub Boxcar_Filter {
    my ($self,$notch_window,$buffer_window) = @_;
    my @data = @{$self->{data_ref}};
    my $total_sample_count=$self->{total_sample_count};
    if (!$notch_window) { $notch_window=1000; }
    if (!$buffer_window) { $buffer_window=100; }
    # running calc to obtain window array for notch_filter_data calc
    # not valid if sample_index<notch_window_size+1
    # this is a running stat calc
    my @notch_filter_data;
    my @lowpass_filter_data;
    my @notch_window_array;
    my $sample_index;
    my $notch_window_sum;
    my $notch_window_avg;
    for $sample_index (0..$total_sample_count-1) {
        my $sample_value = $data[$sample_index];
        if ($sample_index < $notch_window) {
            my $pop = $notch_window_array[0];
            my $i;
            for $i (0..$notch_window-2) {
                $notch_window_array[$i] = $notch_window_array[$i+1];
            }
            $notch_window_array[$notch_window-1] = $sample_value;
            $notch_window_sum += ($sample_value-$pop);
            $notch_window_avg = $notch_window_sum/($sample_index+1);
            $lowpass_filter_data[$sample_index]=$notch_window_avg;
        }
        else {
            my $pop = $data[$sample_index-$notch_window];
            my $push = $data[$sample_index];
            $lowpass_filter_data[$sample_index]=
            $lowpass_filter_data[$sample_index-1]+($push-$pop)/$notch_window;
        }
    }
    for $sample_index (0..$total_sample_count-1) {
        my $sample_value = $data[$sample_index];
        if ($sample_index < $notch_window+$buffer_window) {
            $notch_filter_data[$sample_index]=$data[$sample_index]-
            $lowpass_filter_data[$sample_index+2*$notch_window+$buffer_window];
        }
        else {
            $notch_filter_data[$sample_index]=$data[$sample_index]-
            0.5*($lowpass_filter_data[$sample_index-$notch_window-$buffer_window]
            +$lowpass_filter_data[$sample_index+2*$notch_window+$buffer_window]);
        }
    }
    $self->{notch_filter_data}=\@notch_filter_data;
}
```

Fig. 4.13. Subroutine example for Boxcar filter.

Efficient implementations for statistical tools (O(L))

Working with the native integer encoded binary representation of the data is faster on multiple levels. This would not be of much benefit, however, if the subroutines for the statistical moments (mean, standard deviation, etc.) could not operate at the integer variable level for most of their evaluation. An implementation of statistical methods for evaluating the mean and standard deviation at integer-variable level is shown in Fig. 4.14.

```perl
my $Get_Lowpass = sub {
    my ($ref,$sample_index,$offset,$window_size,$factor) = @_;
    my @data;
    if ($ref) {
        @data = @{$ref}[$sample_index+$offset-$window_size+1..$sample_index+$offset];
    }
    else {
        print "error in passing raw data array\n";
    }
    my $i;
    my $window_sum=0;
    for $i (0..$window_size-1) {
        $window_sum += $data[$i];
    }
    my $mean = $factor*$window_sum/$window_size;
    return $mean;
};
my $Get_Std_Dev = sub {
    my ($ref,$sample_index,$offset,$window_size,$factor) = @_;
    my @data;
    if ($ref) {
        @data = @{$ref}[$sample_index+$offset-$window_size+1..$sample_index+$offset];
    }
    my $mean = $Get_Lowpass->($ref,$sample_index,$offset,$window_size,$factor);
    my $i;
    my $sum_squared_central_moment=0;
    my $factorlessmean = int($mean/$factor+0.5);
    for $i (0..$window_size-1) {
        my $diff = $data[$i]-$factorlessmean;
        $sum_squared_central_moment += $diff*$diff;
    }
    my $variance=$sum_squared_central_moment/$window_size;
    my $std_dev = $factor*sqrt($variance);
    return $std_dev;
};
```

Fig. 4.14. Code examples for non-lossy statistical moment evaluations that are integer-based.

The window-based implementation is more robust with variable, unstable, baseline, but is less precise at identifying falling edges and other sharp transitions. Since the window based method often involves sums over the window, it is sometimes called an integration (or calculus-based) tFSA.

4.7 EE Tools: Signal spectrum: Nyquist Criterion, Gabor Limit, Power Spectrum

In discussions of noise properties spectral analysis plays a large role. The fundamental tool in spectral analysis is the Fourier transform (FT), which gives the frequency decomposition of the transformed signal [185]. For noise fluctuations in an electrical signal, attention is usually focused on the FT of the signal squared. This is because the square of a voltage or current signal is proportional to the power. Depending on the incorporation of that proportionality constant (i.e., impedance value) the spectral densities obtained are known as voltage, current, or power spectral density [157]. Due to properties of FTs, convolution, such as in the definition of the autocorrelation function, transforms to multiplication. This provides a FT relationship between a signal's power spectral density and its autocorrelation function (the Weiner-Khinchine theorem).

4.7.1 Nyquist Sampling Theorem

Let $x(t)$ be a band limited signal with $X(\omega) = 0$ for $|\omega| > \omega_M$. Then $x(t)$ is uniquely determined by its samples $x(nT)$, $n=0, \pm 1, \pm 2, \ldots$ if $\omega_S > 2\omega_M$, where $\omega_S = 2\pi/T$. The frequency $2\omega_M$ is known as the Nyquist rate and must be exceeded by the sampling frequency to satisfy the sampling theorem [185].

4.7.2 Fourier Transforms, and other classic Transforms

The response of a linear (i.e., superposition property) time-invariant system (time-shift in input leads to same output but with that time-shift) to a complex exponential input (a phasor) is the same phasor with a change in amplitude: $e^{i\omega t} \rightarrow H(\omega)e^{i\omega t}$. This motivates phasor reconstruction of a periodic signal $x(t)$, with fundamental period $T = 2\pi/\omega$, using $x(t) = \Sigma_k \, a_k e^{ik\omega t}$, where k summation is over both positive and negative integers. Evaluation of the Fourier series components a_k is via: $a_k = 1/T \int x(t)e^{-ik\omega t}dt$ [185]. (Similar form for continuous time transform.) Other classic transforms include the Laplace, Mellin, Hankel, and Z-transform. There are also a variety of (non-lossy) data-compression methods that can be used as transforms insofar as feature extraction purposes.

4.7.3 Power Spectral Density

The power spectral density, $S(f)$, of a signal, $x(t)$, is a real, even, nonnegative function of frequency. Integration over $S(f)$ gives total average power per ohm: $P = \int S(f)df = <x^2(t)>$ (frequency integration $-\infty$ to ∞ unless specified otherwise). The autocorrelation function, $R(\tau) = <x(t)x(t+\tau)>$, of a power signal, $x(t)$, is defined as the time average $<x(t)x(t+\tau)> = \lim_{T \to \infty} (1/2T) \int x(t)x(t+\tau)dt$. For an ergodic process, $S(f)$ and $R(\tau)$ are a Fourier Transform pair (Weiner-Khinchine theorem): $R(\tau) = \int S(f) \, e^{i2\pi ft} \, df$ and $S(f) = \int R(\tau) \, e^{-i2\pi ft} \, df$ [185].

Power-spectrum based feature extraction

Typical power spectra for captured nine-base-pair DNA hairpins are shown in Fig. 4.1, along with a spectrum for the open channel. Below 10 kHz, the current fluctuation caused by the captured DNA molecule (i.e. the blockade noise) is greater than all other noise sources. Such blockade noise typically arises from changes in transient bonds with the protein channel, DNA conformational changes in molecular structure or overall orientation vis-a-via the channel, changes in DNA conplexation/solvation (involving waters of hydration and salt ions), and changes in internal chemical bonds (terminus fraying, for example). The power spectra for all the signals examined in [45] had approximately Lorentzian profiles, indicative of a predominately two-state switching process (seen as random telegraph noise). Discriminating between the DNA hairpins on the basis of their power spectral (or other Fourier transform properties, or wavelet properties) is possible for small sets of hairpins. For larger sets of hairpins, or for very similar hairpins like here, the HMM-based feature extraction proved critical, due to their strengths at extracting features from aperiodic (stochastic) sequential data. HMMs can be used for classification as well as feature extraction. In what follows HMMs are mainly used for feature extraction in conjunction with a Support Vector Machine (SVM).. The resulting signal processing and pattern recognition architecture enabled real-time single molecule classification on blockade samplings of only 100 milliseconds [45].

4.7.4 Cross-Power Spectral Density

For ergodic processes, time and ensemble averages are interchangeable, in particular, $R(\tau) = \langle x(t)x(t+\tau) \rangle = E\{x(t)x(t+\tau)\}$. If the ergodic processes for two power signals are present the net power signal (noise voltage, for example) is $z(t) = x(t) + y(t)$ and $P_z = E\{z^2(t)\} = P_x + P_y + 2P_{xy}$, where $P_{xy} = E\{x(t)y(t)\}$. The latter quantity is the $\tau=0$ case of the cross-correlation function: $R_{xy}(\tau) = E\{x(t)y(t+\tau)\}$. Similarly, cross-power spectral density, $S_{xy}(f)$, is the Fourier transform of $R_{xy}(\tau)$ [186].

4.7.5 AM/FM/PM Communications Protocol

Amplitude modulation involves addition of a DC bias to a message signal and using this as an amplitude modulation factor on some carrier frequency [186]: $x(t) = [A+m(t)] \cos (\omega t)$, this is often rewritten as:

$$x(t) = A[1 + am_N(t)] \cos (\omega t), \text{ where } m_N(t) = m(t)/|\min m(t),| \text{ and } a = |\min m(t)|/A.$$

The parameter 'a' is the modulation index and envelope detection can only be used if $a < 1$. The total power in the AM modulated signal is proportional to: $\langle x^2(t) \rangle = \langle [A+m(t)]^2 \cos^2(\omega t) \rangle$. If $m(t)$ is more slowly varying than $\cos (\omega t)$, then the latter time integration can be performed to yield a factor of ½: $\langle x^2(t) \rangle = [A^2 + 2A\langle m(t) \rangle + \langle m^2(t) \rangle]/2$, which typically reduces with $\langle m(t) \rangle = 0$ to: $\langle x^2(t) \rangle = [A^2 + \langle m^2(t) \rangle]/2$. Note: the AM signal power with maximum information content is for square-wave signal max $m(t) = 1$ and min $m(t) = -1$, efficiency is 50%. For sinusoidal, efficiency is 33%.

AM does not require a coherent reference for demodulation, this leads to AM radios that are simple and inexpensive. Similarly, this is one point for a branching in the communications theory to other instances where there is not necessarily a coherent reference - such as with the stochastic carrier wave methods to be described in what follows.

4.8 Generalized Hidden Markov Models (gHMMs)

Hidden Markov models have been used in speech recognition since the 1970s [187], and in bioinformatics since the 1990's [188], and have an extensive, and growing, breadth of applications in other areas (especially as more computational resources become available). Other areas of HMM application include gesture recognition [189,190], handwriting and text recognition [191-194], image processing [195,196], computer vision [197], communication [198], climatology [199], and acoustics [200,201]. An HMM is the central method in all of these approaches because they are the simplest, most efficient, modeling approach that is obtained when you combine a Bayesian statistical foundation for Markovian stochastic sequential analysis [202] with the efficient dynamic programming table constructions possible on a computer.

In automated gene finding there are two types of approaches, based on data intrinsic to the genome under study [203], or extrinsic to the genome (e.g., homology, and EST data). Since c.a. 2000 the best gene finders have been based on combined intrinsic/extrinsic statistical modeling [204], [205-207]. The most common intrinsic statistical model is an HMM, so the question naturally arises -- how to optimally incorporate extrinsic side-information into an HMM? We resolve that question in [2] by treating duration distribution information *itself* as side-information and demonstrate a process for incorporating that side-information into an HMM. We thereby bootstrap from an HMM formalism to a HMM-with-duration formulation (more generally, a hidden semi-Markov model or HSMM).

In many applications, the ability to incorporate the state duration into the HMM is very important because the standard, HMM-based, Viterbi and Baum-Welch algorithms are otherwise critically constrained in their modeling ability to distributions on state intervals that are geometric. This can lead to a significant decoding failure in noisy environments when the state-interval distributions are not geometric (or approximately geometric). The starkest contrast occurs for multimodal distributions and heavy-tailed distributions. The hidden Markov model with binned duration (HMMBD) algorithm, presented in [21], eliminates the HMM geometric distribution constraint, as well as the HMMD maximum duration constraint, and offers a significant reduction in computational time for all HMMD-based methods to approximately the computational time of the HMM-process alone. In adopting any model with 'more parameters', such as an HMMD over an HMM, there is potentially a problem with having sufficient data to support the additional modeling. This is generally not a problem in any HMM model that requires thousands of samples of non-self transitions for sensor modeling, such as for the gene-finding that is described in what follows, since knowing the boundary positions allows the regions of self-transitions (the durations) to be extracted with similar sample number as well, which is typically sufficient for effective modeling of the duration distributions in a HMMD.

Critical improvement to overall HMM application rests not only with the aforementioned generalizations to the HMM/HMMD, but also with generalizations to the hidden state model and emission model. This is because standard HMMs are at low Markov order in transitions (first) and in emissions (zeroth), and transitions are decoupled from emissions (which can miss critical structure in the model, such as state transition probabilities that are sequence dependent). This weakness is eliminated if we generalize to the largest state-emission clique possible, fully interpolated on the data set, as is done with the generalized-clique hidden Markov model (HMM) described in [18], where gene finding is performed on the *C. elegans* genome. The objective with the clique generalization is to improve the modeling of the critical signal information at the transitions between exon regions and non-coding regions, e.g., intron and junk regions. In doing this we arrive at a HMM structure identification platform that is novel, and robustly-performing, in a number of ways.

The generalized clique HMM ("meta-HMM") application to gene-finding begins by enlarging the primitive hidden states associated with the individual base labels (as exon, intron, or junk) to substrings of primitive hidden states or *footprint* states. The emissions are likewise expanded to higher order in the fundamental joint probability that is the basis of the generalized-clique, or 'meta-State', HMM. In [18] we show how a meta-state HMM significantly improves the strength of coding/noncoding-transition contributions to gene-structure identification when compared to similar, intrinsic-statistics-only, geometric models. We describe situations where the coding/noncoding -transition modeling can effectively 'recapture' the exon and intron heavy tail distribution modeling capability as well as manage the exon-start 'needle-in-the-haystack' problem. In analysis of the *C. elegans* genome, the sensitivity and specificity (SN,SP) results for both the individual-state and full-exon predictions are greatly enhanced over the standard HMM when using the generalized-clique HMM [18]. These meta-HMMBD developments provide a foundation from which to explore a core new paradigm, the holographic HMM, where generalization to multiple labels are possible at each emission. Holographic HMMs are to be explored theoretically and in implementations, such as alternative-splice gene finding (as described in what follows).

The improved signal resolution possible via a meta-HMMBD signal processing method will allow for reduced signal processing overhead, thereby reducing power usage. This directly impacts satellite communications where a minimal power footprint is critical, and cell phone construction, where a low-power footprint allows for smaller cell phones, or cell phones with smaller battery requirements, or cell phones with less expensive power system methodologies. For real-time signal processing, meta-HMMBD signal processing permits much more accurate signal resolution and signal de-noising than current, HMM-based, methods. This impacts real-time operational systems such as voice recognition hardware implementations, over-the-horizon radar detection systems, sonar detection systems, and receiver systems for streaming low-power digitial signal broadcasts (such an enhancement could improve receiver capabilities on various high-definition radio and TV broadcasts). For batch (off-line) signal resolution, the meta-HMMBD signal processing allows for significantly improved gene-structure resolution in genomic data, and extraction of binding/conformational kinetic feature data from nanopore detector channel current data. For scientific and engineering endeavors in general, where there is any data analysis that can be related to a sequence of measurements or observations, the meta-HMMBD signal processing systems that can be implemented permit improved signal resolution and speed of signal processing.

In a 'holographic' HMM we extend to a multi-track label-sequence (e.g., a multi-state labeling framework at each observation instance). The simplest example of this is to have two label sequnces for one observation sequence, and this is the implementation used in the alternative-splice gene-finding in the *C. elegans* genome effort described in what follows and [12], where it is shown that there is sufficient statistical support for a two-track label model.

In instances of 2-D and higher order dimensional data, such as 2-D images, the data can be reduced to a single-track sequence of measurements via a rastering process, as has been done with HMM methods in the past, *or* the reduction from the rastering could be to a multi-track hidden-label state to better track the local 2-D information, e.g., a 3x3 window with hidden states corresponding to the different 3x3 windows that can be seen, etc., in a self-consistent tiling (the center of the 3x3 grid could be the the former, single sample, datum used in the 1-D reduced rasterization, for example). This can be extended for larger 2-D 'windows', or n-D neighborhoods (the latter reducible to a 2-D representation in an extension of the holographic hypothesis [87,208]). (Another area impacted by the multi-track HMM method is protein folding and conformational analysis. This is because now contemporanious information can be absorbed into the multi-track HMM. This is an extensive application area in its own right.)

All of the HMM generalizations and feature extraction methods discussed in what follows can be optimized for speed with binned durations and through dynamic ("null") binning, distributed table-chunking, and GPU-usage. This allows the limiting speed constraint on the core HMMBD component in the Stochastic Sequential Analysis (SSA) protocol (Fig. 4.15) to be controlled as much as possible. The SSA protocol outlined in what follows is for the discovery, characterization, and classification of localizable, approximately-stationary, statistical signal structures in channel current data, or genomic data, or stochastic sequential data in general, and changes between such structures.

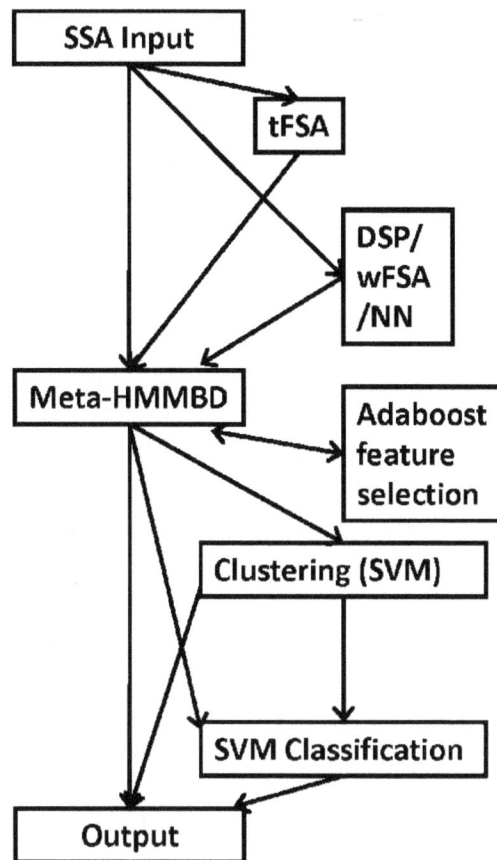

Fig. 4.15. The most common stochastic sequential analysis flow topology. The main signal processing flow is typically Input → tFSA→ Meta-HMMBD → SVM → Output. Notable differences occur in channel current cheminformatics (CCC) where there is use of EVA-projection, or similar method, to achieve a quantization on states, then have Input → tFSA → HMM/EVA → meta-HMMBD-side → SVM → Output. While, in gene-finding just have: Input → meta-HMMBD-side → Output. In gene-finding, however, the HMM internal 'sensors' are sometimes replaced, locally, with profile-HMMs [1] or SVM-based profiling [26], so topology can differ not only in the connections between the boxes shown, but in their ability to embed in other boxes as part of an internal refinement.

The core signal processing stage in Fig. 4.15 is usually the feature extraction and feature selection stage, where central to the signal processing protocol is the Hidden Markov model (HMM). The HMM methods are the central methodology/stage in the CCC protocol in that the other stages can be dropped or merged with the HMM stage in many incarnations. For example, in some data analysis situations the tFSA methods could be totally eliminated in favor of the more accurate HMM-based approach to the problem, with signal states defined/explored in much the same setting, but with the optimized Viterbi path solution taken as the basis for the signal acquisition structure identification. The reason this is not typically done is that the FSA methods are usually only $O(T)$ computational expense, where 'T' is the length of the stochastic sequential data that is to be examined, and '$O(T)$' denotes an order of computation that scales as 'T' (linearly in the length of the sequence). The typical HMM Viterbi algorithm, on the other hand, is $O(TN^2)$, where 'N' is the number of states in the HMM. So, use of the tFSA provides a faster, and often more flexible, means to acquire signal, but it is more hands-on. If the core HMM/Viterbi method can be approximated such that it can run at $O(TN)$ or even $O(T)$ in certain data regimes, for example, then the non-HMM methods can be phased out.

The HMM emission probabilities, transition probabilities, and Viterbi path sampled features, among other things, provide a rich set of data to draw from for feature extraction (to create 'feature vectors'). The choice of features in the SSA Protocol is optimized along with the classification or clustering method that will make use of that feature information. In typical operation of the protocol, the feature vector information is classified using a Support Vector Machine (SVM) [81-84]. Once again, however, the separate classification step could be totally eliminated in favor of the HMM's log likelihood ratio classification capability, for example, when a number of template HMMs are employed (one for each signal class). This classification approach is weaker and slower than the (off-line trained) SVM methodology in many respects, but, depending on the data, there are circumstances where it may provide the best performing implementation of the protocol.

The HMM features, and other features (from neural net, wavelet, or spike profiling, etc.) can be fused and selected via use of various data fusion methods, such as a modified Adaboost selection (from [1,33,37,38], and described in what follows). The HMM-based feature extraction provides a well-focused set of 'eyes' on the data, no matter what its nature, according to the underpinnings of its Bayesian statistical representation. The key is that the HMM not be too limiting in its state definition, while there is the typical engineering trade-off on the choice of number of states, N, which impacts the order of computation via a quadratic factor of N in the various dynamic programming calculations used (comprising the Viterbi and Baum-Welch algorithms among others).

The HMM 'sensor' capabilities can be significantly improved via switching from profile-HMM sensors to pMM/SVM-based sensors, as indicated in [26], where the superior performance and generalization capability of this approach was demonstrated. A martingale feature vector is described in this context in [81-84].

Preliminary work with HMMD binning [21], clique-generalized HMMs [18], and HMMs with side information [19], lays the foundation for an intrinsic-statistics optimized HMM and shows how to optimally incorporate extrinsic side-information (if available). This is described in what follows and may provide a transformative platform for gene-structure identification. The methods described will also show how to collect data for the statistics to validate the strength of a multichannel statistical model. An analysis of the *C. elegans* genome's alternative-splice state-space complexity shows that a two-track alt.-splice modeling is possible. If successful with high accuracy, alternatively-spliced regions could be analyzed with much greater automation, possibly leading to further breakthroughs as new tracts of genomic data are then understood. The HMM speed optimizations could have a profound effect on HMM real-time applications, as well, although such optimizations may have data-dependent complexity, and this is one of the matters that will be explored. In what follows we describe how to 'upgrade' from a pMM sensor to a pMM/SVM sensor. This could significantly boost standard HMM performance, especially when used to lift the general log-liklihood ratio (LLR) terms that arise in the meta-HMM Viterbi-type algorithm into an SVM classifier.

4.8.1 When to use a Hidden Markov Model (HMM)?

Suppose you have a sequence of observations (or measurements, or samplings, etc.) and take '$b_1 b_2 ... b_L$' to denote an observation sequence of length L. Introduce as Bayesian parameter a hidden label associated with each observation, denote the label sequence as '$\lambda_1\lambda_2 ... \lambda_L$'. The joint probability of the observations and a particular label-sequence is given via:

$$P(B;\Lambda) = P(b_1 b_2 ... b_L; \lambda_1\lambda_2 ... \lambda_L) = P(b_1 b_2 ... b_L| \lambda_1\lambda_2 ... \lambda_L) P(\lambda_1\lambda_2 ... \lambda_L).$$

Markov assumptions for a standard 1st-order hidden Markov model then allow reduction to:

$$P(B;\Lambda;1^{st}\text{-order HMM}) = [P(b_1| \lambda_1) ... P(b_L| \lambda_L)] [P(\lambda_1) P(\lambda_2| \lambda_1) ...P(\lambda_L| \lambda_{L-1})]$$

If there are 50 labels and 50 observations (as in the quantized power signal analysis considered in that follows), then the full joint probability has $(50)^{2L}$ possibilities for labeled observation sequence of length L, which is unmanagable for typical sequences of interest, while, after reduction to the 1st-order HMM Markov assumptions on the conditional probabilities we have a set of $2\text{x}(50)^2$ possibilities (independent of sequence length). If n^{th}-order Markov models are employed, the set of possibilities grows as $(50)^{n+1}$, which can be easily enumerated for n up to 3 or 4, and still be accessed via hash-indexing for n>4 (as in the hIMM approach described in [39]).

In the Viterbi algorithm we seek the λ-sequence that maximizes the joint probability above for given observation sequence. In order to consider all of the possible λ-sequences in some direct enumeration for the above N-label case, we would have $(N)^L$ possible L length λ-sequences. Here we employ the classic dynamic programming solution employed by Viterbi instead, to perform a HMM Viterbi table calculation that retains information such that $O(LN^2)$ computations are needed to simultaneously consider all paths.

Hidden Markov models (HMMs) are, thus, an amazing tool at the nexus where Bayesian probability and Markov models meet dynamic programming. To properly define/choose the HMM model in a machine learning context, however, further generalization is usually required. This is because the 'bare-bones' HMM description has critical weaknesses in most applications, which are summarized below. Fortunately, these weaknesses can be addressed, and in computationally efficient ways, as will be described in what follows.

4.8.2 Hidden Markov Models (HMMs) – standard formulation and terms

We define the 1st order HMM as consisting of the following:

- A hidden state alphabet, Λ, with "Prior" Probabilities $P(\lambda)$ for all $\lambda\in\Lambda$, and "Transition" Probabilities $P(\lambda_2|\lambda_1)$ for all $\lambda_1 \lambda_2\in\Lambda$ -- where the standard transition probability is denoted $a_{kl} = P(\lambda_n=l|\lambda_{n-1}=k)$ for a 1st order Markov model on states with homogenous stationary statistics (i.e., no dependence on position 'n').

- An observable alphabet, B, with "Emission" Probabilities $P(b|\lambda)$ for all $\lambda\in\Lambda$ $b\in B$ – where the standard emission probability is $e_k(b) = P(b_n=b|\lambda_n=k)$, i.e., a 0th order Markov model on bases with homogenous stationary statistics.

There are three classes of problems that the HMM can be used to solve [183]:

(1) Evaluation - Determine the probability of occurrence of the observed sequence.
(2) Learning (Baum-Welch) - Determine the most likely emission and transition probabilities for a given set of observational data.
(3) Decoding (Viterbi) - Determine the most probable sequence of states emitting the observed sequence.

Most of the examples focus on the 3rd problem, the Viterbi decoding problem, but full gHMM solutions for both Viterbi and Baum-Welch have been derived and implemented [39].

The probability of a sequence of observables $B = b_0\ b_1\ldots\ b_{n-1}$ being emitted by the sequence of hidden states $\Lambda = \lambda_0\ \lambda_1\ldots\ \lambda_{n-1}$ is solved by using $P(B, \Lambda) = P(B|\Lambda)\ P(\Lambda)$ in the standard factorization, where the two terms in the factorization are described as the *observation model* and the *state model*, respectively. In the 1st order HMM, the state model has the 1st order Markov property and the observation model is such that the current observation, b_n, depends only on the current state, λ_n:

$$P(B|\Lambda)\ P(\Lambda) = P(b_0|\lambda_0)\ P(b_1|\lambda_1)\ldots P(b_{n-1}|\lambda_{n-1})\ \text{x}$$
$$P(\lambda_0)P(\lambda_1|\lambda_0)P(\lambda_2|\lambda_0, \lambda_1)\ldots P(\lambda_{n-1}|\lambda_0\ldots \lambda_{n-2})$$

With first order Markov assumption in the state-model this becomes:

$$P(B|\Lambda)\ P(\Lambda) = P(b_0|\lambda_0)\ P(b_1|\lambda_1)\ldots P(b_{n-1}|\lambda_{n-1})\ \text{x}\ P(\lambda_0)P(\lambda_1|\lambda_0)P(\lambda_2|\lambda_1)\ldots P(\lambda_{n-1}|\lambda_{n-2})$$

In the Viterbi algorithm, a recursive variable is defined (following the notation in [183,184]): $v_k(n) =$ *"the most probable path ending in state 'k' with observation 'b_n' "*. The recursive definition of $v_k(n)$ is then: $v_l(n+1) = e_l(b_{n+1})\ \max_k\ [v_k(n)\ a_{kl}]$. From which the optimal path information is recovered according to the (recursive) trace-back:

$$\Lambda^* = \text{argmax}_\Lambda\ P(B, \Lambda) = (\lambda^*_0, \ldots, \lambda^*_{n-1})$$
$$\lambda^*_n|_{\lambda^*_{n+1}=l} = \text{argmax}_k\ [v_k(n)\ a_{kl}], \text{ and where } \lambda^*_{L-1} = \text{argmax}_k\ [v_k(L-1)], \text{ for}$$
length L sequence

4.8.3 HMMs -- Graphical Model formulation
In this section we make a brief foray into graphical representations of MMs and HMMs such that the linear extension of both graphical models is clear, as well as a better understanding of the HMM in relation to the simpler MM via the graphical models (Fig.s 4.16 & 4.17).

4.8.3.1 Markov Models

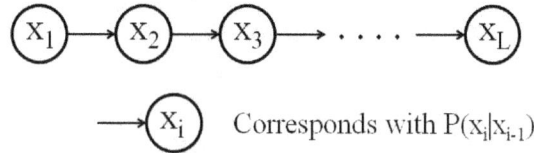

$$P(X) = P(x_1)\, P(x_2|x_1)\, P(x_3|x_2)\, \ldots\ldots\, P(x_L|x_{L-1})$$
$$P(X) = \prod_{i=1}^{L} a_{i-1,i}$$
where $a_{01} = P(x_1)$ and $a_{ij} = P(x_j|x_i)$

Fig. 4.16. Graphical Model for a Markov Model

4.8.3.2 Hidden Markov Models

For (hidden) Markov state sequence $\Pi = $ "$\pi_1\pi_2\pi_3\ldots\ldots\pi_L$", where the states take on values (e.g., $\pi_n = k$) in a finite alphabet, with transition probability $a_{kl} = P(\pi_i = l | \pi_{i-1} = k)$, and with an associated observation sequence $X = $ "$x_1 x_2 x_3 \ldots\ldots x_L$", and 'emission' probability $e_k(b) = P(x_i = b | \pi_i = k)$: (now viewed as an 'emission' outcome of the hidden state, not a transition process in itself):

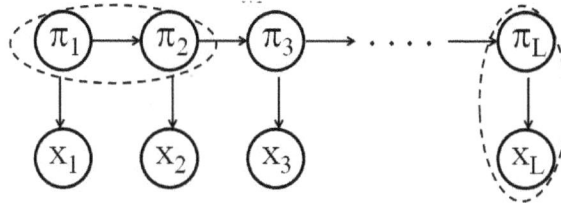

$$P(X; \Pi) = P(X|\Pi) \cdot P(\Pi) = P(X|\Pi) \cdot P(\pi_1)\, P(\pi_2|\pi_1)\, P(\pi_3|\pi_2)\, \ldots\ldots\, P(\pi_L|\pi_{L-1})$$
$$P(X; \Pi) = \prod_{i=1}^{L} P(x_i = x_i | \pi_i = \pi_i) \cdot \prod_{i=1}^{L} P(\pi_i = \pi_i | \pi_{i-1} = \pi_{i-1})$$
$$P(X; \Pi) = a_{0\pi_1} \prod_{i=1}^{L} e_{\pi_i}(xi)\, a_{\pi_i \pi_{i+1}}$$
where $a_{0\pi_1} = P(\pi_1)$ and $a_{\pi_i \pi_{i+1}} = P(\pi_i = \pi_i | \pi_{i-1} = \pi_{i-1})$.

Fig. 4.17. Graphical Model for a Hidden Markov Model

4.8.4 Viterbi Path

In the Viterbi algorithm, a recursive variable is defined: $v_{kn} = v_k(n) = v_k(b_n) = $ *"the most probable path ending in state $\lambda_n = k$ with observation b_n"*. The recursive definition of $v_k(n)$ is then: $v_l(n+1) = e_l(b_{n+1})\, \max_k [v_k(n)\, a_{kl}]$, where $e_l(b_{n+1})$ is the 'emission' probability for the observed b_{n+1} when in state $\lambda_{n+1} = l$, and a_{kl} is the transition probability from state $\lambda_n = k$ to state $\lambda_{n+1} = l$. The optimal path information is recovered according to the (recursive) trace-back:

(1) $\Lambda^* = \text{argmax}_\Lambda\, P(B, \Lambda) = (\lambda^*_0, \ldots, \lambda^*_{L-1});\ \lambda^*_n|_{\lambda^*_{n+1} = l} = \text{argmax}_k [v_k(n)\, a_{kl}]$,

and where

(2) $\lambda^*_{L-1} = \text{argmax}_k [v_k(L-1)]$, for length L sequence.

69

The recursive algorithm for the most likely state path given an observed sequence (the Viterbi algorithm) is expressed in terms of v_{ki} (the probability of the most probable path that ends with observation $b_n = i$, and state $\lambda_n = k$). The recursive relation is lifted directly from the underlying probability definition: $v_{ki} = \max_n\{e_{ki}a_{nk}v_{n(i-1)}\}$, where the $\max_n\{\ldots\}$ operation returns the maximum value of the argument over different values of index n, and the boundary condition on the recursion is $v_{k0} = e_{k0}p_k$. The emission probabilities are the main place where the data is brought into the HMM-EM algorithm. An inversion on the emission probability is possible in this setting because the states and emissions share the same alphabet of states/quantized-emissions. The Viterbi path labelings are, thus, recursively defined by $p(\lambda_i|\lambda_{(i+1)}=n) = \text{argmax}_k\{v_{ki}a_{kn}\}$. The evaluation of sequence probability (and its Viterbi labeling) take the emission and transition probabilities as a given. Estimates on those emission and transition probabilities themselves can be obtained by an Expectation/Maximization (EM) algorithm that is known as the Baum-Welch algorithm in this context. The 50-state generic HMM described above is used extensively in [1,18], and will be described further in the EVA and other methods that follow.

The Most Probable State Sequence

Given an observation sequence $X = $ "$x_1 x_2 x_3 \ldots x_L$", we often want to know the most probable hidden state sequence that might be associated with it:

$$\Pi* = \underset{\Pi}{\text{argmax}}\ P(X; P)$$

aacgcgtagctagttgactctcgaaacgcgtagctagttgactctcgaacgcgtagctagttgactctcgaacgcgtagctagttgactctt

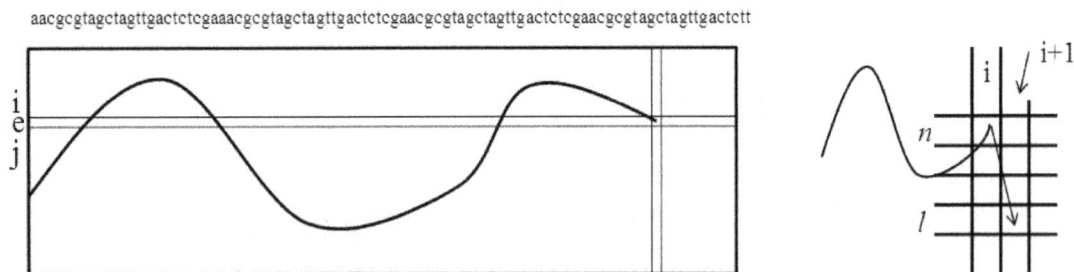

Fig. 4.18. Viterbi path. The path with the most probable state sequence. Viterbi algorithm based on recursive relation, solved via tabulation, seeking *"the most probable path ending in state $\lambda_n=k$ with observation b_n"*.

This can be solved recursively using a dynamic programming algorithm known as the Viterbi algorithm. Let $v_k(i)$ be the probability of the most probable path ending in state "k" with i'th observation (shown in further dtail in Fig. 4.19):

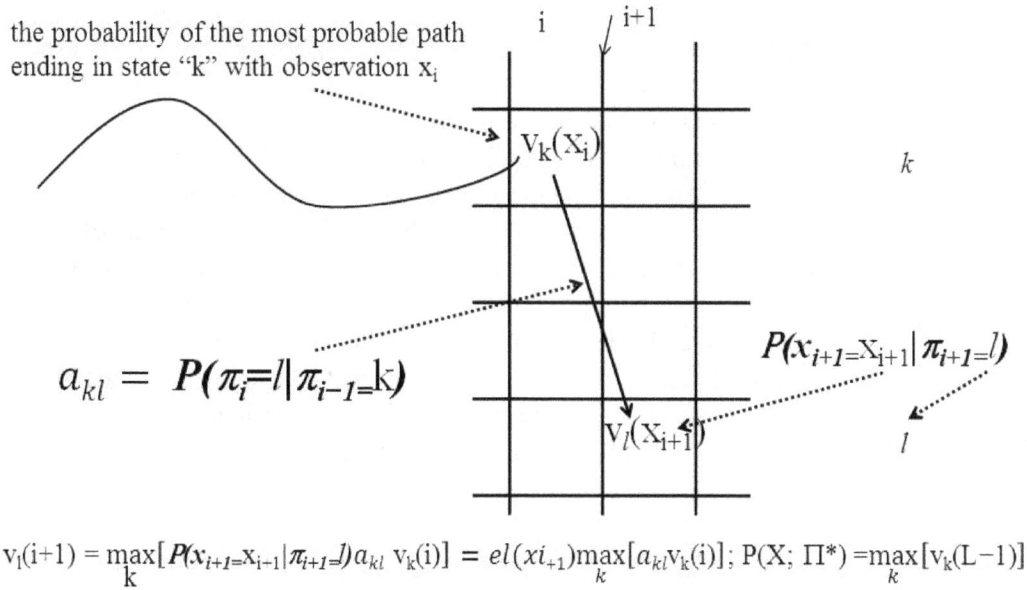

$$v_l(i+1) = \max_{k}[P(x_{i+1}=x_{i-1}|\pi_{i+1}=l)a_{kl} \; v_k(i)] = el(xi_{+1})\max_{k}[a_{kl}v_k(i)]; \; P(X; \Pi^*) = \max_{k}[v_k(L-1)]$$

$$\pi_i^*|(\pi_{i-1}^* = l) = \operatorname*{argmax}_{k}[a_{kl}v_k(i)]; \; v_k(0) = P(x_0|\pi_0=k); \; \pi_{L-1}^* = \operatorname*{argmax}_{k}[v_k(L-1)]$$

Fig. 4.19. Viterbi Path. Optimal Path Identification

4.8.5 gHMMs

Generalized Hidden Markov model (HMM) methods are used for where signal is stationary but not periodic (typical case is stochastic with stationary signal properties, refered to as "stationary statistics"). The generalization to a larger state, the meta-state HMM, is shown in Fig. 4.20. HMM-with-binned-duration, and meta-HMM generalizations, are also used to enable practical stochastic signal processing, where the generalized HMM methods have generalized Viterbi algorithms with all of the inherent benefits of an efficient dynamic programming implementation, as well as Martingale convergence properties when used for filtering and robust feature extraction.

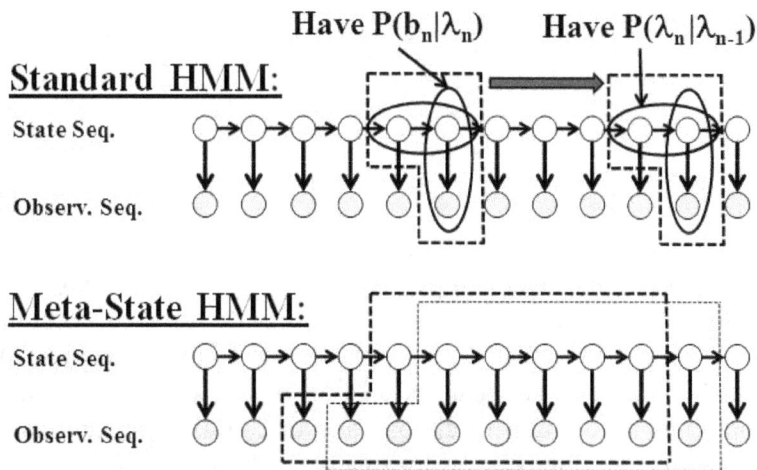

Fig. 4.20. The Meta-State HMM. The meta-HMM entails a massive increase in conditional probability modeling capability, but done within the dynamic programming context (a generalized Viterbi algorithm).

4.9 SVM Overview

A classifier is typically a simple rule whereby a class determination can be made, such as a decision boundary. Learning the decision rule, or a sufficiently good decision rule, especially if simple (and elegant), is the implementation aspect of a classifier, and can be difficult and time consuming. Even so, this is usually manageable because at least you have data to 'learn from', e.g., supervised learning, where you have instances and their classifications (or 'labels'). Learning for classification can be done very effectively using Support Vector Machines (SVMs), as will be described in what follows. With clustering efforts, or unsupervised learning, on the other hand, we don't have the label information during training. In what follows SVMs will also be shown to be incredibly effective at clustering when used with metaheuristics to recover label information in a bootstrap learning process. Also shown will be implementation details for distributed SVM training, and other speedup optimizations, for practical deployment of the powerful SVM classification and clustering methods in real-time operational situations (as will be demonstrated with results on nanopore detector experiments).

Support Vector Machines (SVMs) are variational-calculus based methods that are constrained to have structural risk minimization (maximum margin optimization), unlike neural net classifiers or perceptrons, such that they provide noise tolerant solutions for pattern recognition [1]. An SVM determines a hyperplane that optimally separates one class from another, while the structural risk minimization (SRM) criterion manifests as the hyperplane having a thickness, or "margin," that is made as large as possible in the process of seeking a separating hyperplane (see Fig. 4.21 below).

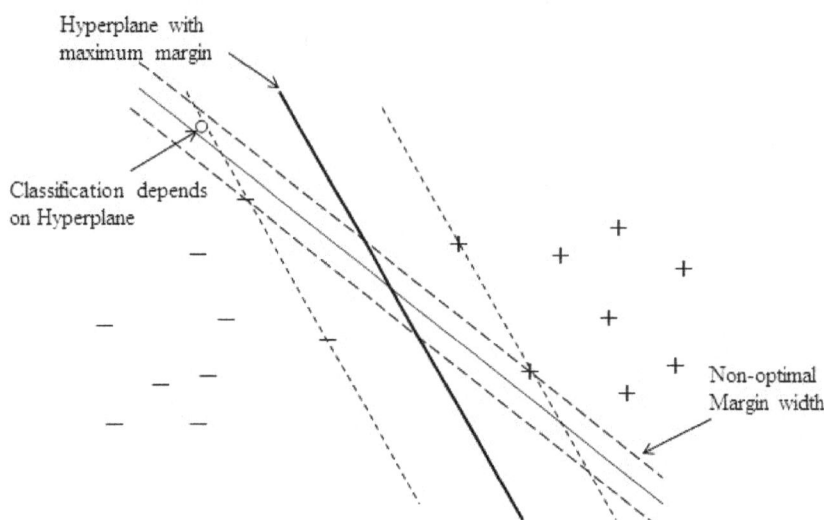

Fig. 4.21. Supervised Learning: Separability and Maximum Margin. Shown in the figure is data separable by a line. With obvious generalization to a plane in 3-D, and in higher dimensions, etc. We have thereby a (manifold) notion of separability on two datasets. Of all of the separating hyperplanes, it would appear to be optimal to choose one that could be made as 'thick' as possible and still be a separating hyperplane. The thickness is known as the margin, so we are seeking a maximal margin hyperplane, as shown. Also shown is a non-optimal margin hyperplane.

A benefit of using SRM is much less complication due to over-fitting. Once learned, the hyperplane allows data to be classified according to the side of the hyperplane in which it resides. The SVM approach encapsulates a significant amount of model-fitting information in its choice of kernel. The SVM kernel also provides a notion of distance in the neighborhood of the decision hyperplane. SVM binary discrimination outperforms other classification methods with or without dropping weak data. SVMs have a built parameter to assess confidence in a signal classification (related to the kernel distance from the separating hyperplane), thus have a built-in notion of weak data. Other classifier methods, if they have a notion of weak data, often introduce it as a separate evaluation that must itself be tuned and analyzed in order to be trusted. SVM multiclass discrimination and SVM-based clustering are also possible [1]. In the SSA protocol SVMs play a central role in performing classification and clustering tasks.

Most SVM uses are restrictive in both training-set size and number of different classes, where most SVM applications involve datasets with fewer than 10,000 training instances and only two classes (the binary SVM). There are SVM implementations, however, that have no such limit on the number of training instances or the number of classes. Efficient new methods have been discovered for multiclass SVM, both internal to the optimization (multi-hyperplane) and external (decision tree and decision forest) [1]. In cases where the SVM training set is much larger than 10,000 instances, or when repeated training over the same training set is needed, significant SVM training computations are necessary. For this reason, distributed/GPU-optimized SVM training processes have been implemented [1].

There is a new approach to unsupervised learning that is based on use of supervised Support Vector Machine (SVM) classifiers. A fundamentally novel aspect of the proposed method is that it provides a non-parametric means for clustering (unsupervised learning) and partially-supervised clustering. In preliminary work the SVM-based clustering method appears to offer prospects for inheriting the very strong performance of standard SVMs from the *supervised* classification setting. This offers a remarkable prospect for knowledge discovery and enhancing the scope of human cognition – the recognition of patterns and clusters without the limitations imposed by explicitly assuming a parametric model, where resolution of the identified clusters can be at an accuracy comparable to a supervised learning setting.

With an SVM 'learning' process, once convergent to solution, what is learned, among other things, is the set of "support vectors" that define the thickness boundary of the separating hyperplane (see Fig. 11.2).

So far we are implicitly describing a two-class problem, a "binary classifier", but what of multiple classes, can this be handled? Yes. Two broad categories of ways to handle this: (i) external refinement, such as via decisions trees (and forests) made from Binary classifier nodes (so more of what we've already got, to be discussed later); and (ii) internal refinement, such as via multiple hyperplanes (now inherently different, so discussed later).

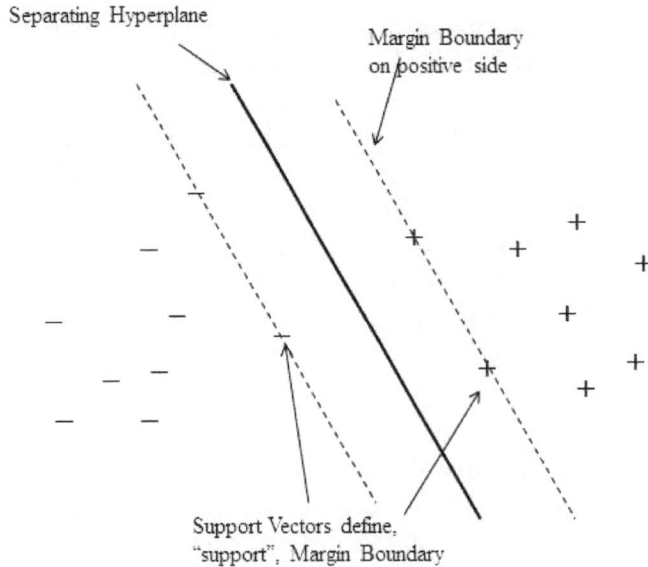

Separating Hyperplane

Margin Boundary
on positive side

Support Vectors define,
"support", Margin Boundary

Fig. 4.22. Support Vectors on Margin boundary.

So far we have been explicitly working with data that was separable. What if it isn't? See Fig. 4.23 Left for example where the data is not linearly separable. It is separable with a curve (or connected line segment as shown in Fig. 4.23 Right).

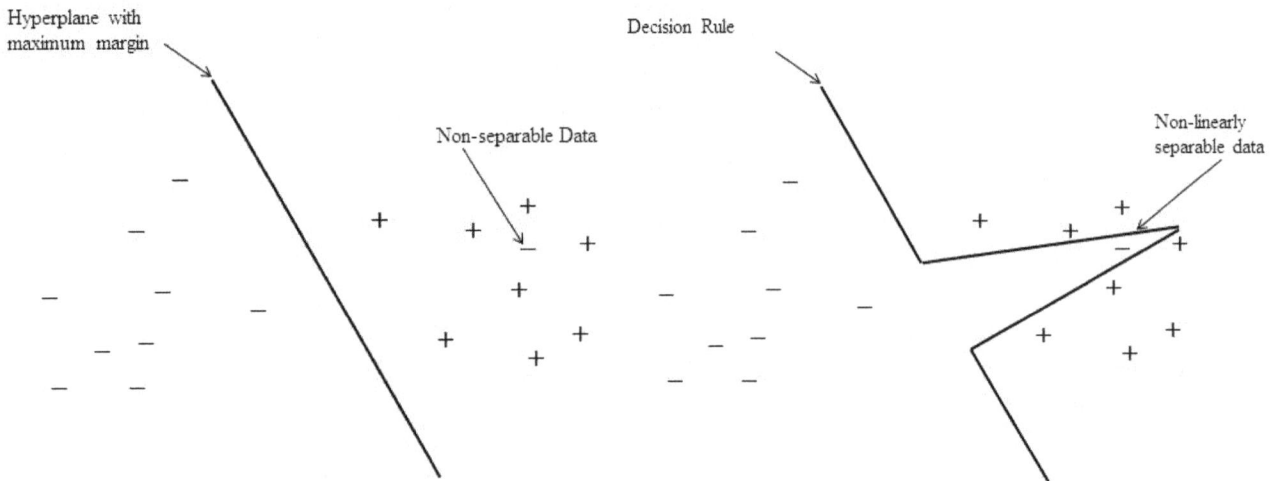

Hyperplane with
maximum margin

Decision Rule

Non-separable Data

Non-linearly
separable data

Fig. 4.23. Left. One of the former positives (central to the positive cluster) is now flipped to have a negative label. The data is now non-separable with a (single) straight line. **Right.** The data is shown separable with a curve that happens to be a connected line segment.

In finding a separable solution (Fig. 4.24, with maximum margin shown) we could also change the problem to separability on *most* of the data, with adjustments to account for those instances not consistent with the decision rule by way of a penalty term (according to how much 'wrong', perhaps). Alternatively, we could establish separability with a non-linear discriminant by mapping the feature vector data to a higher dimensional space (e.g.,

74

introduction of the Kernel map and overall generalization), where linear separability would almost always be possible (almost provably always the case, due to hypersphere shattering in sufficiently high dimensions – something used to obtain an initial convergence in SVM clustering).

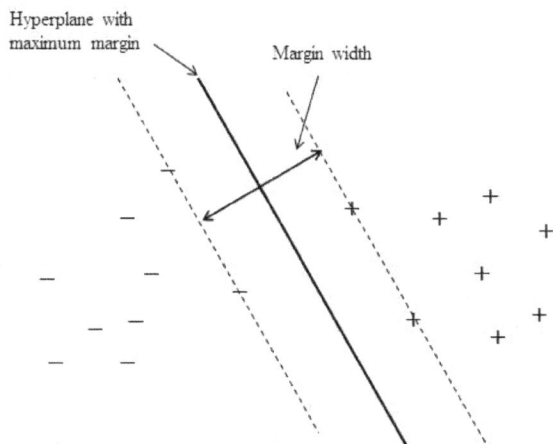

Fig. 4.24. Separable solution with maximum width margin.

Once training is done (have decision rule), we can then do classification (see Fig. 4.25). Using the classifier (training data still shown), we simply classify according to which side of hyperplane (decision surface). Since we know the actual classes of the test data (used in the training/validation), we can score the performance of our classifier – this information, in turn, allows the classifier to be tuned for optimal performance.

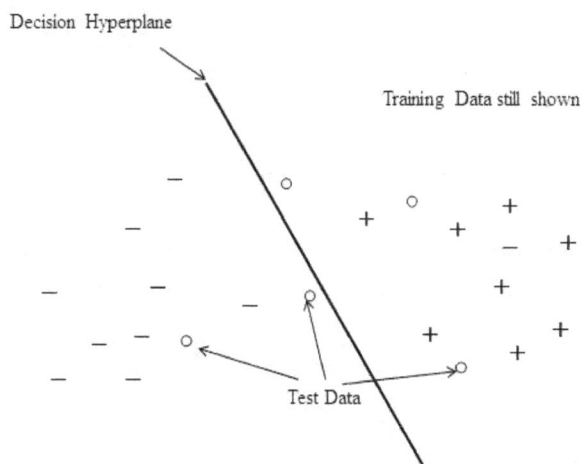

Fig. 4.25. Post-training, have Decision Hyperplane. Now test data (circles in figure) can be introduced and classified according to which side of the hyperplane that they are on. Test data is usually held out from the original training data according to an N-fold cross-validation arrangement. So for analysis on one such fold, the test data used is data for which we "know" the answer ahead of time, but it presented without the knowledge as a test. If the test data is classified positive and it was truly a positively labeled data, then it's a true positive (TP), and similarly for FP, TN, FN.

So far we discussed the problem of learning to classify (with two class data: positive and negative), and we will find that we can solve any of the classification problems mentioned, whether extending to multiple classes or significantly non-separable. We have robust methods to do classification with SVMs and some other methods (carefully managed Neural Nets, for example). So what happens if we take away the label information and want to recover the identification of the positive and negative classes (assuming two classes)? Essentially, we've arrived at the classic clustering problem, where we want to identify clusters in the data. (Clustering is sometimes called unsupervised learning where the lack of label information is related to lack of 'supervision'.) It turns out that clustering is much more difficult than classification since we don't even have a definition for a 'cluster'.

What is a cluster? Vague Rule #1 is indicated in Fig. 4.26, where Inter-cluster distance is required to be greater than intra-cluster distance, and significantly so for good clustering (this fails for two parallel line of dots, for example).

Clustering is qualitative, thus lacks the same level of 'well-definedness' as classification. Having successfully obtained a clustering solution, you can then provide a labeling to the data, and revisit (bootstrap) in a supervised learning scenario to train a classifier....

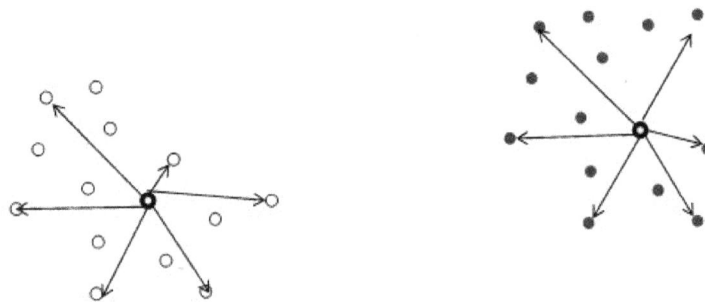

Fig. 4.26. Two clusters shown, with intra-cluster distance to 'centers' shown. The sum of squared error (SSE) score for a cluster is the sum of the distances squared from the "center point" of the cluster (an objective measure of clustering performance if not already the basis for the clustering method, .e.g. k-means).

Instead of bootstrapping from clustering into classification (with addition of label information), can we bootstrap from classification into clustering? By throwing any, random, label info, and having a learning process on the labels? That way we play to our strength, since classification as a learning process is on very form ground, both theoretically and in terms of efficient implementation. This would seem to indicate that a classification → clustering bootstrap solution could be done, but it would also require beginning the initial classification bootstrap run with randomly data. Is this possible for any classifier to manage? The answer is yes, for the SVM, but only for certain choices of SVM kernel. A schematic for the SVM-classifier based clustering method is shown in Fig. 4.27.

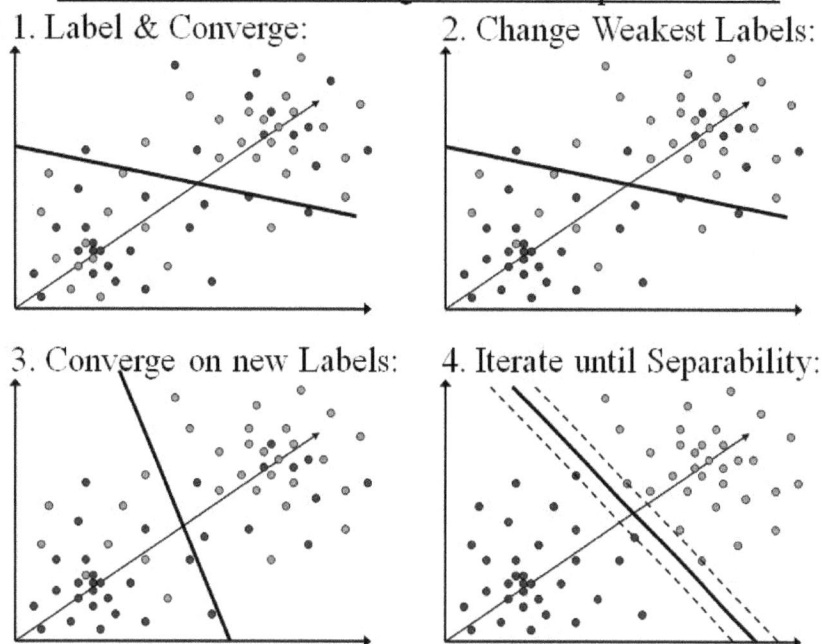

SVM-based Clustering (via multi-pass SVM)

1. Label & Converge:

2. Change Weakest Labels:

3. Converge on new Labels:

4. Iterate until Separability:

Fig. 4.27. Schematic for SVM-based clustering, starting with randomly labeled data for which an SVM 'learning' solution is sought (randomly labeled means extremely non-separable, typically). Remarkably, for certain kernels, convergence is possible.

SVM Multiclass from Decision Tree with SVM Binary Classifiers

The SVM binary discriminator offers high performance and is very robust in the presence of noise. This allows a variety of reductionist multiclass approaches, where each reduction is a binary classification. The SVM Decision Tree is one such approach used extensively with the datasets examined in [37,45], where a collection of SVM Decision Trees (a SVM Decision Forest) can be used to avoid problems with throughput biasing. Alternatively, the variational formalism can be modified to perform a multi-hyperplane optimization situation for a direct multiclass solution [1,37,45], as is described next.

The SVM Decision Tree shown in Fig. 4.28 obtained nearly perfect sensitivity and specificity, with a high data rejection rate, and a highly non-uniform class signal-calling throughput. In Fig. 4.29, the Percentage Data Rejection vs SN+SP curves are shown for test data classification runs with a binary classifier with one molecule (the positive, given by label) versus the rest (the negative). Since the signal calling wasn't passed through a Decision Tree, the way these curves were generated, they don't accurately reflect total throughput, and they don't benefit from the "shielding" shown in the Decision Tree in Fig. 4.28 prototype. In the SVM Decision Tree implementation described in Fig. 4.28 [45], this is managed more comprehensively, to arrive at a five-way signal-calling throughput at the furthest node of 16% (in Fig. 4.28, 9CG and 9AT have to pass to the furthest node to be classified), while the best throughput, for signal calling on the 8GC molecules, is 75%.

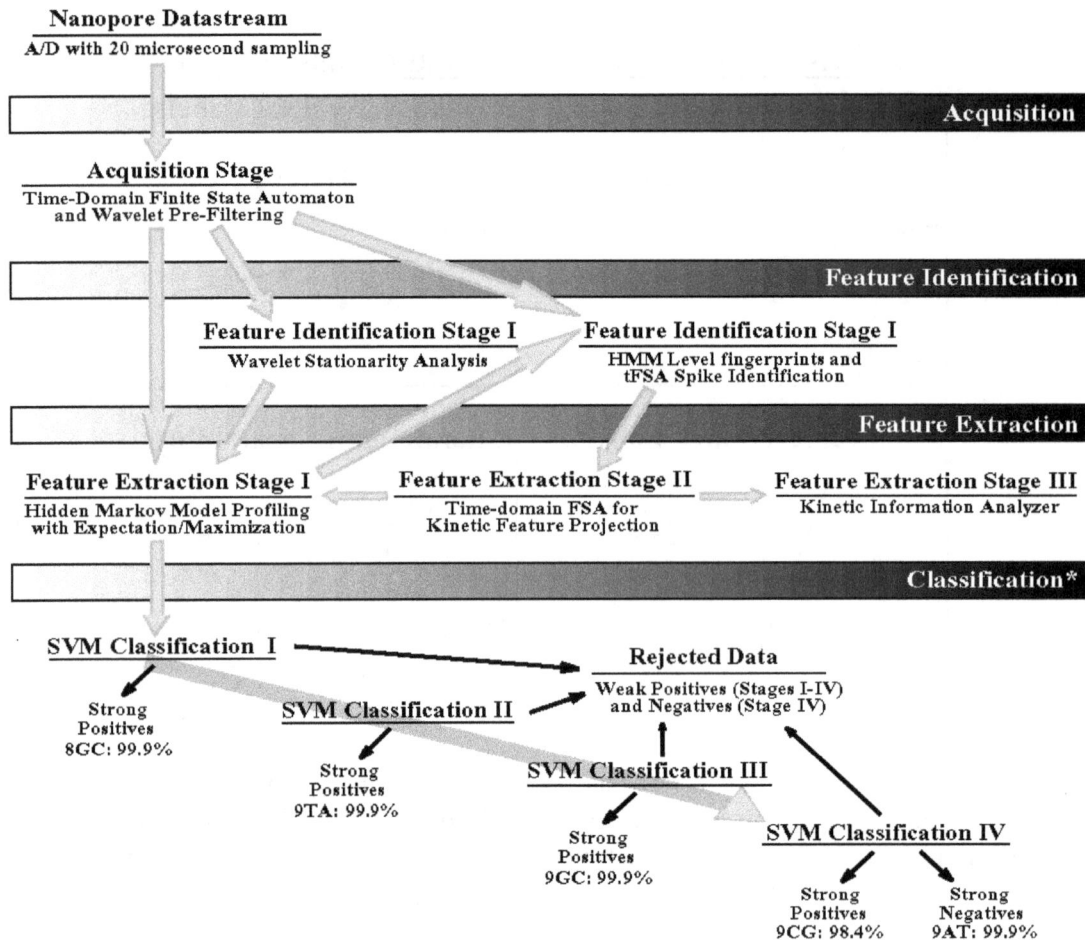

Fig. 4.28. Nanopore Detector signal analysis architecture, with use of an SVM Decision Tree for classification.

The SVM Decision Tree classifier's high, non-uniform, rejection can be managed by generalizing to a collection of Decision Trees (with different species at the furthest node). The problem is that tuning and optimizing a single decision tree is already a large task, even for five species (as in [45]). With a collection of trees, this problem is seemingly compounded, but can actually be lessened in some ways in that now each individual tree need not be so well-tuned/optimized. Although more complicated to implement than an SVM-External method, the SVM-Internal multiclass methods are not similarly fraught with tuning/optimization complications. Fig. 4.30 shows the Percentage Data Rejection vs SN+SP curves on the same train/test data splits as used for Fig. 4.29, except now the drop curves are to be understood as *simultaneous* curves (not sequential application of such curves as in Fig. 4.29). Thus, comparable, or better, performance is obtained with the multiclass-internal approach and with far less effort, since there is no managing and tuning of Decision Trees. Another surprising, and even stronger argument for the SVM-Internal approach to the problem, for many situations, is that a natural drop zone is indicated by the margin. The drawback of the significantly easier use, s shown in Fig. 4.30, is that a much more complicated Lagrangian derivation and implementation computationally is needed (see [54] for complete details).

78

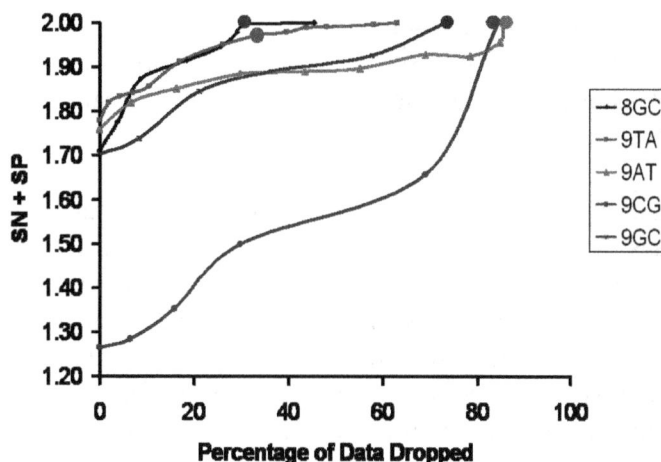

Fig. 4.29. The Percentage Data Rejection vs SN+SP curves are shown for test data classification runs with a binary classifier with one molecule (the positive, given by label) versus the rest (the negative). Since the signal calling wasn't passed through a Decision Tree, it doesn't accurately reflect total throughput, and they don't benefit from the "shielding" shown in the Decision Tree in Fig. 4.28 prototype. The Relative Entropy Kernel is shown because it provided the best results (over Gaussian and Absdiff).

Suppose we define the criteria for dropping weak data as the margin: For any data point x_i; let $\max_m\{f_m(x_i)\} = f_{yi}$, and Let $f_m = \max_m\{f_m(x_i)\}$ for all $m \neq yi$, then we define the margin as: $(f_{yi} - f_m)$, hence data point x_i is dropped if $(f_{yi} - f_m) \leq$ Confidence Parameter. (For this data set using Gaussian, AbsDiff & Sentropic kernel, a confidence parameter of at least $(0.00001)*C$ was required to achieve 100% accuracy.) Using the margin drop approach, there is even less tuning, and there is improved throughput (approximately 75% for *all* species) [1,54].

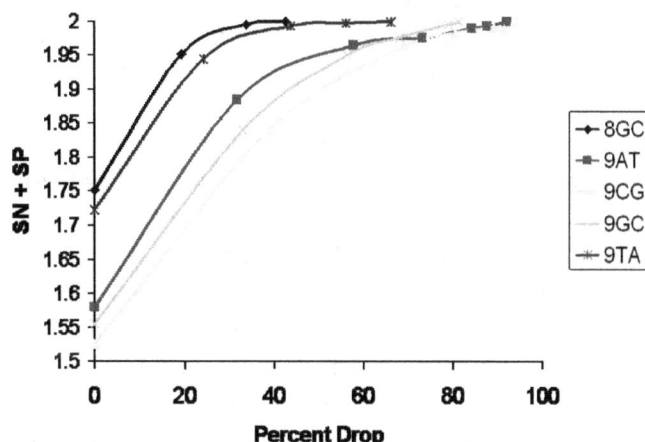

Fig. 4.30. The Percentage Data Rejection vs SN+SP curves are shown for test data classification runs with a *multiclass* discriminator. The following criterion is used for dropping weak data: for any data point x_i; if $\max_m\{f_m(x_i)\} \leq$ Confidence Parameter, then the data point x_i is dropped. For this data set using AbsDiff kernel ($\sigma^2 = 0.2$) performed best, and a confidence parameter of 0.8 achieve 100% accuracy.

79

4.10 Stochastic Sequential Analysis (SSA)

A protocol has been developed for the discovery, characterization, and classification of localizable, approximately-stationary, statistical signal structures in stochastic sequential data, such as channel current data.

The stochasric sequential analysis (SSA) methods described in what follows, provide a robust and efficient means to make a device or process as smart as it can usefully be, with possible enhancement to device (or process) sensitivity and productivity and efficiency, as well as possibly enabling new capabilities for the device or process (via transduction coupling, for example, as with the nanopore transduction detector (NTD) platform). The SSA Protocol can work with existing device or process information flows, or can work with additional information induced via modulation or introduction via transduction couplings (comprising carrier references [95,96], among other things). Hardware device-awakening and process-enabling may be possible via introduction of modulations or transduction couplings, when used in conjunction with the SSA Protocol implemented to operate on the appropriate timescales to enable real-time experimental or operationalcontrol.

4.10.1 Channel Current Cheminformatics (CCC) implementation of the Stochastic Sequential Analysis (SSA) protocol

The components for a stochastic signal analysis (SSA) protocol and a stochastic carrier wave (SCW) communications protocol are described in what follows. NTD, with the channel current cheminformatics (CCC) implementation of the SSA protocol, provides proof-of-concept examples of the SSA methods utilization, and can be used as a platform for finite state communication. From the CCC/NTD starting point it is easier to convey the unique signal boosting capabilities when working with real-time capable HMMBD signal processing [21,90] and other SSA methods. In the larger sense, recognition of stationary statistics transitions allows one to generalize to full-scale encoding/decoding in terms of stationary statistics 'phases', i.e., stochastic phase modulation, a form of stochastic carrier-wave (SCW) communications. Many of the Proof-of-concept experiments described in what follows involve SSA applications, in a CCC implementation or a context for the NTD platform. The SSA Protocol is a general signal processing paradigm for characterizing stochastic sequential data.

4.10.2 NTD 'Binary' event communication, a precursor to stochastic 'phase' modulation (SPM)

In the Nanopore Transduction Detector (NTD) experiments the molecular dynamics of a (single) captured transducer molecule provides a unique stochastic reference signal with stable statistics on the observed, single-molecule blockade, channel current, somewhat analogous to a carrier signal in standard electrical engineering signal analysis. Changes in transient blockade statistics, coupled to SSA signal processing protocols, enables the means for a highly detailed characterization of the interactions of the transducer molecule with binding cognates in the surrounding (extra-channel) environment.

The transducer molecule is specifically engineered to generate distinct signals depending on its interaction with the target molecule. Statistical models are trained for each binding mode, bound and unbound, for example, by exposing the transducer molecule to zero or high (excess) concentrations of the target molecule. The transducer molecule is engineered so that these different binding states generate distinct signals with high resolution. Once the signals are characterized, the information can be used in a real-time setting to determine if trace amounts of the target are present in a sample through a serial, high-frequency sampling process.

Thus, in NTD applications of the SSA Protocol, due to the molecular dynamics of the captured transducer molecule, a unique reference signal with stationary (or approximately stationary) statistics is engineered to be generated during transducer blockade, analogous to a carrier signal in standard electrical engineering signal analysis. The adaptive machine learning algorithms for real-time analysis of the stochastic signal generated by the transducer molecule offer a "lock and key" level of signal discrimination. The heart of the signal processing algorithm is an adaptive Hidden Markov Model (AHMM) based feature extraction method, implemented on a distributed processing platform for real-time operation. For real-time processing, the AHMM is used for feature extraction on channel blockade current data while classification and clustering analysis are implemented using a Support Vector Machine (SVM). The machine learning software has been integrated into the nanopore detector for "real-time" pattern-recognition informed (PRI) feedback [34,20]. The methods used to implement the PRI feedback include *distributed* HMM and SVM implementations, which enable the processing speedup that is needed.

A mixture of two DNA hairpin species {9TA, 9GC} is examined in an experimental test of the PRI system. In separate experiments, data is gathered for the 9TA and 9GC blockades in order to have known examples to train the SVM pattern recognition software. A nanopore experiment is then run with a 1:70 mix of 9GC:9TA, with the goal to eject 9TA signals as soon as they are identified, while keeping the 9GC's for a full 5 seconds (when possible, sometimes a channel-dissociation or melting event can occur in less than that time). The results showing the successful operation of the PRI system is shown in Fig. 4.31 as a 4D plot, where the radius of the event 'points' corresponds to the duration of the signal blockade (the 4th dimension). The result in Fig. 4.31 demonstrates an approximately 50-fold speedup on data acquisition of the desired minority species.

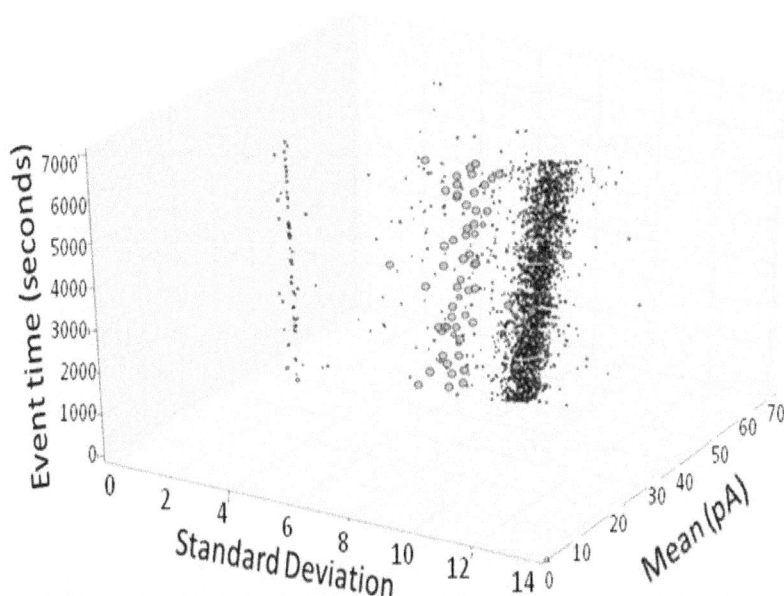

Figure 4.31. PRI Mixture Clustering Test with 4D plot [20]. The vertical axis is the event observation time, and the plotted points correspond to the

standard deviation and mean values for the event observed at the indicated event time. The radius of the points correspond to the duration of the corresponding signal blockade (the 4th dimension). Three blockade clusters appear as the three vertical trajectories. The abundant 9TA events appear as the thick band of small-diameter (short duration, ~100ms) blockade events. The 1:70 rarer 9GC events appear as the band of large-diameter (long duration, ~ 5s) blockade events. The third, very small, blockade class corresponds to blockades that partially thread and almost entirely blockade the channel.

To provide enhanced, autonomous reliability, the NTD is self-calibrating: the signals are normalized computationally with respect to physical parameters (e.g. temperature, ph, salt concentration, etc.) eliminating the need for physical feedback systems to stabilize the device. In addition, specially engineered calibration probes have been designed to enable real-time self-calibration by generating a standard "carrier signal." These probes are added to samples being analyzed to provide a run-by-run self-calibration. These redundant, self-calibration capabilities result in a device which may be operated by an entry level lab technician.

Although the nanopore transduction detector can be a self-contained 'device' in a lab, external information can be used, for example, to update and broaden the operational information on control molecules ('carrier references'). For the general 'kit' user, carrier reference signals and other systemically-engineered constructs can be used, for example, for a wide range of thin-client arrangements (where they typically have minimal local computational resource and knowledge resource). The paradigm for both device and kit implementations involve system-oriented interactions, where the kit implementation may operate on more of a data service/data repository level and thus need 'real-time' (high bandwidth) system processing of data-service requests or data-analysis requests. Although not as system-dependent on database-server linkages, the more self-contained 'device' implementation will still typically have, for example, local networked (parallelized) data-warehousing, and fast-access, for distributed processing speedup on real-time experimental operations.

4.10.3 The SSA Protocol
The SSA Protocol is for the discovery, characterization, and classification of localizable, approximately-stationary, statistical signal structures in stochastic sequential data (such as channel current data), and changes between such structures. The SSA (and CCC) protocol is usually decomposed into the following stages (see Fig.s 4.32-4.34):

(Stage 1) primitive feature identification: this stage is typically finite-state automaton based, with feature identification comprising identification of signal regions (critically, their beginnings and ends), and, as-needed, identification of sharply localizable 'spike' behavior in any parameter of the 'complete' (non-lossy, reversibly transformable) classic EE signal representation domains: raw time-domain, Fourier transform domain, wavelet domain, etc. (The methodology for spike detection is shown applied to the time-domain in [33].) Primitive feature extraction can be operated in two modes: off-line, typically for batch learning and tuning on signal features and acquisition; and on-line, typically for the overall signal acquisition (with acquisition parameters set – e.g., no tuning), and, if needed, 'spike' feature acquisition(s).

The FSA method that is primarily used in the channel current cheminformatics (CCC) signal discovery and acquisition is to identify signal-regions in terms of their having a valid 'start' and a valid 'end', with internal information to the hypothesized signal region consisting, minimally, of the duration of that signal (e.g., the duration between the hypothesized valid 'end and hypothesized valid 'start'). One approach along these lines is a signal 'fishing' protocol " …constraints on valid 'starts' that are weak (with prominent use of 'OR' conjugation) and constraints on valid 'ends' that are strong (with prominent use of 'AND' conjugation)." We underpin our approach to signal analysis in a fundamentally different way, however, although the signal fishing method indicated above is still used as needed. The FSA signal analysis methodology used here, for example, involves identifying anomalously long-duration regions. Identification of anomalously-long duration regions in the more sophisticated Hidden Markov model (HMM) representation would require use of a HMM-with-duration to not lose the information on the anomalous durations, which is one of the application areas for the HMMBD method.

Once identification rules, often threshold-based, are established for the signal start's and signal end's, then those definitions can be explored/used in signal acquisition. As those definitions are tuned over, by exploring the different signal acquisition results obtained with different parameter settings, the signal acquisition counts can undergo radical phase transitions, providing the most rudimentary of the holistic tuning methods on the primitive feature acquisition FSA. By examining those phase transitions, and the stable regimes in the signal counts (and other attributes in more involved holistic tuning), the recognition of good parameter regimes for accurate acquisition of signal can be obtained. As more internal signal structure is modeled by the FSA, the holistic tuning can involve more sophisticated tuning recognition of emergent grammars on the signal sub-states. The end-result of the tuning is a signal acquisition FSA that can operate in an on-line setting, and very efficiently (computation on the same order as simply reading the sequence) in performing acquisition on the class of signals it has been 'trained' to recognize. On-line learning is possible via periodic updates on the batch learning state/tuning process. For typical SSA (and CCC) applications, the tFSA is used to recognize and acquire 'blockade' events (which have clearly defined start and stop transitions).

(Stage 2a) feature identification and feature selection: this stage in the signal processing protocol is typically Hidden Markov model (HMM) based, where identified signal regions are examined using a fixed state HMM feature extractor or a template-HMM (states not fixed during template learning process where they learn to 'fit' to arrive at the best recognition on their train-data, the states then become fixed when the HMM-template is used on test data). The Stage 2 HMM methods are the central methodology/stage in the CCC protocol in that the other stages can be dropped or merged with the Stage 2 HMM in many incarnations. For example, in some data analysis situations the Stage 1 methods could be totally eliminated in favor of the more accurate HMM-based approach to the problem, with signal states defined/explored in much the same setting, but with the optimized Viterbi path solution taken as the basis for the signal acquisition structure identification. The reason this is not typically done is that the FSA methods sought in Stage 1 are usually only $O(T)$ computational expense, where 'T' is the length of the stochastic sequential data that is to be examined, and '$O(T)$' denotes an order of computation that scales as 'T' (linearly in the length of the sequence). The typical HMM Viterbi algorithm, on the other hand, is $O(TN^2)$, where 'N' is the number of states in the HMM. Stage 1 provides a faster, and often more flexible, means to acquire signal, but it is more hands-on. If the core HMM/Viterbi method can be approximated such that it can run at O(TN) or even O(T) in certain data regimes, for example, then the non-HMM methods in stage 1 could be phased out. Such HMM approximation methods present a data-dependent branching in the most efficient implementation of the protocol. If the data is sufficiently regular, direct tuning and regional approximation with HMM's may allow Stage 1 FSA methods to be avoided entirely. For general data, however, some tuning and signal acquisition according to Stage 1 will be needed (possibly off-line) if only to then bootstrap (accelerate) the learning task of the HMM approximation methods.

The HMM emission probabilities, transition probabilities, and Viterbi path sampled features, among other things, provide a rich set of data to draw from for feature extraction (to create 'feature vectors'). The choice of features is optimized according to the classification or clustering method that will make use of that feature information. In typical operation of the protocol, the feature vector information is classified using a Support Vector Machine (SVM). This is described in Stage 3 to follow. Once again, however, the Stage 3 classification could be totally eliminated in favor of the HMM's log likelihood ratio classification capability at Stage 2, for example, when a number of template HMMs are employed (one for each signal class). This classification approach is inherently weaker and slower than the (off-line trained) SVM methodology in many respects, but, depending on the data, there are circumstances where it may provide the best performing implementation of the protocol.

(Stage 2b) Stochastic carrier wave encoding/decoding
Using HMMBD we have an efficient means to establish a new form of carrier-based communications where the carrier is not periodic but is stochastic, with stationary statistics. The HMMBD algorithmic methodology, [90], enables practical stochastic carrier wave (SCW) encoding/decoding with this method.

Stochastic carrier wave (SCW) signal processing is also encountered at the forefront of a number of efforts in nanotechnology, where it can result from establishing or injecting signal modulations so as to boost device sensitivity. The notion of modulations for effectively larger bandwidth and increased sensitivity is also described in [95,96]. Here we choose modulations that specifically evoke a signal type that can be modeled well with a HMMD but not with a HMM. This is a generally applicable approach where conventional, periodic, signal analysis methods will often fail. Nature at the single-molecule scale may not provide a periodic signal source, or allow for such, but may allow for a signal modulation that is stochastic with stationary statistics, as in the case of the nanopore transduction detector (NTD).

(Stage 3) classification: this stage is typically SVM based. SVMs are a robust classification method. If there are more classes to discern than two, the SVM can either be applied in a Decision Tree construction with binary-SVM classifiers at each node, or the SVM can internally represent the multiple classes, as done, for example, in proof-of-concept experiments. Depending on the noise attributes of the data, one or the other approach may be optimal (or even achievable). Both methods are typically explored in tuning, for example, where a variety of kernels and kernel parameters are also chosen, as well as tuning on internal KKT handling protocols. Simulated annealing and genetic algorithms have been found to be useful in doing the tuning in an orderly, efficient, manner. If the feature vectors produced correspond to complete data information/profiling in some manner, such is explicitly the case in a probability feature vector representation on a complete set of signal event frequencies (where all the feature 'components' are positive and sum to 1), then kernels can be chosen that conform to evaluating a measure of distance between feature vectors in accordance with that notion of completeness (or internal constraint, such as with the probability vectors). Use of divergence kernels with probability feature vectors in proof-of-concept experiments have been found to work well with channel blockade analysis and is thought to convey the benefit of having a better pairing of kernel and feature vector, here the kernels have probability distribution measures (divergences), for example, and the feature vectors are (discrete) probability distributions.

(Stage 4) clustering: this stage is often not performed in the 'real-time' operational signal processing task as it is more for knowledge discovery, structure identification, etc., although there are notable exceptions, one such comprising the jack-knife transition detection via clustering consistency with a causal boundary that is described in what follows. This stage can involve any standard clustering method, in a number of applications, but the best performing in the channel current analysis setting is often found to be an SVM-based external clustering approach (see [31]), which is doubly convenient when the learning phase ends because the SVM-based clustering solution can then be fixed as the supervised learning set for a SVM-based classifier (that is then used at the operational level).

A computationally 'expensive' HMM signal acquisition at Stage 1 may be necessary for very weak signals, for example, if the typical Stage 1 methods fail. In this situation the HMM will probably have a very weak signal differential on the different signal classes if it were to attempt direct classification (and eliminate the need for a separate Stage 3). In this setting, the HMM would probably be run in the finest grayscale generic-state mode, with a number of passes with different window sample sizes to 'step through' the sequence to be analyzed. Then, there are two ways to proceed: (1) with a supervised learning 'bias', where windows on one side of a 'cut' are one class, and those on the other side the other class, can a the SVM classify at high accuracy on train/test with the labeled data so indicated? If so, a transition is identified. In (2) the idea is to use an unsupervised learning SVM-based clustering method where we look for a strong knife-edge split on clustered populations along the sequence of window samples. When this occurs, there is a strong identification of a transition. Since regions are identified (delineated) by their transition boundaries, we arrive at a minimally-informed means for state and state-transition discovery in stochastic sequential data involving HMM/SVM based channel current signal processing.

(All Stages) Database/data-warehouse System specification:
The adaptive HMM (AHMM) and modified SVM systems require implementation-specific data schema designs, for both input and output. The signal processing algorithms depend on information, represented structurally in the data, the algorithms are both process driven and data driven - these components impact the implementation of the algorithms.

The data schemas are typically implemented for optimal read time and ease of re-use and deployment, and have system dependencies that can be very significant, such as with client data-services involving distributed data access. The data schemas are typically implemented using flat files, low level operating system specific system calls to map data onto virtual memory, Relational Database Management Systems (RDBMS), and Object Database Management Systems (ODBMS). The database schemas are defined in two system contexts, 1) real time data acquisition, which includes feature recognition (AHMM) and classification (SVM), and, 2) data warehousing for client data-service, and for further analysis that can be computationally intensive and requires substantial data processing.

The real-time data acquisition systems associated with the signal processing can be implemented using flat file systems and operating system specific virtual memory management interfaces. These interfaces are optimized to be scalable and high-bandwidth, to meet the requirements of high speed, real-time, data acquisition and storage. The data schemas allow for real-time signal processing such as feature recognition and classification, as well as local storage for subsequent export to a data warehouse, which can be implemented using industry standard RDBMS and ODBMS systems.

(All Stages) Server-based data analysis System specification:
The data warehouse data schemas are optimized for applications-specific analysis of the signal processing tools in a distributed, scalable environment where substantial computing power can extend the analysis beyond what is possible in real-time. The local data acquisition systems produce and identify structure in real-time, storing the data locally, while another process can stream the data transparently to an off-site data warehouse for subsequent analysis. The database uses data modeling tools to identify data schemas that work in tandem with the signal processing algorithms. The structure of the data schemas are typically integral to efficient implementation of the algorithms. Substantial off-line data pre-processing, for example, is used to create data structures based on inherent structure identified in the data. An internet-based user interface allows for access to the stored data and provides a suite of server-based, application-specific analysis and data mining tools.

I. Channel Current Cheminformatics (CCC) Protocol

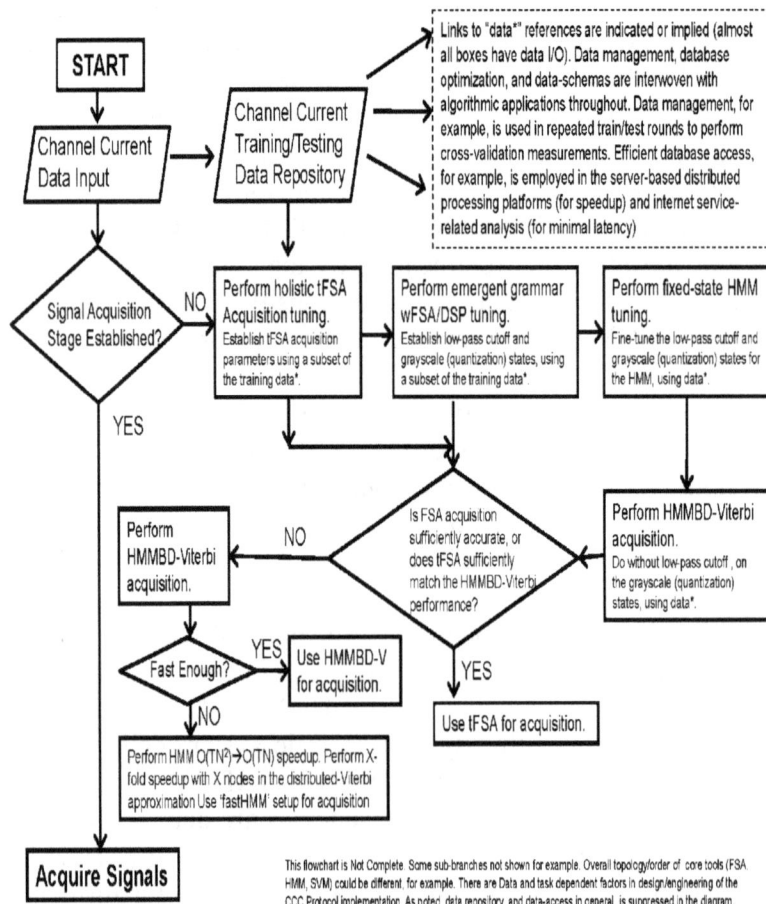

Fig. 4.32. CCC Protocol Flowchart (part 1)

II. Channel Current Cheminformatics (CCC)Protocol

This flowchart is Not Complete. Some sub-branches not shown for example. Overall topology/order of core tools (FSA, HMM, SVM) could be different, for example. There are Data and task dependent factors in design/engineering of the CCC Protocol implementation. As noted, data repository, and data-access in general is suppressed in the diagram.

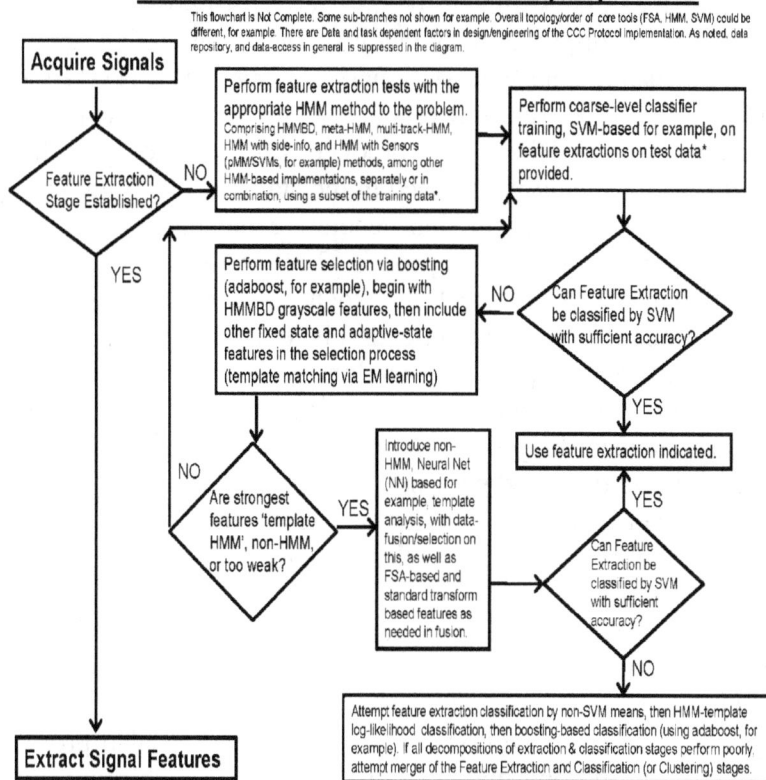

Fig. 4.33. CCC Protocol Flowchart (part 2)

III. Channel Current Cheminformatics (CCC)Protocol

This flowchart is Not Complete. Some sub-branches not shown for example. Overall topology/order of core tools (FSA, HMM, SVM) could be different, for example. There are Data and task dependent factors in design/engineering of the CCC Protocol implementation. As noted, data repository, and data-access in general, is suppressed in the diagram.

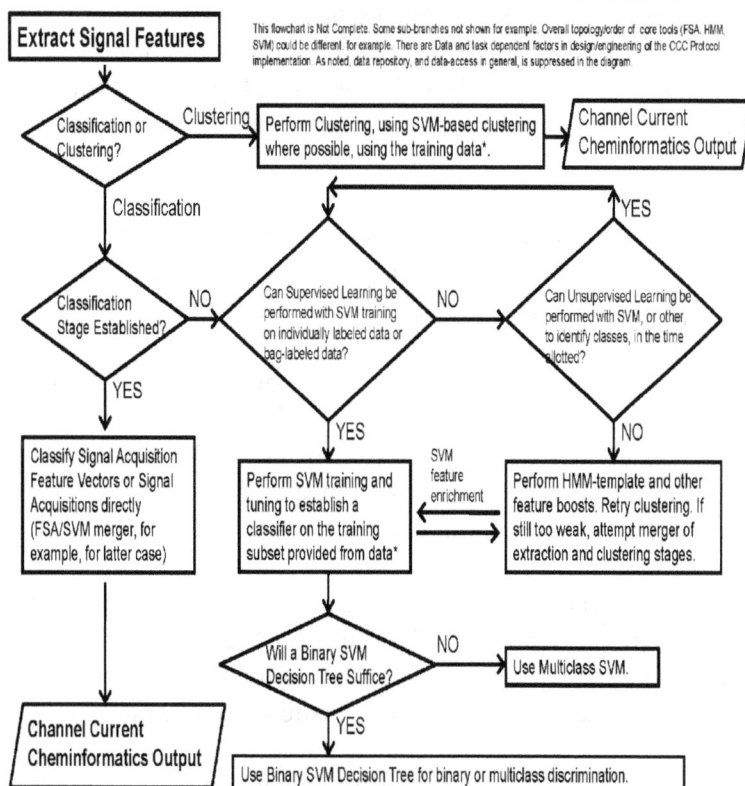

Fig. 4.34. CCC Protocol Flowchart (part 3)

4.10.4 SCW for detector sensitivity boosting

Since an HMM template match can be weighted by its Viterbi path probability, the HMM's generative projection might offer improved efficiency in the compressive sensor method [209] when working with stochastic data with stationary statistics. The HMM's generative capability derives from its ability to be 'run in reverse', e.g., given the learned parameterization of a particular state model, an HMM can produce stochastic data with the same stationary statistics as that which originally was used in the statistical learning process. (The terminology 'emission probability' relates to this generative perspective.) The SCW approach is similar to the compressive sensing approach [209] in that is involves a random (stochastic) projection, but in the SCW approach the stochastic projection is constrained to have stationary statistics and the main area of application is to stochastic sequential analysis. Compressive sensing makes use of a sequence of random projections, while in SCW a sequence of generative projections are done by using template HMMs and compressing stationary statistical parameters according to the generalized Viterbi algorithms. Stationary statistical signals that are truly modulatory, e.g., not simply at a fixed level with Gaussian noise fluctuations about that level, will have a sparse data representation in terms of generative feature sets (which is used by SVMs for some classification tasks).

New HMM implementations, allow a new form of carrier-based communications where the carrier is not periodic but is stochastic, with stationary statistics. The "stochastic carrier wave" (SCW) approach is not only a means to understand the messages Nature provides (in near-equilibrium flow phenomenologies with stationary statistics), but also provides a hidden carrier method, enabling security and making signal jamming much more difficult. An algorithmic methodology allows for 100-fold, or faster, implementation of a Hidden Markov model with duration (the HMMBD algorithm [21]), is critical to this encoding/decoding method.SCW communications are found at the forefront of a number of efforts in nanotechnology. This is because nature at the single-molecule scale has a signal modulation that is stochastic, sometimes with stationary statistics. Such is the case with the signal analysis in a nanopore transduction detector (NTD). Thus, further developments with stochastic carrier wave methods would serve both communications efforts and biosensing efforts.

If states have self-transitions with a notably non-geometric distribution on their self-transition 'durations', then a fit to a geometric distribution in this capacity, as will be forced by the standard HMM, will be weak, and HMMD modeling will serve better. In engineered communications protocols, or in engineered, modulated, nanopore transduction detector (NTD) signals, highly non-geometric distributions can be sought. One encoding scheme that is strongly non-geometric in same-state duration distribution is the familiar *long* open-reading-frame (ORF) encoding found in genomic data. This suggests a similar ORF-like encoding scheme to establish a carrier duration peak in the self-transition distribution's tail region, e.g., a second peak in the duration distribution (perhaps one even more skewed from the geometric distribution than the heavy-tail distributions found for ORFs).

The NTD signal analysis demonstrates the simplest stochastic carrier wave utilization, in a biophysics experimental setting. A minor elaboration on the signal analysis, to go from a simple two-state (bound/unbound) signal recognition to a lengthy two-state telegraph signal, then yields the rudimentary implementation for stochastic carrier communications purposes.

The adaptive machine learning algorithms for real-time analysis of the stochastic signal generated by the NTD transducer molecule are critical to realizing the increased sensitivity of the NTD and offer a "lock and key" level of signal discrimination. The transducer molecule is specifically engineered to generate distinct signals depending on its interaction with the target molecule. Statistical models are trained for each binding mode, bound and unbound, by exposing the transducer molecule to high concentrations of the target molecule. The transducer molecule has been engineered so that these different binding states generate distinct signals with high resolution. In operation, the NTD-biosensing process is analogous to giving a bloodhound a distinct memory of a human target by having it sniff a piece of clothing. Once the signals are characterized, the information is used in a real-time setting to determine if trace amounts of the target are present in a sample through a serial, high frequency sampling process.

The algorithms which describe the stochastic channel current modulations are a pure form of Machine Learning (a branch of Artificial Intelligence) in that there is no assumption of an underlying probability distribution; the statistical representation is directly generated from the data being produced by the molecular dynamics of the transducer molecule. A statistical model that is based on direct observation of the transducer molecule's dynamics eliminates the need for a parameterized statistical model, resulting in a higher resolution of discrimination.

General methods are proposed for (i) stochastic sequential analysis; (ii) stochastic carrier-wave communications; (iii) holographic HMM extensions; and, (iv) distributed HMM implementations. In method (ii), in particular, we establish a new type of communication process where the carrier wave is a stochastic observation sequence that obeys stationary statistics. In standard periodic carrier wave signal processing convolving with the carrier frequency allows the signal modulations of that carrier to be obtained. Here we have something analogous, but we have a carrier with stationary statistics, not fixed frequency, and can recognize different phases of stationary statistics via HMM methods for class-independent feature extraction, with Support Vector Machines (SVMs) for sparse data classification, or via HMM methods for class-dependent HMM generative projection (as mentioned in earlier comments).

In standard signal analysis with periodic waveforms, sampling is done at the Nyquist rate and the data compressed and transmitted (a 'smart' encoder). The received data is then decompressed and reconstructed (by simply summing wave components, e.g., a 'simple' decoder). If the signal is sparse or compressible, then compressive sensing [209] can be used, where sampling and compression are combined into one efficient step to obtain compressive measurements (refered to as 'dumb' encoding in [209] since a set of random projections are employed), which are then transmitted. On the receiving end, the decompression and reconstruction steps are, likewise, combined using an asymmetric 'smart' decoding step. This progression towards asymmetric compressive signal processing can be taken a step further if we consider signal sequences to be equivalent if they have the same stationary statistics. What is obtained is a method similar to compressive sensing, but involving stationary-statistics generative-projection sensing, where the signal processing is non-lossy at the level of stationary statistics equivalence. In the SCW signal analysis the signal source is generative in that it is describable via use of a hidden Markov model, and the HMM's Viterbi-derived generative projections are used to describe the sparse components contributing to the signal source. In SCW encoding the modulation of stationary statistics can be man-made or natural, with the latter in many experimental situations that involve flow phenomologies that have stationary statistics. If the signal is man-made, usually the underlying stochastic process is still a natural source, where it is the changes in the stationary statistics that is under the control of the man-made encoding scheme. Transmission and reception are then followed by generative projection via Viterbi-HMM template matching or via Viterbi-HMM feature extraction followed by separate classification (using SVM). So in the SCW approach the encoding is even 'dumber' in that it can be any noise source with stationary statistics (the case for many experimental observations), with stationary statistics phase modulation for encoding. The decoding must be even 'smarter', on the other hand, in that generalized Viterbi algorithms are used to perform a generative projection (and possibly other machine learning methods as well, SVMs in particular). An example of the stationary statistics sensing with a machine learning based decoder is described in application to channel current cheminformatics studies in what follows.

In the standard HMM [187], when a state 'i' is entered, that state is occupied for a period of time, via self-transitions (with self-transition probability denoted as a_{ii}), until transiting to another state 'j' (with probability a_{ij}). If the same-state interval is given as 'd', the standard HMM description of the probability distribution on state intervals is implicitly given by a geometric distribution. The best-fit geometric distribution, however, is inappropriate in many cases. The standard HMMD replaces the above equation with a $p_i(d)$ that models the real duration distribution of state i. In this way explicit knowledge about the duration of states is incorporated into the HMM.

The original description of an explicit HMMD required computation of order $O(TN^2+TND^2)$ [210] (where T is the sequence length to be examined, N is the number of states in the HMM/HMMD model, and D is the maximum duration length allowed in the HMMD model). The 'D^2' term made the original approach prohibitively computationally expensive in practical, real-time, operations, and introduced a severe maximum-duration constraint on the duration-distribution model. Improvements via hidden semi-Markov models to computations of order $O(TN^2+TND)$ are described in [211,212], where the maximum-interval constraint is still employed, and comparisons of these methods were subsequently detailed in [213]. In [21] we show that $O(TN^2+TND^*)$ is possible with the HMMBD algorithm, where D* is the number of binned length states. The HMMBD implementation brings the HMMD modeling within the range of computational viability for many applications. In the HMMBD approach we also eliminate the maximum-duration constraint. We can often reduce to a bin representation with D*<10, such that D*<<N in many situations, in which case that the HMMBD requires computations of order $O(TN^2)$, the same as for the HMM alone.

One important application of the HMM-with-duration (HMMD) method used in [21] includes kinetic feature extraction from EVA projected channel current data (the HMM-with-Duration is shown to offer a critical stabilizing capability in an example in[21]). The EVA-projected/HMMD processing offers a hands-off (minimal tuning) method for extracting the mean dwell times for various blockade states (the core kinetic information on the blockading molecule's channel interactions).

The HMM-with-Duration implementation, described in [21], is being explored in terms of its performance at parsing synthetic blockade signals. In the [21] experiment the synthetic data was designed to have two levels, with lifetime in each level determined by a governing distribution (Poisson and Gaussian distributions with a range of mean values were considered). The results clearly demonstrate the superior performance of the HMMD over the simpler HMM formulation on data with non-geometrically distributed same-state interval durations. With use of the EVA-projection method this affords a robust means to obtain kinetic feature extraction. The HMM with duration is critical for accurate kinetic feature extraction, and the results in [21] suggest that this problem can be elegantly solved with a pairing of the HMM-with-Duration stabilization with EVA-projection.

In Fig. 4.35 we show state-decoding on synthetic data that is representative of a biological-channel two-state ion-current decoding problem, or an encode/decode software radio signal. For this problem 120 data sequences were generated that have two states with channel blockade levels set at 30 and 40 pA (a typical scenario in practice). Every data sequence has 10,000 samples. Each state has emitted values in a range from 0 to 49 pA. The maximum duration of states is set at 500. The mean duration of the 40 pA state is given as 200 samples (typically have one sample every 20 microseconds in actual experiments), while the 30 pA level has mean duration set at 300 samples. The task is to train using 100 of the generated data sequences and attempt state-decoding on the remaining 20 data sequences. Example sequences are shown in Fig. 4.35, along with their decoding when an HMM or an HMMD is employed. The performance difference is stark: the exact and adaptive HMMD decodings are 97.1% correct, while the HMM decoding is only correct 61% of the time (where random guessing would accomplish 50%, on average, in a two-state system). Three parameterized distributions were examined: geometric, Gaussian, and Poisson. Distributions that were segmented and "messy" were also examined. In all cased the HMMD performed robustly, similar to the above, and in all cases the adaptive HMMD optimization performed comparably to the more computationally expensive exact HMMD.

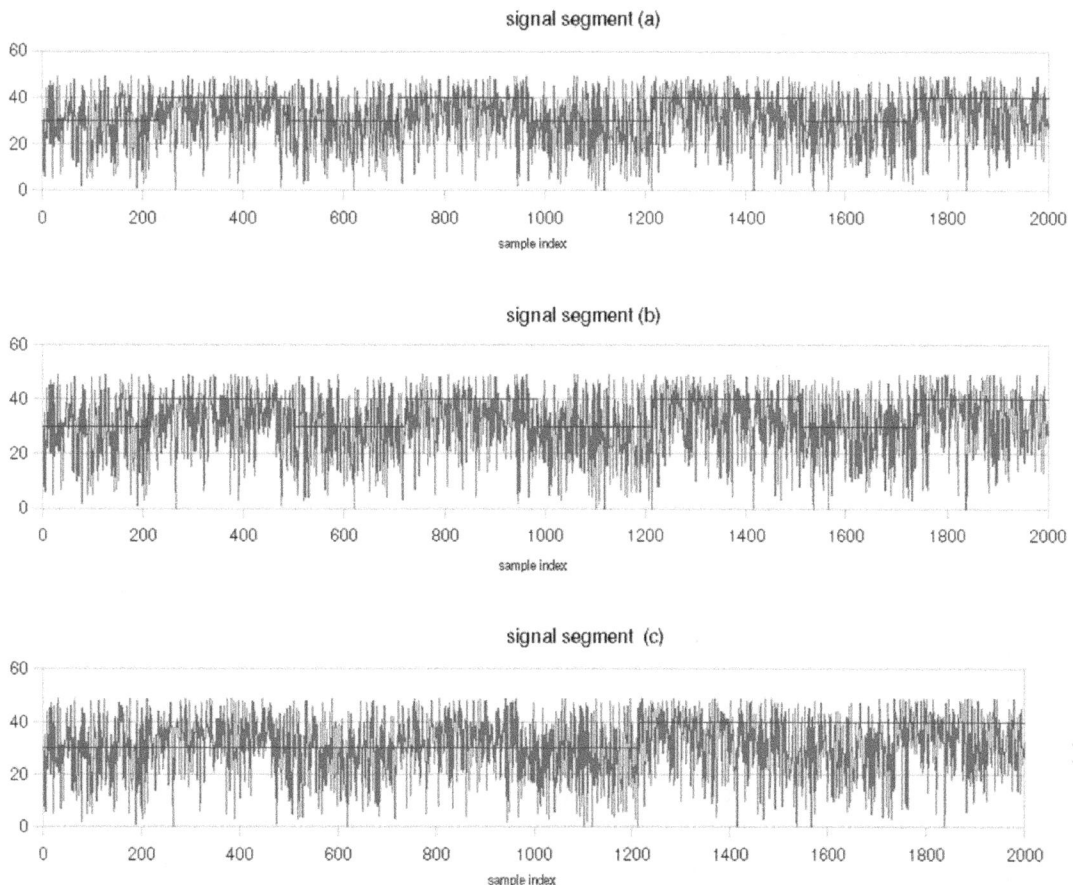

Fig. 4.35. In the figure we show state-decoding results on synthetic data that is representative of a biological-channel two-state ion-current decoding problem.

91

Signal segment (a) (at the top) shows the original two-level signal as the dark line, while the noised version of the signal is shown in red. Signal segment (b) (in the middle) shows the noised signal in red and the two-state denoised signal according to the HMMD decoding process (whether exact or adaptive), which is stable (97.1% accurate) allowing for state-lifetime extraction (with the concomitant chemical kinetics information that is thereby obtained in this channel current analysis setting). Signal segment (c) (at the bottom) shows the standard HMM signal resolution, and its failure to properly resolve the desired level-lifetime information.

Stochastic Carrier Wave (SCW) signal processing occurs in both natural and engineered situations. Whenever Nature is observed with a sequence of observations that have stationary statistics (associated with equilibrium and near-equilibrium flow situations, for example), then the basis for SCW signal processing arises. SCW also parallels all electrical engineering carrier-wave methodologies where periodic wave methods are used in some modulation scheme, thus the number of engineering applications is enormous. AM heterodyning, for example, can be replaced with stochastic carrier wave with pattern recognition informed (PRI) heterodyning . Also have phase modulation equivalence: the standard periodic carrier wave approach has a coherent phase reference, while SCW introduces a stochastic carrier wave with stationary statistics 'phase'. Have similar capabilities as with phased-locked loop (PLL), for example, where the phase tracking is done on SCW encoded information.

Chapter 5

Channel current modulation detection (*transduction* detection)

5.1 Channel-based detection mechanisms
5.1.1 Partitioning and translocation-based ND biosensing methods

The standard Nanopore Detector (ND) detection paradigm, that is predominantly translocation (or dwell-time) based, is shown in Fig. 5.1 side-by-side with the Nanopore *transduction* detector paradigm. Fig. 5.2 elaborates on the possible ND detection platform topologies possible with translocation-based approaches, where the difference in translocation times is often the critical information that is used. The difference in dwell times can depend on the off-binding time of the target binding entity (possibly in a high strain environment), where binding failure allows polymer (ssDNA) translocation to complete (and the channel blockade to end). By this mechanism, and its variants, bound probes can be distinguished from unbound. There are specificity limits on the melting-based detection, however, that are not a problem in the NTD approach.

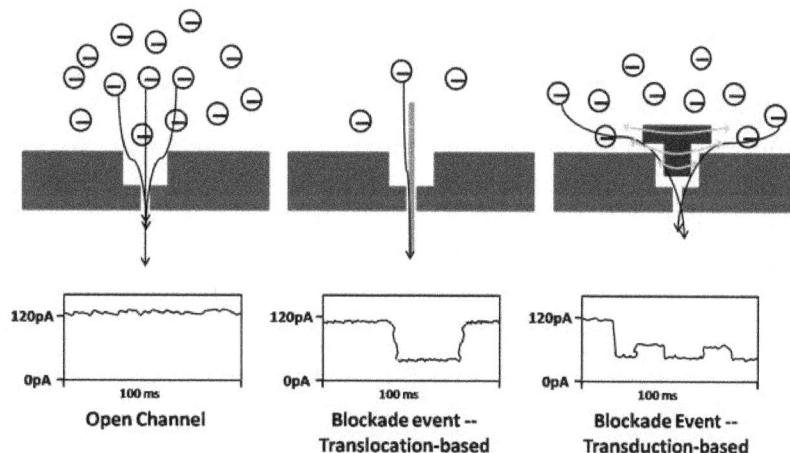

Fig. 5.1. Translocation Information and Transduction Information . Left. Open Channel . **Center.** A channel blockade event with feature extraction that is typically dwell-time based. A Single-molecule coulter counter. **Right.** Sin-

93

gle-molecule transduction detection is shown with a transduction molecule modulating current flow (typically switching between a few dominant levels of blockade, dwell time of the overall blockade is not typically a feature -- many blockade durations will not translocate in the time-scale of the experiment, for example, active ejection control is often involved).

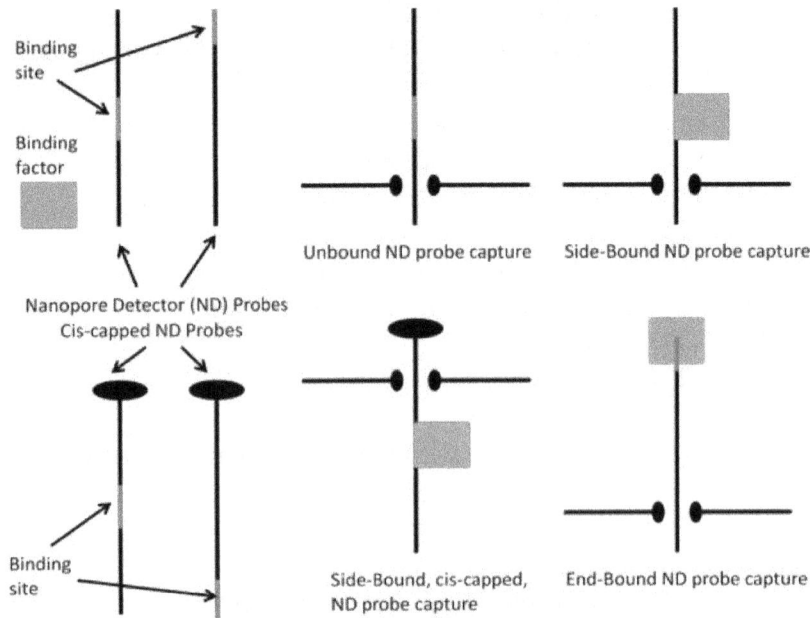

Fig. 5.2. Nanopore Detector detection topologies involving polymer translocation or threading. The detection event is given by polymer (ssDNA) translocations that are delayed if bound (side and end configurations shown). If bound entity is on the trans side (with cis-side capped, or vice versa), and bound entity is a processive DNA enzyme, then sequencing may be possible.

5.1.2 Transduction *vs.* Translation

There are two ways to functionalize measurements of the flow (of something) through a 'hole' : (1) translocation sensing; and (2) transduction sensing. The translocation methods in the literature are typically a form of a 'Coulter Counter', with a wide range of channel dimensions allowable, that typically measures molecules non-specifically via pulses in the current flow through a channel as each molecule translocates. The transduction biosensing method, on the other hand, requires nanopore sizes that are much more restricted, to the 1-10nm inner diameters that might capture, and not translocate, most biomolecules. Transduction functionalization uses a channel flow modulator (see Table 5.1 below) that also has a specific binding moiety, the transducer molecule. In transduction, the transducer molecule is used to measure molecular characteristics *indirectly*, by using a transducer/reporter molecule that binds to certain molecules, with subsequent distinctive blockade by the bound, or unbound, molecule complex. One such transducer, among many studied, was a channel-captured dsDNA "gauge" that was covalently bound to an antibody. The transducer was designed to provide a blockade shift upon antigen binding to its exposed antibody binding sites. In turn, the dsDNA-antibody transducer platform then provides a means for directly observing the *single molecule* antigen-binding affinities of any antibody in single-molecule focused assays, in addition to detecting the presence of binding target in biosensing applications.

94

There are two approaches to utilizing a nanopore for detection purposes: translocation /dwell-time (T/DT) based approaches, which strongly rely on blockade dwell-times, and nanopore transduction detection (NTD) based approaches, which functionalizes the nanopore by utilizing an engineered blockade molecule with blockade features typically not including dwell-time.

Translocation/dwell-time methods introduce different states to the channel via use of the frequency of channel blockades, from a series of individual molecular blockades (often during their translocation). The strongest feature employed in translocation/dwell-time discrimination, and often the only feature, is the blockade dwell-time where the dwell-time is typically engineered to be associated with the lifetime until a specific bond failure occurs. Other feature variations include time *until* a bond-formation occurs, or simply measuring the approximate length of a polymer according to its translocation 'dwell'-time.

Feature Space.	The T/DT approach typically has a single feature, the dwell time. Sometimes a second feature, the fixed blockade level observed, is also considered, but usually not more features sought (or engineered) than that.	The NTD approach has multiple features, e.g., blockade HMM parameters, etc., with number and type according to modulator design objectives.
Versatility.	T/DT: highly engineered/pre-processed for detection application to a particular target.	NTD: requires minimal preparation/augmentation to the transduction platform via use of separately provided binding moieties (antibody or aptamer, for example) for particular target or biomarker (which are then simply linked to modulator)
Speed.	T/DT: Slow: entire detection "process" is at the channel, and typically restricted on processing speed to the average time-scale feature (dwell-time) for the longest-lived blockade signal class.	NTD: Fast: feature extraction not dependent on dwell-time. Very low probability to get a false positive.
Multichannel.	T/DT: Method not amenable to multichannel gain with single-potential platform (can't resolve single-channel blockade signal with multichannel noise).	NTD: Have multichannel gain due to rich signal resolution capabilities of an engineered modulator molecule.
Feature Refinement/Engineering.	T/DT: No buffer modifications or off-channel detection extensions via introduction of substrates; the weak feature set limited to dwell-time doesn't allow such methods to be utilized.	NTD: Have "lock-and-key" level signal resolution. The introduction of off-channel substrates in the buffer solution can increase sensitivity.
Multiplex capabilities.	T/DT: Each modified channel is limited to detect a single analyte or single bond-change-event detection, so no multiplexing without brute force production of arrays of T/DT detectors in a semiconductor production setting.	NTD: Supports multi-transducer, multi-analyte detection from a single sample. Supports multichannel with a single aperture.

Table 5.1. Comparative analysis of the Translocation/Dwell-Time (T/TD) and Nanopore Transduction Detection (NTD) approaches.

5.1.3 Single-molecule *vs.* Ensemble

When the extra-channel states correspond to bound or unbound, there are two protocols for how to set up the Nanopore Transduction Detection (NTD) platform: (1) observe a sampling of bound/unbound states, each sample only held for the length of time necessary for a high accuracy classification. Or, (2), hold and observe *a single* bound/unbound system and track its history of bound/unbound states. The single molecule binding history in (2) has significant utility in its own right, especially for observation of critical conformational change information not observable by any other methods (critical information for understanding antibodies, allosteric proteins, and many enzymes). The ensemble measurement approach in (1), however, is able to benefit from numerous further augmentations, and can be used with general transducer states, not just those that correspond to a bound/unbound extra-channel states.

Fundamentally, the weaknesses of the standard ensemble-based binding analysis methods are directly addressed with the single-molecule approach, even if only to do a more informed type of ensemble analysis. The role of conformational change during binding, in particular, could potentially be directly explored in this setting. This approach also offers advantages over other single-molecule translation-based nanopore detection approaches in that the transduction-based apparatus introduces two strong mechanisms for boosting sensitivity on single-molecule observation: (i) engineered enhancement to the device sensitivity via the transduction molecule itself; and (ii) machine learning based signal stabilization with highly sensitive state resolution. NTD used in conjunction with recently developed pattern recognition informed sampling capabilities [20] greatly extends the usage of the single-channel apparatus. For medicine and biology, NTD and machine learning methods may aid in understanding multi-component interactions (with co-factors), and aid in designing co-factors according to their ability to result in desired binding or modified state.

In ensemble *single-molecule* measurements (via serial detection process), the pattern recognition informed (PRI) sampling on molecular populations provides a means to accelerate the accumulation of kinetic information. PRI sampling over a population of molecules is also the basis for introducing a number of gain factors. In the ensemble detection with PRI approach [20], in particular, one can make use of antibody capture matrix and ELISA-like methods [1], to introduce two-state NTD modulators that have concentration-gain (in an antibody capture matrix) or concentration-with-enzyme-boost-gain (ELISA-like system, with production of NTD modulators by enzyme cleavage instead of activated fluorophore). In the latter systems the NTD modulator can have as 'two-states', cleaved and uncleaved binding moieties. UV- and enzyme-based cleavage methods on immobilized probe-target can be designed to produce a high-electrophoretic-contrast, non-immobilized, NTD modulator, that is strongly drawn to the channel to provide a 'burst' NTD detection signal [1].

5.1.4 Biosensing with high sensitivity in presence of interference

Clinical studies have shown an abundance of protein-based disease markers that accumulate in the blood of patients suffering from chronic kidney disease. In the case of the Bioscience PXRF01marker the stage of kidney disease is linearly correlated ($r=.83$) indicating that the more severe the disease, the greater the accumulation of the marker in the bloodstream of patients. The NTD biosensing platform provides a tool for quantifying the relationship between PXRF01 and its biosystem interactants with an unparalleled fidelity. With higher quantification of PXRF01 a more accurate characterization of the disease biomarker and kidney disease progression can be established. Greater sensitivity translates directly to earlier diagnosis and improved outcomes. The electrophoretic nature of the biosensing platform also allows for significant advantage in dealing with interference agents, whether in the blood sample itself, say, or due to contaminants, since the reporter molecule can be designed to have a charge that easily separates it from the interference agents. (This is why blood can be scraped off the dirty floor at a crime scene and still accurately report on the identity or identities of those present.)

5.1.5 Introductory Nanopore Transduction Detection

Transduction methods introduce different states to the channel via observations of changes in blockade statistics on a single molecular blockade event that is modulatory. This is a specially engineered arrangement involving a partially-captured, single-molecule, channel modulator, typically with a binding moiety for a specific target of interest linked to the modulator's extra-channel portion. The modulator's 'state' changes according to whether its binding moiety is bound or unbound. For further comparative analysis, see Table 5.1:

The nanopore transduction detection (NTD) platform (Fig. 5.3, in Sec. 5.2) involves functionalizing a nanopore detector platform in a new way that is cognizant of signal processing and machine learning capabilities and advantages, such that a highly sensitive biosensing capability is achieved. The core idea in the NTD functionalization of the nanopore detector is to design a molecule that can be drawn into the channel (by an applied potential) but be too big to translocate, instead becoming stuck in a bistable 'capture' such that it modulates the ion-flow in a distinctive way. An approximately two-state 'telegraph signal' has been engineered for a number of NTD modulators. If the channel modulator is bifunctional in that one end is meant to be captured and modulate while the other end is linked to an aptamer or antibody for specific binding, then we have the basis for a remarkably sensitive and specific biosensing capability. The biosensing task is reduced to the channel-based recognition of bound or unbound NTD modulators (or formed/unformed NTD modulators if target is ssDNA).

In order to have a *capture* state in the channel with a *single* molecule, a true nanopore is needed, not a micropore, and to establish a coherent capture-signal exhibiting non-trivial stationary signal statistics, which is the modulating-blockade desired, the nanopore's limiting inner diameter typically needs to be sized at approximately 1.5nm for duplex DNA channel modulators (precisely what is found for the alpha-hemolysin channel). The modulating-blockader is captured at the channel for the time-interval of interest by electrophoretic means, which is established by the applied potential that also establishes the observed current flow through the nanopore.

The NTD molecule providing the channel blockade has a second functionality, typically to specifically bind to some target of interest, with blockade modulation discernibly different according to binding state. NTD modulators are engineered to be bifunctional: one end is meant to be captured and modulate the channel current, while the other, extra-channel-exposed end, is engineered to have different states according to the event detection. Examples include extra-channel ends linked to binding moieties such as antibodies, antibody fragments, or aptamers. Examples also include 'reporter transducer' molecules with cleaved/uncleaved extra-channel-exposed ends, with cleavage by, for example, UV or enzymatic means. By using signal processing with pattern recognition to manage the streaming channel current blockade modulations, and thereby track the molecular states engineered into the transducer molecules, a biosensor or assayer is enabled.

Nanopore transduction detection (NTD) works at a scale where physics, chemistry, and biomedicine methodologies intersect. In some applications the NTD platform functions like a biosensor, or an artificial nose, at the single-molecule scale, e.g., a transducer molecule rattles around in a single protein channel, making transient bonds to its surroundings, and the binding kinetics of those transient bonds is directly imprinted on a surrounding, electrophoretically driven, flow of ions. The observed channel current blockade patterns are engineered or selected to have distinctive stationary statistics, and changes in the channel blockade stationary statistics are found to occur for a transducer molecule's interaction moiety upon introduction of its interaction target. In other applications the NTD functions like a 'nanoscope', e.g., a device that can observe the states of a single molecule or molecular complex. With the NTD apparatus the observation is not in the optical realm, like with the microscope, but in the molecular-state classification realm. NTD, thus, provides an unprecedented new technology for characterization of transient complexes. The nanopore detection method uses the stochastic carrier wave signal processing methods developed and described in prior work [5,7], and comprises machine learning methods for pattern recognition that can be implemented on a distributed network of computers for real-time experimental feedback and sampling control [20].

5.1.5.1 Things to 'contact' with the channel: Aptamers
Aptamers are synthetically-derived, single-stranded, RNA or DNA molecules up to ~80 oligonucleotides in length with a high affinity towards bonding to specific targets. In 1990, a new method dubbed SELEX (Systematic Evolution of Ligands by EXponential Enrichment) provided a process of producing aptamers from random DNA or RNA libraries ([214] and [215]). Application of real-time PCR in the production of aptamers has contributed to the growing effectiveness of aptamers in a variety of research areas today [216,217].The main advantages of aptamers over antibodies are that aptamers are more durable (i.e., longer shelf life, do not require in vivo conditions, can sustain high immune response and toxins), are more obtainable (i.e., cost effective, quicker to make, easily modified, uniformity due to synthetic origin), and have greater specificity and sensitivity (i.e., the degree of binding target recognition, lack of cross-species overlap) [216,217]. Aptamers may bind with anything from dyes, drugs, peptides, proteins, metal ions, antibodies, and enzymes. The values of Kd range between ~pML^{-1} to ~nML^{-1}, better than that of antibodies [216,218,219]. Aptamers are now replacing antibodies as detection reagents, in particular, due to having several advantages over antibodies: versatility, the creation of a lab-on-a-chip to process, low detection limits, simpler reactions to perform, diversity and specificity of aptamer-target binding properties [217]. The use of aptamer beacons has been used in flow cytometry [219], in place of antibody-based assays [217,220] and most abundantly in studies of specific proteins [218,221,222].

5.1.5.2 Things to 'contact' with the channel: Immunoglobulins
The immunoglobulin molecule IgG is often described as a bifunctional molecule: one region for binding to target antigen, the other region for mediating effector function. Effector functions include binding of the antibody to host tissues, to various cells of the immune system, to some phagocytic cells, and to the first component (C1q) of the classical complement system. Activation of the immune system in response to a specific antigen is an amazing example of how a series of protein phosphorylation and dephosphorylation reactions convert a cell surface event to changes in DNA transcription and cell replication.

The structure of the IgG antibody forms three globular regions that are attached to each other in the middle of its grouping. The overall shape of the structure forms a Y configuration. At the base of this structure is the Constant (Fc) region where the effector functions take place and at the tips of the two arms, both referred to as the variable region (Fab), are the antigen binding sites. These variable regions are tethered to the trunk of the Y shaped molecule by a flexible hinge which allows for a high degree of arm movement. The relative size of the antibody is about three times the size of the alpha-hemolysin channel. Its length from base (Fc) to arm tip (Fab) is 25 nm and the width of each globular arm ranges from 6-10nm.

The forces binding antigen to antibody are an important and difficult area of study. Hydrophobic bonds, in particular, are very difficult to characterize by existing crystallographic and other means, and often contribute half of the overall binding strength of the antigen-antibody bond. Hydrophobic groups of the biomolecules exclude water while forming lock and key complementary shapes. The importance of the hydrophobic bonds in protein-protein interactions, and of critically placed waters of hydration, and the complex conformational negotiation whereby they are established, may be accessible to direct study using nanopore detection methods in future developments of this technology.

5.2 The NTD Nanoscope

Nanopore event transduction involves using single-molecule biophysics, engineered information flows, and nanopore cheminformatics. Nanopore transduction detection (NTD) is a unique platform, or 'nanoscope', for detection and analysis of single molecules. Proof-of-Concept experiments shown in what follows indicate a promising approach for single nucleotide polymorphism (SNP) detection, and other biosensing, for clinical diagnostics. This is accomplished via use of the channel-blockade signals produced by engineered event-transducers or by channel modulators in general. The transducer molecule central to te approach is a bi-functional molecule: one end is captured in the nanopore channel while the other end is outside the channel. This extra-channel end is typically engineered to bond to a specific target: the analyte being measured. When the outside portion is bound to the target, the molecular changes (conformational and charge) and environmental changes (current flow obstruction geometry and electro-osmotic flow) result in a change in the channel-binding kinetics of the portion that is captured in the channel. The change in channel interaction kinetics generates a change in the channel blockade current (which is engineered to have a signal unique to the target molecule). The transducer molecule is, thus, a bi-functional molecule which is engineered to produce a change in its stationary-statistics channel-blockade profile upon binding to cognate. For detection of DNA molecules, the binding can itself lead to NTD modulator *formation*, including formation of the modulator function itself (a duplex DNA molecule annealed to form a Y-branching, for example).

NTD Methods for SNP detection alone offers the tantalizing prospect of medical diagnostics and cancer screening by highly accurate assaying of targeted genomic regions. Common methods for SNP detection are typically PCR-based, thus inherit the PCR error rate (0.1% in some situations). The percentages of minority SNP population might be 0.1%, or less, in instances of clinical interest, thus the PCR error rate is critically limiting in the standard approach. Although standard methods for SNP detection have high sensitivity, they typically lack high specificity and versatility. As will be shown, the Nanopore Transduction Detector is a unique platform with both high sensitivity and high specificity.

An interdisciplinary perspective is important to understanding the experimental approach, so initial background describes nanopore electrochemistry and single-molecule biophysics and how the biophysics information flows can result in stationary statistics observations. Then details are provided on the use of engineered stationary statistics signal processing in device enhancement, and communication, as inferred from the selected nanopore detector (ND) blockade sensing experimental results that are shown.

5.2.1 Nanopore Transduction Detection (NTD)

The nanopore transduction detection (NTD) platform [17,27] includes a single nanometer scale channel and an engineered, or selected, channel blockading molecule. The channel blockading molecule is engineered to provide a current modulating blockade in the detector channel when drawn into the channel, and held, by electrophoretic means. The channel has inner diameter at the scale of that molecule. For most biomolecular analysis implementations this leads to a choice of channel that has inner diameter in the range 0.1-10 *nanometers* to encompass small and large biomolecules, where the inner diameter is 1.5 nm in the alpha-hemolysin protein based channel used in the results that follow (see Fig. 5.3). Given the channel's size it is referred to as a nanopore in what follows. In efforts by others 'nanopore' is sometimes used to describe 100-1000 nm range channels, which are here referred to here as micropores.

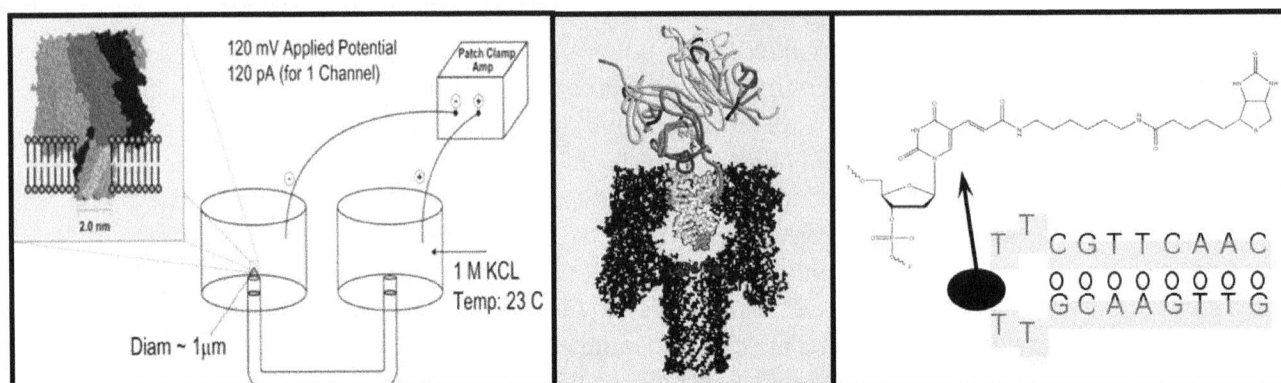

Figure 5.3. Schematic diagram of the Nanopore Transduction Detector. Reprinted with permission of [17,27]. **Left**: shows the nanopore detector consists of a single pore in a lipid bilayer which is created by the oligomerization of the staphylococcal alpha-hemolysin toxin in the left chamber, and a patch clamp amplifier capable of measuring pico Ampere channel currents located in the upper right-hand corner. **Center**: shows a biotinylated DNA hairpin molecule captured in the channel's cis-vestibule, with streptavidin bound to the biotin linkage that is attached to the loop of the DNA hairpin. **Right**: shows the biotinylated DNA hairpin molecule (Bt-8gc).

In order to have a *capture* state in the channel with a *single* molecule, a nanopore is needed. In order to establish a coherent capture-signal exhibiting non-trivial stationary signal statistics the nanopore's limiting inner diameter typically needs to be sized at approximately 1.5nm for duplex DNA channel modulators (precisely what is found for the alpha-hemolysin channel). The modulating-blockader is captured at the channel for the time-interval of interest by electrophoretic means.

The NTD molecule providing the modulating blockade in what follows has a second functionality, to specifically bind to some target of interest such that its blockade modulation is discernibly different according to binding state. Thus, the NTD modulators are engineered to be bifunctional in that one end is meant to modulate the channel current, while the other end is engineered to have different states according to the event detection, or event-reporting, of interest. Examples include extra-channel ends linked to binding moieties such as antibodies or aptamers. Examples also include 'reporter transducer' molecules with cleaved/uncleaved extra-channel-exposed ends, with cleavage by UV or enzymatic means. By using pattern recognition to process the channel current blockade modulations, and thereby track the molecular states, a biosensor is thereby enabled.

With the NTD apparatus the observation is not in the optical realm, like with the microscope, but in the molecular-state classification realm. NTD, thus, provides a technology for characterization of transient complexes. The nanopore detection method uses the stochastic carrier wave signal processing methods, and comprises machine learning methods for pattern recognition that can be implemented on a distributed network of computers for real-time experimental feedback and sampling control [20].

In assaying applications the nanopore detector offers two types of analysis: (1) direct glycoform assaying according to blockade modulation produced directly by the analyte interacting with the nanopore detector, which works on negatively charged glycosylation and glycation profiling best ; and (2) indirect isomer assaying by means of surface feature measurements using a specifically binding intermediary, such as with the antibody used in HbA1c testing. A mixture of the direct and indirect assaying methods may be necessary for complex problems of interest.

One of the most challenging nanopore assaying applications is for discriminating between isomers, approximately mass equivalent molecular variants, or aptamers. Other nanopore-based efforts include DNA sequencing applications, and nanopore device physics studies in general, including with channels other than alpha-hemolysin.

Pattern recognition informed sampling capabilities greatly extends the usage of the single-channel apparatus, including learning the avoidance of blockades associated with channel failure when contaminants necessitate, and nanomanipulation of a single-molecule under active control in a nanofluidics-controlled environment.

The nanopore transduction detection (NTD) system, deployed as a biosensor platform (Fig. 5.3), possesses highly beneficial characteristics from multiple technologies: (i) the specificity of antibody binding, aptamer binding, or nucleic acid annealing; (ii) the sensitivity of an engineered channel modulator to specific environmental change; and (iii) the robustness of the electrophoresis platform in handling biological samples [1].

A NTD transducer can often be constructed by covalently tethering a molecule of interest to a nanopore channel modulator. In previous work, using inexpensive (commoditized) biomolecular components, such as DNA hairpins, as channel-modulators, and antibodies as specific binding moieties (with inexpensive immuno-PCR linkages to DNA), experiments were done to analyze individual antibodies and DNA molecules, their conformations, glycosylations, and their binding properties. It was found that in many applications the DNA-based transducers worked well, but in efforts to extend the methodology to biosensing and glycosylation profiling the DNA modulators often had too short a lifetime until melting. To make matters worse, the DNA-based modulators often had internal conformational freedom of their own that complicated analysis of any linked molecule's conformational changes. Worst of all, sometimes the DNA modulators only modulated when unbound (and the NTD method works best with clearly different modulatory states). Efforts to fix the non-modulatory aspect were partly solved by using a laser-tweezer apparatus to drive distinctive stochastic modulatory blockades in the DNA modulator. This was accomplished by introducing a periodic laser-tweezer 'tugging' on channel-modulator variants that had a biotinylated portion that was bound to a streptavidin-coated magnetic bead (another commoditized component). With modulations 'reawakened', however, the number of types of blockade signal appeared to proliferate significantly, and it wasn't clear if an automated signal analysis could be implemented as had been done previously.

At the nanometer-scale of the nanopore experiment the Reynold's number of the flow is incredibly small (10^{-10}). Thus the flow environment is not fluid-like in a familiar sense. The fluid strongly damps transverse vibrations, for example, so no string-like-motion on polymers. The motions are strongly driven by electrostatic forces and steric constraints and have significant thermal energy contributions, such that a stochastic process is effectively obtained in typical measurements.

5.2.2 NTD: a versatile platform for biosensing

The use of a channel modulator introduces significant, engineered, signal analysis complexity, that we resolve using artificial intelligence (machine learning) methods. The benefit of this complication is a significant gain in sensitivity over T/TD, that uses a 'sensing' moiety covalently attached to the channel itself, where they have a T/TD-type blockade 'lifetime' event, with minimal or no internal blockade structure engineered. The NTD approach, on the other hand, has significant improvement in versatility, e.g., we can 'swap out' modulators on a given channel, in a variety of ways, since they are not covalently attached to the channel. The improvements in sensitivity derive from the measurable stationary statistics of the channel blockades (and how this can be used to classify state with very high accuracy). The overall improvement in versatility is because all that needs to be redesigned for a different NTD experiment (or binding assay) is the linkage-interaction moiety portion of the bifunctional molecules involved. There is also the versatility that *mixtures* of different types of transducers can be used, a method that can't be employed in single-channel devices that use covalently bound binding moieties (or that discriminate by dwell-time in the channel).

At the nanopore channel one can observe a sampling of bound/unbound states, each sample only held for the length of time necessary for a high accuracy classification. Or, one could hold and observe a single bound/unbound system and track its history of bound/unbound states or conformational states. The *single* molecule detection, thus, allows measurement of molecular characteristics that are obscured in ensemble-based measurements. Ensemble averages, for example, lose information about the true diversity of behavior of individual molecules. For complex *bio*molecules there is likely to be a tremendous diversity in behavior, and in many cases this diversity may be the basis for their function. There can also be a great deal of diversity via post-translational modifications, as well, such as with heterogeneous mixtures of protein glycoforms that typically occur in living organisms (e.g., for TSH and hemoglobin proteins in blood serum and red blood cells, respectively). The hemoglobin 'A1c' glycoprotein, for example, is a disease diagnostic (diabetes), and for TSH, glycation is critical component in the TSH-based regulation of the endocrine axis. Multi-component regulatory systems and their variations (often sources of disease) could also be studied much more directly using the NTD approach, as could multi-component (or multi-cofactor) enzyme systems. Glycoform assays, characterization of single-molecule conformational variants, and multi-component assays are significant capabilities to be developed further with the NTD approach, further details on NTD assaying will follow in a later section.

In NTD applications we seek DNA modulators with specific, non-linear, topologies, such as Y-shaped DNA duplexes, to obtain molecules whose non-translocating blockades modulate the channel . We include shorter nucleic acids, with channel modulating and simple, DNA-complement, annealing properties, in the collection of DNA-based 'NTD aptamers' described in the NTD biosensor applications that follows. This is because the detection of ssDNA can enable the NTD-transducer's channel-modulatory formation, for direct signal validation, as will be described in what follows.

Nanopore transduction detection provides an inexpensive, quick, accurate, and versatile method for performing medical diagnostics. It is hypothesized that NTD biomarkers can be developed for early stage disease detection with femtomolar to attomolar sensitivity (see Table 5.2) for doing the standard clinical tests of the future. The potentially incredible sensitivity of the NTD targeting on biomarkers also provides a significant new tool for public health and biodefense in general.

In the preliminary results shown in what follows, we first demonstrate a 0.17 µM streptavidin sensitivity in the presence of a 0.5 µM concentration of detection probes with a 100 second detection window. The detection probe is a biotinylated DNA-hairpin transducer molecule (Bt-8gc). In repeated experiments we see the sensitivity limit ranging inversely to the concentration of detection probes. If taken to its limits, with established PRI sampling capabilities [20], and with stock Bt-8gc at 1mM concentration conveniently available, we believe it is possible to boost probe concentration almost three magnitudes. In doing so, we would boost sensitivity by similar measure, until the minimal observation time needed to reject limits this gain mechanism (see Table 5.2).

METHOD	SN
Low-probe concentration, 100s obs.	100 nM
High probe conc, 100s observation	100 pM
High probe conc, <u>long observation (~1dy)</u>	100 fM *
TARISA (conc. gain), 100s observation	100 fM
TERISA (enzyme gain), 100s obs.	100 aM **
Electrophoretic contrast gain, 100 s	1.0 aM

Table 5.2. Sensitivity limits for detection in the streptavidin-biosensor model system. *We have done 1 -1.5 day long experiments in other contexts, but not longer. Thus, current capabilities, with no modifications to the NTD platform for specialization for biosensing, can achieve close to 100 fM sensitivity by pushing the device limits and the observation window. **Only a slow enzyme turnover of 10 per second is assumed.

Detection in the attomolar regime (see Table 5.2) is critical for *early* discovery of type I diabetes destructive processes and for early detection of Hepatitis B. Early PSA detection currently has a 500 aM sensitivity. For some toxins, their potency, even at trace amounts, precludes their usage in the typical antibody-generation procedures (for mAb's that target that toxin). In this instance, however, aptamer-based NTD probes can still be obtained.

5.2.3 NTD Platform
The components comprising the NTD platform include an engineered molecule that can be drawn, by electrophoretic means (using an applied potential), into a channel that has inner diameter at the scale of that molecule, or one of its molecular-complexes, a means to estab-

lish a current flow through that nanopore (such as an ion flow under an applied potential), a means to establish the molecular capture for the timescale of interest (electrophoresis, for example), and the computational means to perform signal processing and pattern recognition (see Fig. 5.4 and 5.5).

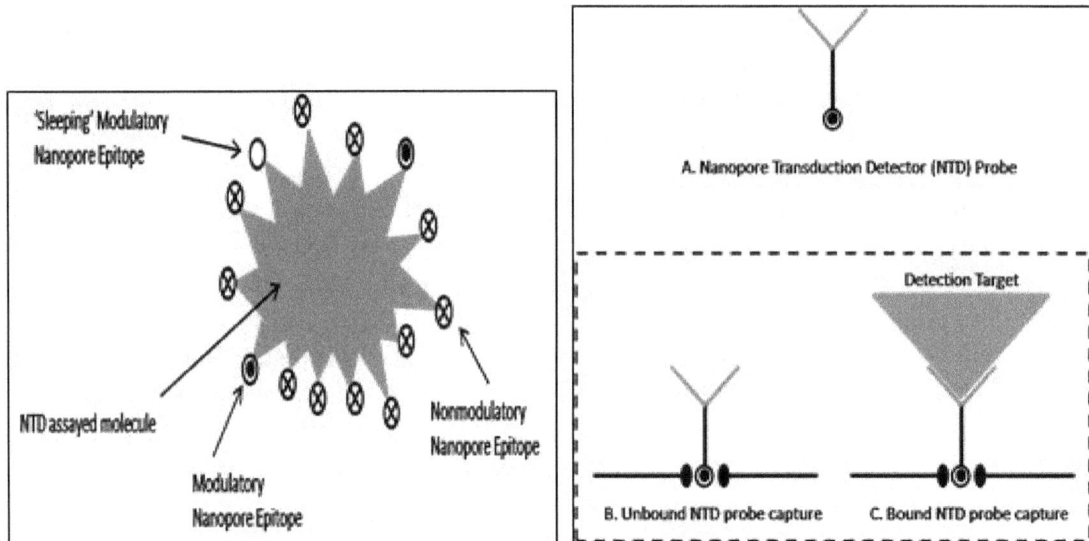

Fig. 5.4. Right. Nanopore Transduction Detector (NTD) Probe – a bifunctional molecule (A), one end channel-modulatory upon channel-capture (and typically long-lived), the other end multi-state according to the event detection of interest, such as the binding moieties (antibody and aptamer, schematically indicated in bound and unbound configurations in (B) and (C)), to enable a biosensing and assaying capability. **Left.** NTD assayed molecule (a protein, or other biomolecule, for example) Antibodies (proteins) are NTD assayed in the Proof-of-Concept Experiments, for example. Nanopore epitopes may arise from glyocprotein modifications and provide a means to measure surface features on heterogeneities mixture of protein glycoforms (such mixtures occur in blood chemistry, commercially available test on HbA1c glycosylation common, for example). A molecule may be examined via NTD sampling assay upon exposure to nanopore detector, (or molecular complex including molecule of interest).

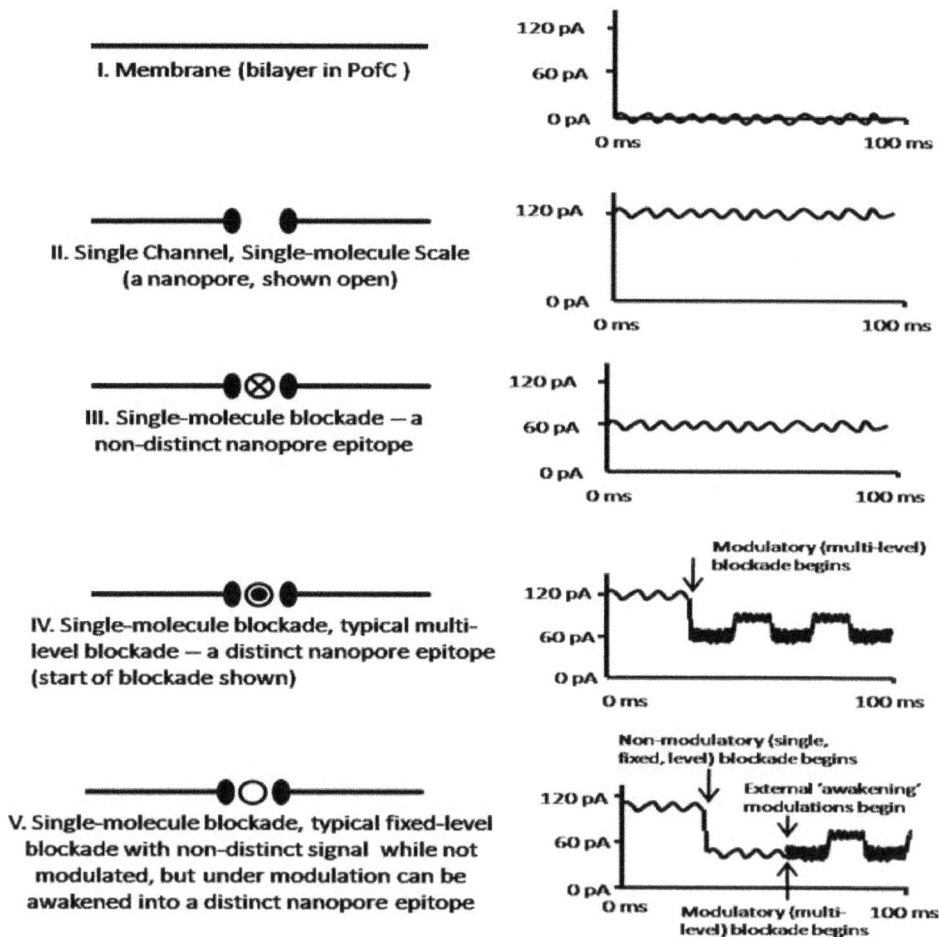

I. Membrane (bilayer in PofC)

II. Single Channel, Single-molecule Scale (a nanopore, shown open)

III. Single-molecule blockade — a non-distinct nanopore epitope

IV. Single-molecule blockade, typical multi-level blockade — a distinct nanopore epitope (start of blockade shown)

V. Single-molecule blockade, typical fixed-level blockade with non-distinct signal while not modulated, but under modulation can be awakened into a distinct nanopore epitope

Fig. 5.5. The various modes of channel blockade are shown: I. No channel – e.g., a Membrane (bilayer). II. Single Channel, Single-molecule Scale (a nanopore, shown open). III. Single-molecule blockade, a brief interaction or blockade with fixed-level with non-distinct signal -- a non-modulatory nanopore epitope. IV. Single-molecule blockade, typical multi-level blockade with distinct signal modulations (typically obeying stationary statistics or shifts between phases of such). V. Single-molecule blockade, typical fixed-level blockade with non-distinct signal while not modulated, but under modulation can be awakened into distinct signal, with distinct modulations.

The channel is sized such that a transducer molecule, or transducer-complex, is too big to translocate, instead the transducer molecule is designed to get stuck in a 'capture' configuration that modulates the ion-flow in a distinctive way.

The NTD modulators are engineered to be bifunctional in that one end is meant to be captured, and modulate the channel current, while the other, extra-channel-exposed end, is engineered to have different states according to the event detection, or event-reporting, of interest. Examples include extra-channel ends linked to binding moieties such as antibodies, antibody fragments, or aptamers. Examples also include 'reporter transducer' molecules with cleaved/uncleaved extra-channel-exposed ends, with cleavage by, for example, UV or enzymatic means. By using signal processing to track the molecular states engineered into the transducer molecules, a biosensor or assayer is thereby enabled. By tracking transduced states of a coupled molecule undergoing conformational changes, such as an antibody, or a protein with a folding-pathway associated with disease, direct examination of co-factor, and other, influences on conformation can also be assayed at the single-molecule level.The channel blockade modes in an NTD experiment thus make special use of channel current modulation scenarios (with stationary statistics).

5.2.4 NTD Operation

When the extra-channel states correspond to bound/unbound, there are two protocols for how to set up the NTD platform: (1) observe a sampling of bound/unbound states, each sample only held for the length of time necessary for a high accuracy classification. Or, (2), hold and observe a single bound/unbound system and track its history of bound/unbound states. The single molecule binding history in (2) has significant utility in its own right, especially for observation of critical conformational change information not observable by any other methods. The ensemble measurement approach in (1), however, is able to benefit from numerous further augmentations, and can be used with general transducer states (see Fig. 5.6-5.9), not just those that correspond to a bound/unbound extra-channel states.

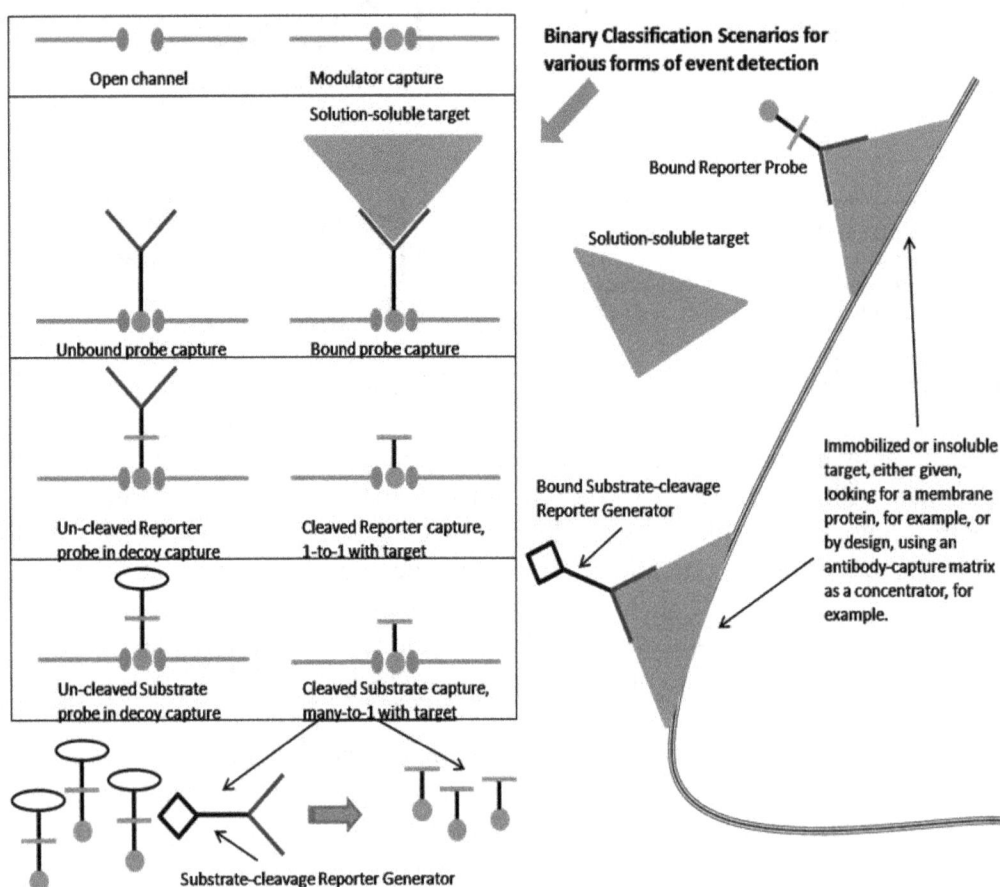

Fig. 5.6. Probes shown: bound/unbound type and uncleaved/cleaved type.

The pattern recognition informed (PRI) sampling 'acceleration', in ensemble-based measurements, for example, provides a means to accelerate the accumulation of kinetic information in most situations [22]. Furthermore, the sampling over a population of molecules is the key to a number of other gain factors that may be realized. In the ensemble detection with PRI approach [22], in particular, one can make use of antibody capture matrix and ELISA-like methods [88], to introduce two-state NTD modulators that have concentration-gain (in an antibody capture matrix) or concentration-with-enzyme-boost-gain (ELISA-like system, with production of NTD modulators by enzyme cleavage instead of activated fluorophore production). (Note that in the latter systems the NTD modulator is simply specified as 'two-state', where here we typically don't have bound/unbound, but cleaved/uncleaved instead.) In the ensemble evaluations, with the aforementioned off-channel-engineered event gain factors, we can introduce a NTD probe substrate that thoroughly probes the sample presented if some element of the probe-target system is immobilized, or significantly reduced in mobilization (see Fig. 5.6). In this circumstance, UV- and enzyme-based cleavage methods on immobilized probe-target can be designed to produce a high concentration, or concentration burst, of NTD modulators, that will be strongly drawn to the channel and provide a UV-event correlated 'burst' concentration detection signal.

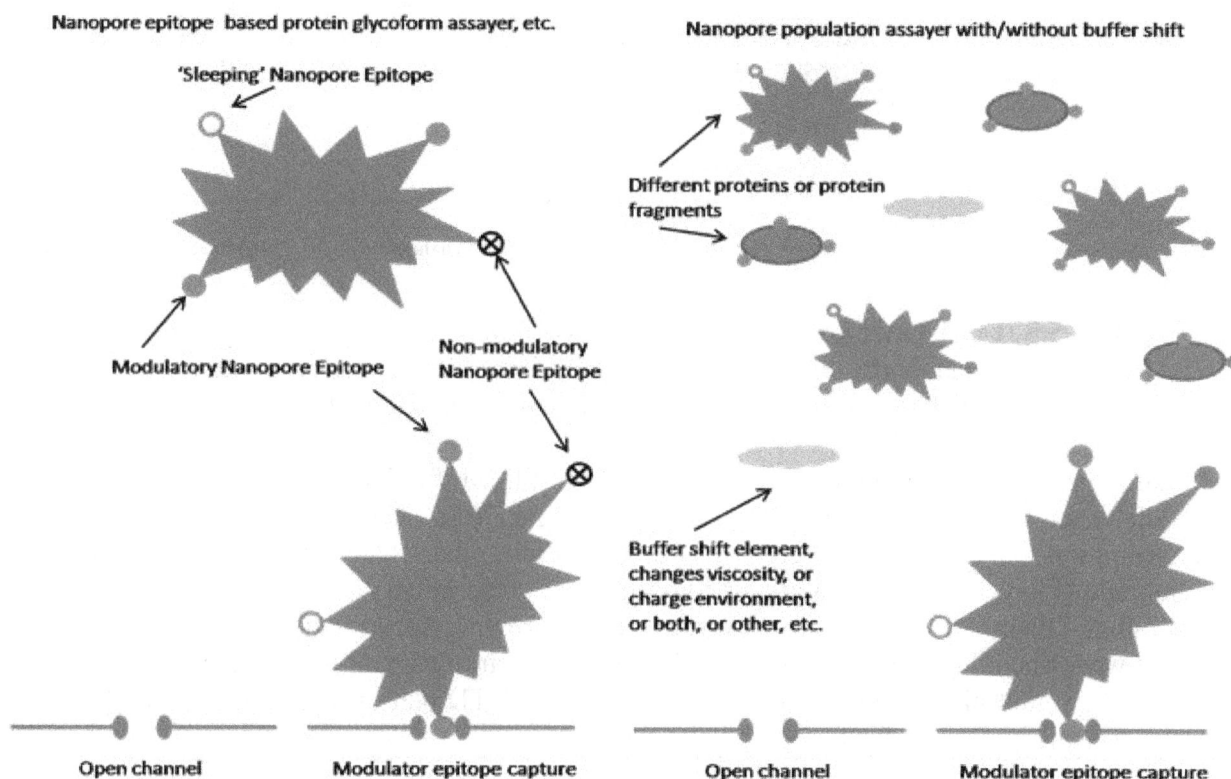

Fig. 5.7 (Left). Nanopore epitope assay (of a protein, or a heterogenous mixture of related glycoprotein, for example, via glycosilation that need not be enzynatically driven, as occurs in blood, for example). **(Right). Gel-shift mechanism.** Electrophoretically draw molecules across a diffusionally resistive buffer, gel, or matrix (PEG-shift experiments). If medium in buffer, gel, or matrix is endowed with a charge gradient, or a fixed charge, or pH gradient, etc., isoelectric focusing effects, for example, might be discernable.

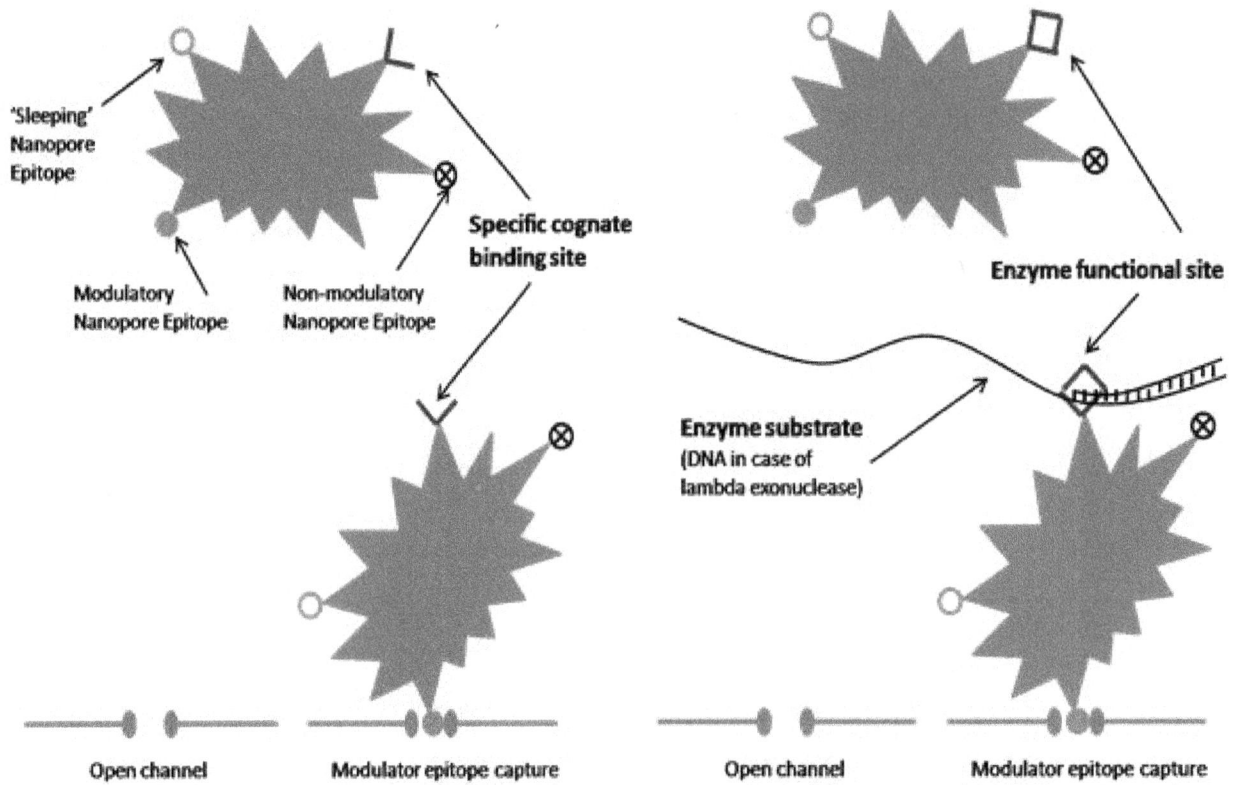

Fig. 5.8 (Left). Oriented modulator capture on protein (or other) with specific binding (an antibody for example). **(Right).** Oriented modulator capture on protein (or other) with enzymatic activity (lambda exonuclease for example).

A multi-channel implementation of the NTD can be utilized if a distinctive-signature NTD-modulator on one of those channels can be discerned (the scenario for trace, or low-concentration, biosensing, see Fig. 5.9). In this situation, other channels bridging the same membrane (bilayer in case of alpha-hemolysin based experiment) are in parallel with the first (single) channel, with overall background noise growing accordingly. In the stochastic carrier wave encoding/decoding with HMMD, we retain strong signal-to-noise, such that the benefits of a multiple-receptor gain in the multi-channel NTD platforms can be realized (see Proof-of-Concept Results for further details).

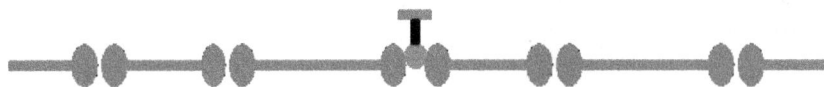

Fig. 5.9. Multichannel scenario, with only one blockade present (at low concentration, for example).

5.2.5 Driven modulations

It is possible to probe higher frequency realms than those directly accessible at the operational bandwidth of the channel current based device (~200 kHz), or due to the time-scale of the particular analyte interaction kinetics, by introducing modulated excitations. This can be accomplished by chemically linking the analyte or channel to an excitable object, such as a magnetic bead, under the influence of laser pulsations [41]. In one configuration, the excitable object can be chemically linked to the analyte molecule to modulate its blockade current by modulating the molecule during its blockade. In another configuration, the excitable object is chemically linked to the channel, to provide a means to modulate the passage of ions through that channel. In a third experimental variant, the membrane is itself modulated (using sound, for example) in order to effect modulation of the channel environment and the ionic current flowing though that channel. Studies involving the first, analyte modulated, configuration (Fig.s 5.10 & 5.11), indicate that this approach can be successfully employed to keep the end of a long strand of duplex DNA from permanently residing in a single blockade state. Similar study of magnetic beads linked to antigen may be used in the nanopore/antibody experiments if similar single blockade level, "stuck," states occur with the captured antibody (at physiological conditions, for example). Likewise, this approach can be considered for increasing the antibody-antigen dissociation rate if it does not occur within the time-scale of the experiment. It may be possible, with appropriate laser pulsing, or some other modulation, to drive a captured DNA molecule in an informative way even when not linked to a bead, or other macroscopic entity (Fig. 5.12).

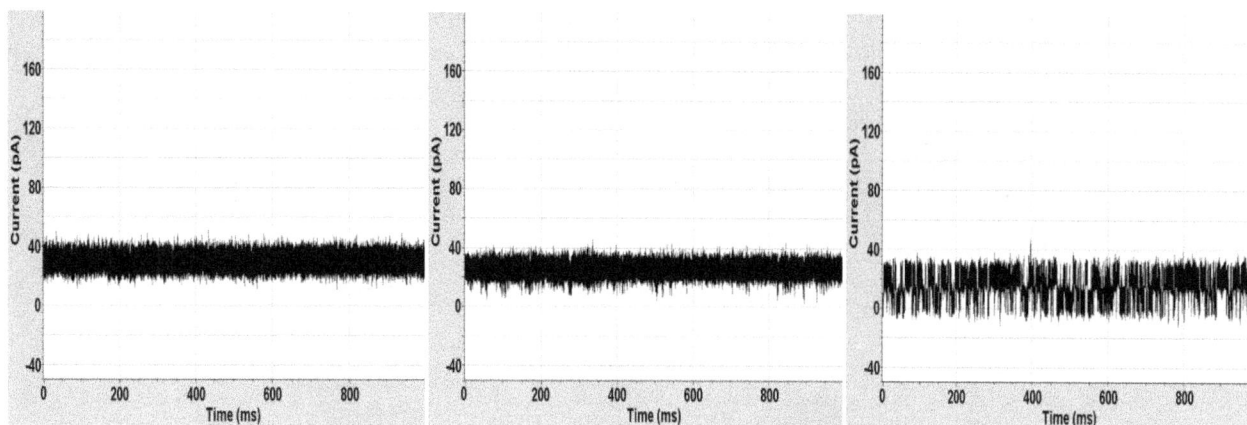

Fig. 5.10. A (Left) Channel current blockade signal where the blockade is produced by 9GC DNA hairpin with 20 bp stem. (Center) Channel current blockade signal where the blockade is produced by 9GC 20 bp stem with magnetic bead attached. (Right) Channel current blockade signal where the blockade is produced by c9GC 20 bp stem with magnetic bead attached and driven by a laser beam chopped at 4 Hz. Each graph shows the level of current in picoamps over time in milliseconds.

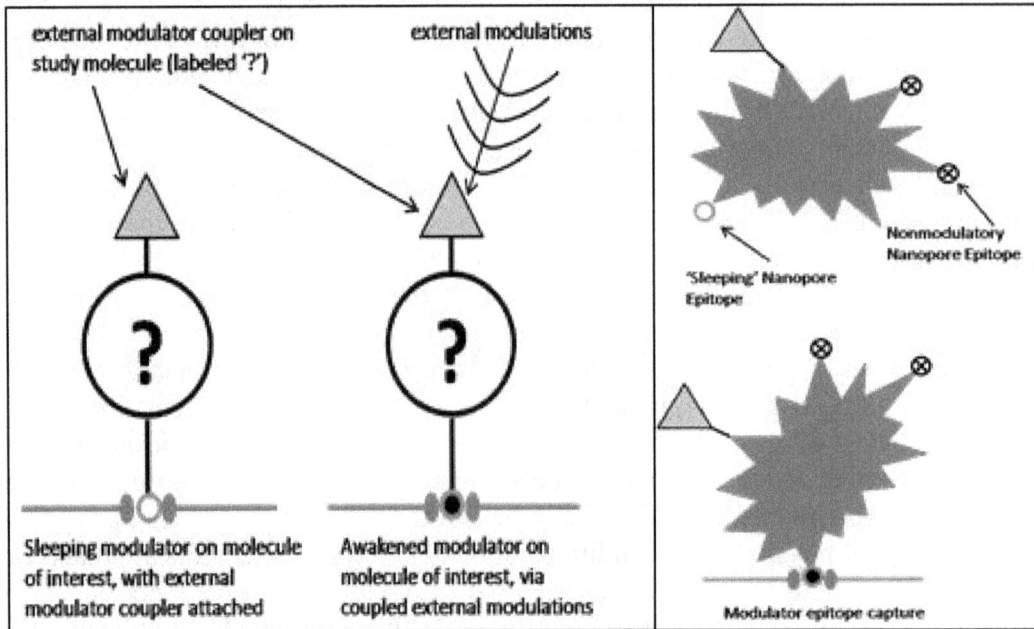

Fig. 5.11 (Left). Study molecule with externally-driven modulator linkage to awaken modulator signal. **(Right).** Study molecule with externally-driven modulator linkage to awaken modulator signal, with epitope-selection to obtain sleeping epitope , then determine its identity, and based on known modulator-activation driving signals, proceed with driving the system to obtain a modulator capture linkage.

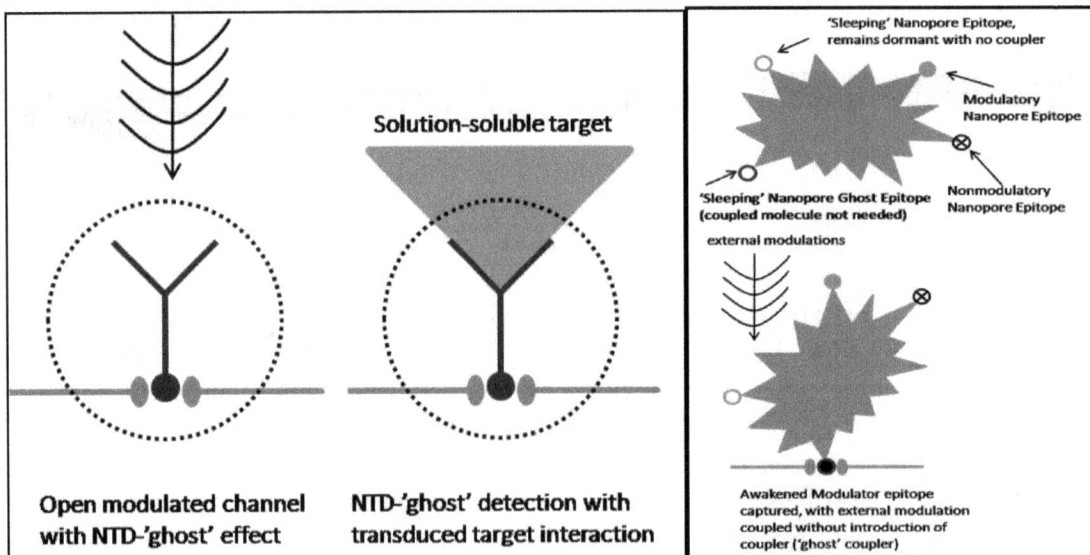

Fig. 5.12 (Left). Same situation as in cases with linked-modulator, but more extensive range of external modulations explored, such that, in some situations, a sleeping nanopore epitope is 'awakened' (modulatory channel blockades produced), and the target molecule does not require a coupler attachment., e.g., using external modulations with no coupler, may be able to obtain 'ghost' transducers in some situations. **(Right).** 'Sleeping' Nanopore Ghost Epitope (coupled molecule not needed).

110

5.2.6 Driven modulations with multichannel augmentation

The *S. aureus* alpha-hemolysin pore-forming toxin that is used to produce our single-channel nanopore-detector self- oligomerizes to derive the energetics necessary to create a channel through the bi-layer membrane. In the nanopore construction protocol, the process is limited to the creation of a single channel. It is possible to allow the process to continue unabated to create 100 channels or more. The 100 channel scenario has the potential to increase the sensitivity of the NTD, but the signal analysis becomes more challenging since there are 100 parallel noise sources. The recognition of a transducer signal is possible by the introduction of 'time integration' to the signal analysis akin to heterodyning a radio signal with a periodic carrier in classic electrical engineering. In order to introduce a 'time integration' benefit in the transducer signal, periodic (or stationary stochastic) modulations can be introduced to the transducer environment. In a high noise background, modulations can be introduced such that some of the transducer level lifetimes have heavy-tailed distributions. With these modifications to the signal processing software a single transducer molecule signal could be recognizable in the presence of 100 channels or more. Increasing the number of channels by 100 and retaining the capability of recognizing a single transducer blockading one of those channels provides a direct gain in sensitivity according to the number of channels (e.g., 100 channels would provide a sensitivity boost of two orders of magnitude). It is important to note that this type of increase in sensitivity is mostly implemented computationally and does not add complexity or cost to the NTD device.

The single-channel biosensing methods used here can be generalized to where many channels are present, where each channel offers parallel conductance paths for the ionic current, and where each channel is augmented with antibody (or aptamer) to establish a background collection of channel/antibody signals that is modifiable in the presence of antigen. Such 'passive' multi-channel methods offer similar capabilities to surface plasmon resonance approaches for characterizing binding affinity. Multiple antibody (aptamer) species can be present in this multi-channel operation. Anything that can evoke an antibody response (or SELEX selection, for aptamers) can be taken as the antigen or collection of antigens for which the bio-sensing is designed.

Multichannel with modulation is shown in Fig. 5.13, where modulation forces a population inversion, such that state durations are strongly non-geometrically distributed. Even without such modulations, however, there may be a strong enough signal recognitionwith the HMM methods without duration modeling enhancements.

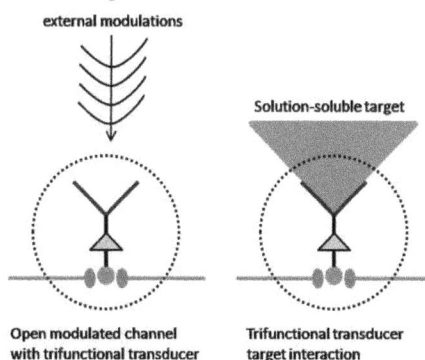

Fig. 5.13. External modulations with transducer with coupler, a trifunctional molecule.

Chapter 6
NTD Biosensing Methods

NTD biosensing is possible with any channel modulator. In this chapter, however, we mainly focus on examples where the NTD transducer has a DNA-based modulator portion. NTD biosensing methods typically involve a DNA modulator given their establishedproduction of reproducible and distinct channel modulations (according to sequence and secondary structure). The modulator typically has linkage to an aptamer, antibody, or some other binding moiety, including simply a ssDNA overhang. The linkages needed to connect a DNA-based channel-modulator to a DNA-based aptamer involve a trivial join of the underlying ssDNA sequences involved. The linkage needed to connect a DNA-based channel-modulator to an antibody *could* involve use of linker technology, and this has been used in the past with dsDNA hairpins [30], but another, more commoditized route to be discussed, easily accessible with use of the NADIR refined Y-shaped DNA channel modulators [27,28], is that the antibody need merely be 'tagged' with the appropriate ssDNA strand, e.g., where the DNA sequence is complement to part of the 'Y' shaped DNA channel modulator, and antibody tagging with DNA is a standard service for use in immuno-PCR. Proof-of-Concept Biosensing Experiments are described in what follows for the streptavidin-biotin and DNA annealing model systems, a pathogen/SNP detection prototype, and for aptamer and antibody based detection.

6.1   Model biosensor based on streptavidin and biotin

A biotinylated DNA-hairpin that is engineered to generate two signals depending on whether or not a streptavidin molecule is bound to the biotin (see Fig. 5.3). Results in Fig. 6.1 (Right) suggest that the new signal class on binding is actually a racemic mixture of two hairpin-loop twist states. At T=4000 urea is introduced at 2.0 M and gradually increased to 3.5 M at T=8,100.

The transducer molecule in the NTD "Streptavidin Toxin Biosensor" configuration (shown in Fig. 5.3) consists of a bi-functional molecule: one end is captured in the nanopore channel while the other end is outside the channel. This exterior-channel end is engineered to bond to a specific target: the analyte being measured. When the outside portion is bound to the target, the molecular changes (conformational and charge) and environmental changes (current flow obstruction geometry and electro-osmotic flow) result in a change in the channel-binding kinetics of the portion that is captured in the channel. This change of kinetics generates a change in the channel blockade current which represents a signal unique to the target molecule.

Some of the transducer molecule results from [17] are shown in Fig. 6.1, for a biotinylated DNA-hairpin that is engineered to generate two unique signals depending on whether or not a streptavidin molecule is bound.

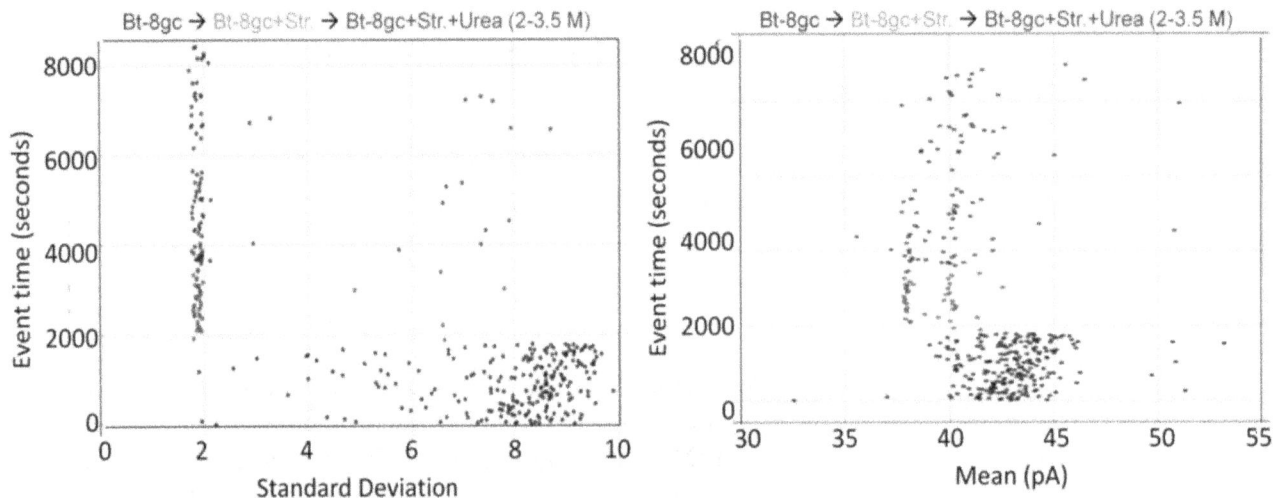

Figure 6.1. Left. Observations of individual blockade events are shown in terms of their blockade standard deviation (x-axis) and labeled by their observation time (y-axis) [20]. The standard deviation provides a good discriminatory parameter in this instance since the transducer molecules are engineered to have a notably higher standard deviation than typical noise or contaminant signals. At T=0 seconds, 1.0 μM Bt-8gc is introduced and event tracking is shown on the horizontal axis via the individual blockade standard deviation values about their means. At T=2000 seconds, 1.0 μM Streptavidin is introduced. Immediately thereafter, there is a shift in blockade signal classes observed to a quiescent blockade signal, as can be visually discerned. The new signal class is hypothesized to be due to (Streptavidin)-(Bt-8gc) bound-complex captures. **Right.** As with the Left Panel on the same data, a marked change in the Bt-8gc blockade observations is shown immediately upon introducing streptavidin at T=2000 seconds, but with the mean feature we clearly see two distinctive and equally frequented (racemic) event categories. Introduction of chaotropic agents degrades first one, then both, of the event categories, as 2.0 M urea is introduced at T=4000 seconds and steadily increased to 3.5 M urea at T=8100 seconds.

In the NTD platform, sensitivity increases with observation time [1] in contrast to translocation technologies where the observation window is fixed to the time it takes for a molecule to move through the channel. Part of the sensitivity and versatility of the NTD platform derives from the ability to couple real-time adaptive signal processing algorithms to the complex blockade current signals generated by the captured transducer molecule. If used with the appropriately designed NTD transducers, NTD can provide excellent sensitivity and specificity and can be deployed in many applications where trace level detection is desired. The monoclonal antibody-based NTD system, deployed as a biosensor platform, possesses highly beneficial characteristics from multiple technologies: the specificity of monoclonal antibody binding, the sensitivity of an engineered channel modulator to specific environmental change, and the robustness of the electrophoresis platform in handling biological samples. In combination, the NTD platform can provide trace level detection for early diagnosis of disease as well as quantify the concentration of a target analyte or the presence and relative concentrations of multiple distinct analytes in a single sample.

In [20] a 0.17 μM streptavidin sensitivity is demonstrated in the presence of a 0.5 μM concentration of detection probes, with only a 100 second detection window. The detection probe is the biotinylated DNA-hairpin transducer molecule (Bt-8gc) described in Fig. 1. In repeated experiments, the sensitivity limit ranges inversely to the concentration of detection probes (with PRI sampling) or the duration of detection window. The stock Bt-8gc has 1mM concentration, so a 1.0 mM probe concentration is easily introduced. (Note: The higher concentrations of transducer probes need not be expensive on the nanopore platform because the working volume can be very small: *cis* chamber volume is 70 μL, and could be reduced to 1.0 μL with use of microfluidics.) In [20] the selectivity of the detector in the presence of interference agents, such as albumin and sucrose and a variety of antibodies (without specific binding to biotin or the channel) was also examined, and a control transducer molecule with the same six-carbon linker arm from the DNA hairpin, but without the biotin 'fishing lure' binding site, was introduced, where it was shown that no interaction (via change of blockade signal) was observed upon introduction of streptavidin, as expected.

6.2 Model system based on DNA annealing
6.2.1 Linear DNA annealing test
Proof of Concept experiments for DNA annealing were initially tested for detection of a specific 5-base ssDNA molecule, where we have a linear molecule with a bulge in the center. To one side of the bulge is the blunt-ended stem sequence like that used in one of our DNA hairpin controls, where the bulge is now in the position of the hairpin's loop. To the other side of the bulge is a cap-section of base-pairs followed by an overhang section of length five bases. A similar set of experiments is performed with the "Y-aptamer", a Y-shaped DNA complex with one arm of the Y with an overhang (6Ts), while the other arm is capped with a 4dT loop. The base of the Y is a stem of 10 base-pairs length, prior to the Y-nexus of the molecule. Here the Y-nexus is in the place of the bulge, or the hairpin loop. Nine or ten base-pairs is approximately the length in dsDNA from the mouth of the channel to its limiting aperture. The significance of this length in the modeling is due to its delicate placement of the end of the captured molecule over the high electrophoretic field strength zone near the limiting aperture of the channel, permitting operation in transduction model. The overhang's binding strength can be adjusted by tailoring its length in both of these experiments, and in future work this will also permit a highly precise study of DNA annealing.

The linear duplex DNA molecule, with bulge, and ssDNA overhang, is given below. Examples of the signals that occur when a properly annealed duplex is captured are shown in Fig. 6.2. Fig. 6.2 compares signal traces before/after in terms of their standard 150-component feature set. The linear aptamer with bulge consists of annealing the following two ssDNA strands:

(1) 5'-GAGGCTTGG TTT CAATAGGTA-3'
(2) 5'-ATTG TTT CCAAGCCTC-3'
The complementary 5 nucleotide ssDNA sequence (3):
(3) 5'-TACCT-3'

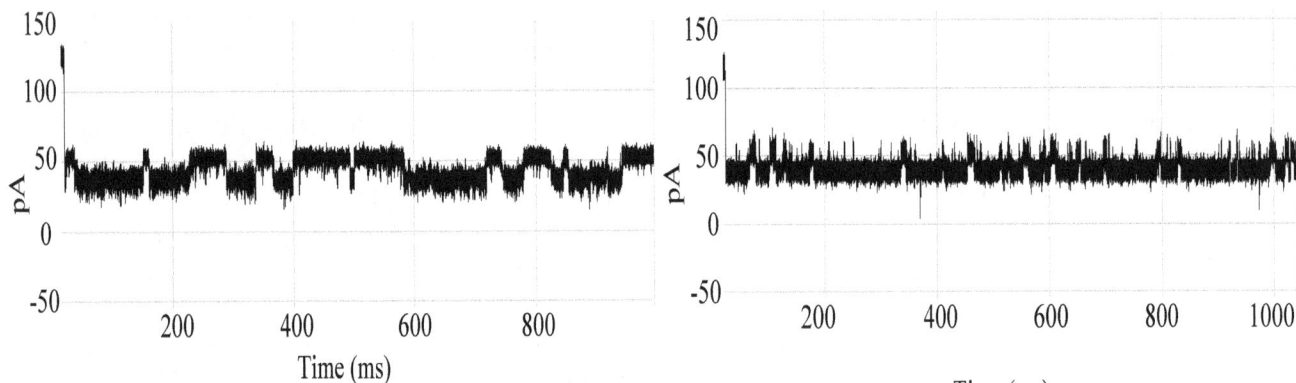

Fig. 6.2. Pseudo-aptamer: DNA overhang binding complement – signal blockades. Left: Before introduction of 5-base ssDNA complement. **Right:** After introduction of complement.

6.2.2 'Y' DNA annealing test

The Y-aptamer DNA molecule consists of a three-way DNA junction created by annealing two DNA molecules:

(1) 5'-CTCCGTCGAC GAGTTTATAGAC TTTTTT-3'
(2) 5'-GTCTATAAACTC GCAGTCATGCTTTTGCATGACTGC GTCGACGGAG-3'

For the resulting Y-aptamer, one of the junctions' arms terminate in a 4dT-loop and the other arm has a 6T overhang in place of a 4dT-loop. Preliminary results are shown in Fig. 6.3. The blunt ended arm has to be carefully designed such that when it is captured by the nanopore it produces a toggling blockade. One of the arms of the Y-shaped aptamer (Y-aptamer) has a TATA sequence, and is meant to be a binding target for TBP. In general, any transcription factor binding site could be studied (or verified) in this manner. Similarly, transcription factor could be verified, or the efficacy of a synthetic transcription factor could be examined.

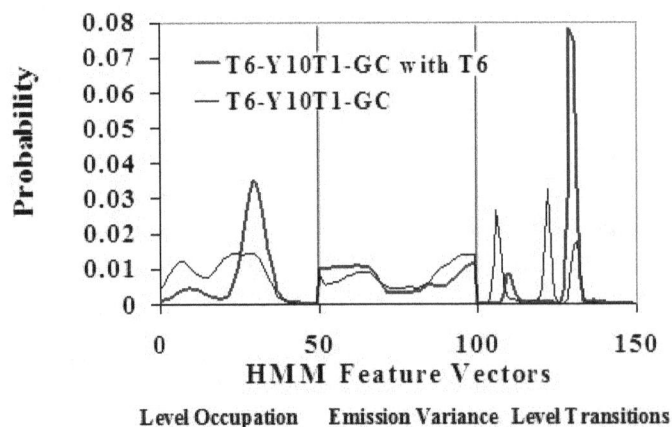

Level Occupation   Emission Variance   Level Transitions

Fig. 6.3. Y-aptamer with DNA overhang that binds complement. Left: signal profiles before and after binding. **Right:** the dwell-time distributions on

116

the three dominant levels indicated in the *unbound* blockade signal. The profiles aresurprisingly different, the bound case, with annealed complement, appears to be more "stable", with only two dominant blockade levels. This is consistent with it being a molecule with fewer degrees of freedom (with 6T overhang now annealed to 6A complement).

6.2.3 Y-shaped NTD-aptamer

A unique, Y-shaped, NTD-aptamer is described in Fig. 6.4. In this experiment a stable modulator is established using a Y-shaped molecule, where one arm is loop terminated such that it can't be captured in the channel, leaving one arm with a ssDNA extension for annealing to complement target.

A preliminary test of DNA annealing has been performed with the Y-shaped DNA transduction molecule indicated, where the molecule is engineered to have an eight-base overhang for annealing studies. A DNA hairpin with complementary 8 base overhang is used as the binding partner. Fig. 6.4 shows the binding results at the population-level (where numerous single-molecule events are sampled and identified), where the effects of binding are discernible, as are potential isoforms, and the introduction of urea at 2.0 M concentration is easily tolerated (a mild chaotrope) and actually helps in discerning collective binding interactions such as with the DNA annealing.

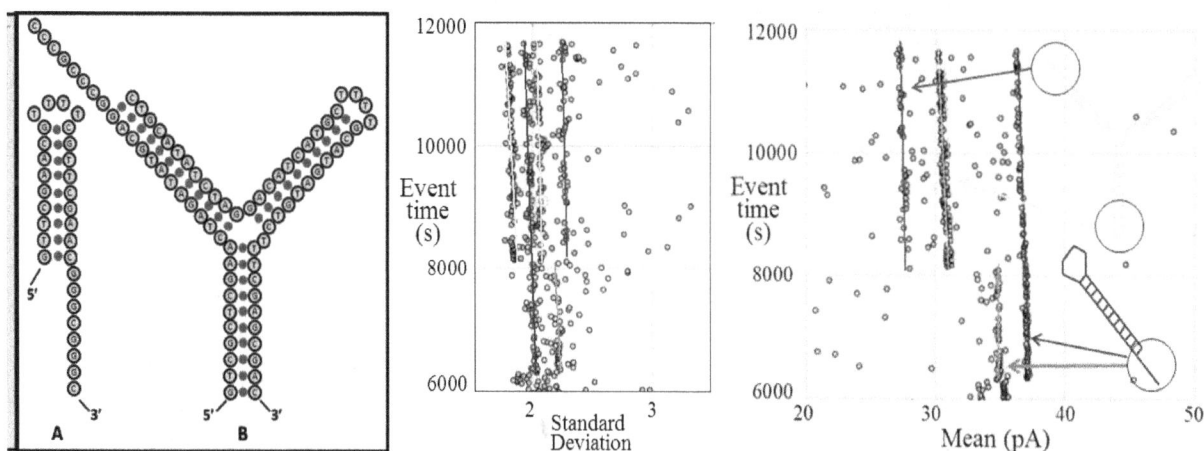

Figure 6.4. Eight-base annealing using a NTD Y-transducer [17]. **Left:** The DNA hairpin and DNA Y-nexus transducer secondary structures with sequence information shown. **Center and Right: Y-shaped DNA transducer with overhang binding to DNA hairpin with complementary overhang.**

The ability of the of the NTD apparatus to tolerate high chaotrope concentration, up to 5M urea, was demonstrated in [3]. DNA hairpin control molecules have demonstrated a manageable amount of isoform variation even at 5M urea.

In Fig. 6.4, only a portion of a repetitive validation experiment is shown, thus time indexing starts at the 6000th second. From time 6000 to 6300 seconds (the first 5 minutes of data shown) only the DNA hairpin (sequence details in [17,27]) is introduced into the analyte chamber, where each point in the plots corresponds to an individual molecular blockade

117

measurement. At time 6300 seconds urea is introduced into the analyte chamber at a concentration of 2.0 M. The DNA hairpin with overhang is found to have two capture states (clearly identified at 2 M urea). The two hairpin channel-capture states are marked with the green and red lines, in both the plot of signal means and signal standard deviations. After 30 minutes of sampling on the hairpin+urea mixture (from 6300 to 8100 seconds), the Y-shaped DNA molecule is introduced at time 8100. Observations are shown for an hour (8100 to 11700 seconds). A number of changes and new signals now are observed: (i) the DNA hairpin signal class identified with the green line is no longer observed – this class is hypothesized to be no longer free, but annealed to its Y-shaped DNA partner; (ii) the Y-shaped DNA molecule is found to have a bifurcation in its class identified with the yellow lines, a bifurcation clearly discernible in the plots of the signal standard deviations. (iii) the hairpin class with the red line appears to be unable to bind to its Y-shaped DNA partner, an inhibition currently thought to be due to G-quadruplex formation in its G-rich overhang. (iv) The Y-shaped DNA molecule also exhibits a signal class (blue line) associated with capture of the arm of the 'Y' that is meant for annealing, rather than the base of the 'Y' that is designed for channel capture.

6.3 Pathogen Detection, miRNA detection, and miRNA haplotyping

In clinical diagnostics, as well as in biodefense testing, patient blood samples can be drawn for the purpose of assaying the DNA content. Obviously there will be a preponderance of human DNA in such a sample, but if there is infection then trace amounts of the associated viral or bacterial DNA will be present as well. The question then arises as to how to detect unique elements of bacterial DNA sequence that are singled-out for detection, with very high sensitivity and specificity. This may be possible in the NTD approach, with annealing-based detection along the lines described earlier and in [17,27], where ssDNA sequences are targeted for detection of approximate length 22 base sub-sequences. A 22-mer is shown in Fig. 6.4, 'B'-labeled secondary structure, in the leftmost, linear, ssDNA segment. The Y-shaped secondary structure in Fig. 6.4 ('B') shows the blueprint for a NTD ssDNA probe for any targeted ssDNA segment, upon 'recognition' (annealing-based), a Y-shaped channel modulator is engineered to occur. If not the correct modulator, due to a few mis-matches or inserts (particularly at the Y-nexus), then the difference can be discerned with high discrimination. All that is needed is a specific set of enzyme digestion steps on the DNA sample to 'chop' it into shorter segments, and leave targeted regions at the ends of (some) of the resulting ssDNA digests → so as to obtain-dsDNA annealed targets with probe match as in Fig. 6.4('B'), where the excess ssDNA length (beyond the 22-mer match template) is left to dangle off of one end, as shown for the eight-base segment shown in Fig. 6.4('B').

In clinical diagnostics patient blood samples can also be drawn for the purpose of assaying glycoprotein contents. In the case of DNA there will be a preponderance of the individual's own genomic DNA in such a sample, but if there is infection then trace amounts of the associated viral or bacterial DNA will be present as well. One of the questions that then arises is how to detect unique elements of bacterial DNA sequence with very high sensitivity and specificity. In [1] annealing-based detection is explored, where Y-shaped NTD transducer results are shown for tests involving an eight base ssDNA target [17]. The method can be extended to other lengths of targeted ssDNA, using annealing-based recognition. For longer lengths we can arrive at interesting detection scenarios for pathogens or for miRNA's (some possibly pathogenic). The known pathogen ssDNA targets could be longer, 15-25 bases say, to enable unique identifiers respective to a particular pathogen. For miRNA detection probes could be designed for ssDNA target annealing that is in the 7-15 base range.

MicroRNA detection follows a similar approach to the pathogen detection problem, but now typically working with a much shorter length nucleic acid detection target, a miRNA sequence based annealing target. In this setting often have similar 'informed' analysis to pathogen detection analysis.

Y-DNA modulator platforms for biosensing can also provide a simple linker platform for use with antibody binding moieties, where a 'linker' aptamer can be used that is covalently linked to the common base of the antibody (IgG) molecule (using a DNA tagged antibody approach). Aptamer tuning can also be enhanced in the nanopore setting using nanopore directed SELEX (referred to as NADIR in [1]), where binding strength can be selected to be not too strong or weak according to the desired tuning on the observed binding lifetimes, as seen in the state durations of the observed state noise.

Linkage of ssDNA to antibody is commonly done in immuno-PCR preparations, so another path with rapid deployment is to make use of a linkage technology that is already commoditized, e.g., a good NTD signal can then be produced with immuno-PCR tagged antibodies that are designed to anneal to another DNA molecule to form an NTD 'Y-transducer'.

The explosive geographic expansion of the Zika virus provides another reminder that rapid diagnostic tools for new viral infections are an ever increasing need. The rapid deployment of a fast diagnostic tool in the example of the Zika virus is all the more pertinent given that the virus has been shown to be the cause of microcephaly in the fetuses of exposed pregnant women, along with results indicating possible brain damage (Guillain-Barre reaction) to a significant fraction of those exposed. A rapid development, deployment, and evaluation of a Zika virus diagnostic would afford the patient the critical time needed to undergo aggressive prophylactic measures. Similarly, certain fungal infections need to be diagnosed as early as possible (cryptococcus neoformans, for example, can disrupt and cross the blood-brain barrier). The treatments for many fungal infections are highly toxic, however, such that they will only be undertaken if infection is highly likely.

Pathogens that are suspected can potentially be probed in a matter of hours using an NTD platform with the methods described here using probes designed according to the pathogen's genomic profile. Unknown pathogens would first need to either have their genomes sequenced (less than a day) if sufficient DNA already available, or a sample directly measured via a test assay template (same procedure as for biomarker discovery) for assay-level fingerprint determination, then testing for that pathogen fingerprint in the patient.

The NTD platform can be enhanced to be a rapid annealing-based detection platform due a recently established ability [14] to operate under high chaotropic conditions (up to 5M urea), which allows measurement of collective binding interactions such as nucleic acid annealing with other simpler binding and related complexes thereby eliminated and effectively filtered from the analysis task. What remains to be done is to establish a general production method for creating a NTD transducer for the sequence of interest, and this is described in [9].

6.4   SNP Detection

The proposed test of DNA SNP annealing is with the Y-shaped DNA transduction molecule shown in Fig. 6.4 ('B') that is minimally altered, and such that the SNP variant occurs in the Y-nexus region. For the case where digestion can't conveniently provide extension only to one-side, a Y-shaped annealed dsDNA molecule can still be obtained, but such that the ssDNA extensions outside the annealed region are now free to extend on both arms of the Y-molecule.

SNP variant detection is reduced to resolving the signals of two Y-shaped duplex DNA molecules, one with mismatch at SNP, one with Watson-Crick base-pairing match at SNP. In preliminary studies of Y-shaped DNA molecules, numerous Y-shaped DNA molecules were considered. Three variants that successfully demonstrated the easily discernible, modulatory, channel blockade signals are shown in Fig. 6.5 [17]. In those variants we considered the Y-nexus with and without an extra base (that is not base-paired). And if an extra base is inserted we explore the three positions at the Y (left and middle inserts shown in the left and center Y-molecules shown in Fig. 6.5. Fig. 6.9 diagrams the NADIR nanopore directed SELEX refinement.

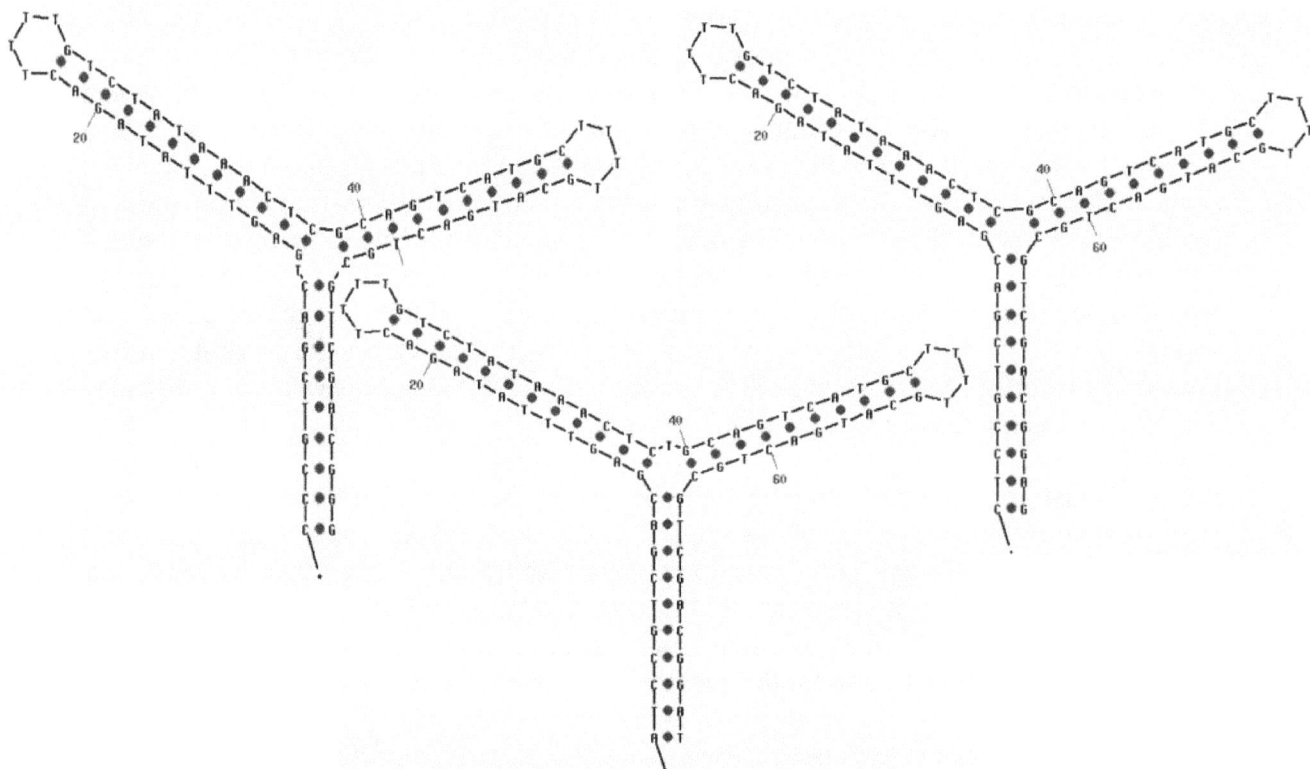

Fig. 6.5. Shown are Y-shaped aptamers that have shown they have capture states with the desired blockaded toggling.

The DNA molecule design we are currently using consists of a three-way DNA junction created: 5'-CTCCGTCGAC   GAGTTTATAGAC   TTTT   GTCTATAAACTC GCAGTCATGC TTTT GCATGACTGC GTCGACGGAG-3'. Two of the junctions' arms terminate in a 4T-loop and the remaining arm, of length 10 base-pairs, is usually designed to be blunt ended (sometimes shorter with an overhang). The blunt ended arm has to be carefully designed such that when it is captured by the nanopore it produces a toggling blockade. One of the arms of the Y-shaped aptamer (Y-aptamer) has a TATA sequence, and is meant to be a binding target for TBP. In general, any transcription factor binding site could be studied (or verified) in this manner. Similarly, transcription factor could be verified by such constructions, or the efficacy of a synthetic transcription factor could be examined. The other Y-aptamer, used in the integrase binding analysis, is shown later.

A preliminary test of DNA SNP annealing can be done with the Y-shaped DNA transduction molecule shown in Fig. 6.6, which is minimally altered (e.g., mostly common sequence identity) from the Y-annealing transducer introduced in Fig. 6.5.

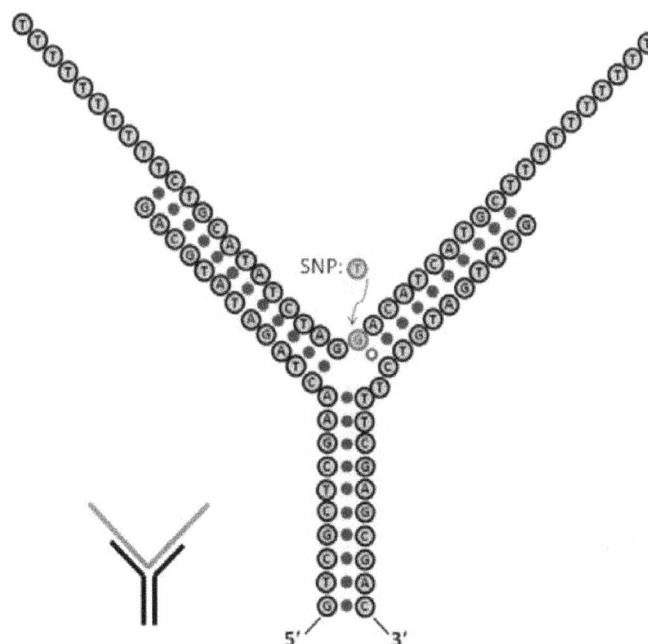

Fig. 6.6. The Y-SNP with test complex is shown at the base-level specification and at the diagrammatic level, where a SNP base is as indicated. If the SNP is its variant form (typically only one other base possibility is common), then a base-pairing will not occur at the nexus of the Y-SNP shown (with the red base becoming a 'T' in the variant as indicated). This allows discrimination between the annealed forms with high accuracy, while also discerning from the signals produced by the non-annealed Y-SNP, where there is no target-bound, or only non-specific molecular interactions imparting much less conformational structure as occurs with the matching (or mostly matching) annealing interaction.

Once the Y-SNP transducer has been tested on a single-species of short overhang length test molecules the next experimental challenge will be to detect SNP variants using the Y-SNP transducer probe in the presence of a heterogeneous length mixture (some with target SNP region of interest), with overhang as shown in Fig. 6.7.

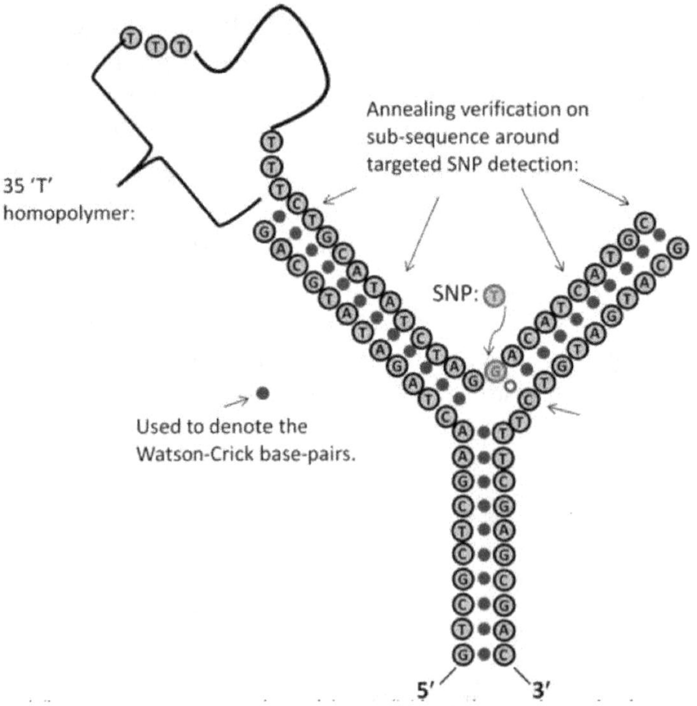

Fig. 6.7. The Y-SNP test complex with 35 dT length overhang is shown at the base-level specification, where a SNP base is as shown. If the SNP is its variant form (typically only one other base possibility is common), then a base-pairing will not occur at the nexus of the Y-SNP shown. This allows discrimination between the annealed forms with high accuracy, while also discerning from the signals produced by the non-annealed Y-SNP, where there is no target-bound, or only non-specific molecular interactions imparting much less conformational structure as occurs with the matching (or mostly matching) annealing interaction.

The value of 35 'T's on the extension is to also match the approximate extension, with same 'Y'-sequence (except for a 4 dT cap) as the previously 'blunt-ended' annealed conformation. SNP variant detection is reduced to resolving the signals of two Y-shaped duplex DNA molecules, one with mismatch at SNP, one with Watson-Crick base-pairing match at SNP. From the above it is clear that the NTD method provides a viable prospect for SNP variant detection to very high accuracy (possibly the accuracy with which the NTD can discern DNA control hairpins that only differ in terminal base-pair, greater than 99.999%). SNP detection via *translocation-based* methods, on the other hand, must discern between two SNP variants according to the different dwell times of the complement-template annealed SNPs, until dissociation from the template allows translocation of the blockading dsDNA annealed conformation.

6.5   Aptamer-based Detection

Aptamers are especially appropriate for study by nanopore detection due to the fact they can be designed with an end to be captured and modulate a nanopore (i.e., the captured end is dsDNA) while other parts of the aptamer are intended to bind a specific target. This directly provides a NTD transducer if one or both of the bound/unbound states (captured in the channel at the dsDNA end) provides distinctive channel modulations. The binding statistics derived from the study of aptamers in a nanopore detector can also be used in the design of the aptamer itself, e.g., NADIR selection instead of further SELEX-based selection [1,7]. In Fig. 6.8 we see the first aptamer test case to be considered, where we seek to detect thrombin [224] in one case, and IgG [223] in another. We use the thrombin aptamer found by Ikebukuro et al [224], it is selected via SELEX and EMA and is a 31-mer, linked by a 4 dT spacer to link to the Y-transducer (see Fig. 6.8).

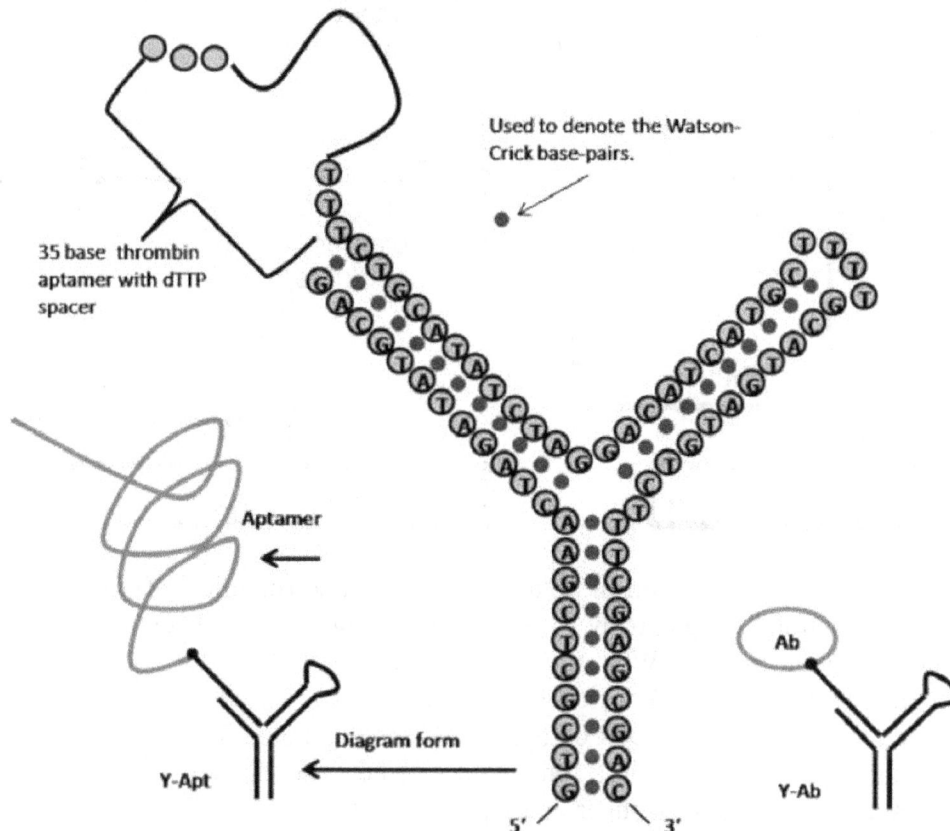

Fig. 6.8 The thrombin aptamer from [224] is 5'-CACTGGTAGGTTGGTGTGGTTGGGGCCAGTG-3'.

6.5.1 NaDir SELEX

In using the NADIR refinement process to arrive at the Y-transducer used in the DNA annealing test [1,7], we have demonstrated how *single-base insertions or modifications at the nexus of the Y-shaped molecule can have clearly discernible changes in channel-blockade signal*. Y-molecules as DNA probes with single point mutations discernible at the Y-nexus are explored in [1,7] (see Fig.s 6.5 and 6.9). What is described in [1,7] is a linkage to a *na*nopore-detector *dir*ected (NADIR) search for aptamers that is based on bound-state lifetime measurements (or some other selection criterion of interest). NADIR complements and augments SELEX in usage.

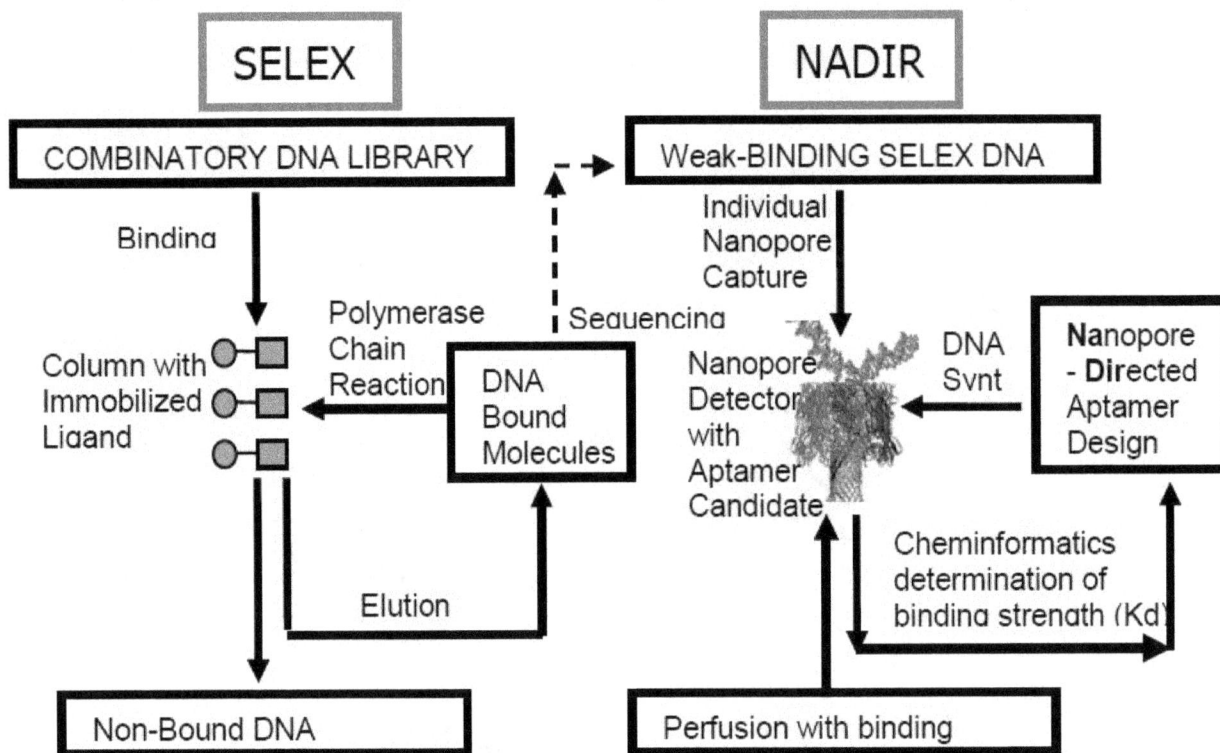

Fig. 7.18. The determination of aptamers can be done (or initiated) via Systematic Evolution of Ligands by Exponential Enrichment (SELEX), as shown schematically on the left. What is proposed here is a linkage to a *na*nopore-detector *dir*ected (NADIR) search for aptamers that is based on bound-state lifetime measurements. NADIR complements and augments SELEX in usage: SELEX can be used to obtain a functional aptamer, and NADIR used for directed modifications (for stronger binding affinity, for example).

6.6   Antibody-based Detection

Linkage of ssDNA to antibody is commonly done in immuno-PCR preparations, so another path with rapid deployment is to make use of a linkage technology that is already commoditized, e.g., the molecules required for the antibody-based biosensing with this approach are simple (non-specialty) molecular components. The core issue to be tested here is whether a good NTD signal can be produced with immuno-PCR tagged antibodies that are designed to anneal to another DNA molecule to form an NTD 'Y-transducer' (see Fig. 6.8, lower right). From previous efforts [30], with more complicated EDC linkages between a modified thymine and an antibody (see Fig. 6.12), it is clear that there are strong prospects for success with this method. What is sought is not just further validation of the method, however, but a less expensive, accessible, platform from which to refine and develop NTD-based systems.

Some mAb blockades produce a very clean toggling between two levels (shown in figures that follow). The mAb interference modulatory signals are easily discerned from a modulatory signal of interest, however, especially with increased observation time as needed. Aside from being an interference agent, antibodies offer a direct means for having a NTD transducer since their modulatory blockade signals are observed to change upon introduction of antigen. The problem with using an antibody directly as a transducer in a biosensor arrangement is that the antibody produces multiple blockade signal types (a dozen or more)

124

just by itself (without binding). This weakness for use directly as a biosensor (they can still be linked indirectly as in [30]) is because the antibody is a glycoprotein that has numerous heterogeneous glycosylations and glycations, with many molecular side-groups that might be captured by the nanopore detector to produce modulatory blockades. If the purpose is to study the post-translational modifications (PTMs) themselves, a glyco-profile of the antibody in other words, then the numerous signal types seen are precisely the information desired. A more complete analysis of antibody blockades on the nanopore detector is beyond the scope of this paper, and will be in a separate paper. Some further details on the Antibody structure and its direct glyco-profiling is still given next, however, since similar PTMs can be analyzed on other proteins of critical biomedical interest.

Managing antibodies as easily identifiable interference or transducer

Antibodies are the secreted form of a B-cell receptor, where the difference between forms is in the C-terminus of the heavy chain region. Fig. 6.10 shows the standard antibody schematic. Standard notation is shown for the constant heavy chain sequence ('CH', 'H', and 'S' parts), variable heavy chain region ('VH' part), the variable light chain region ('VL' part), and constant light chain region ('CL' part). The equine IGHD gene for the constant portion of the heavy chain has exons corresponding with each of the sections CH1,H1,H2,CH2,CH3,CH4(S), and for the membrane-bound form of IGHD, there are two additional exons, M1 and M2 for the transmembrane part, thus, CH1, H1, H2, CH2, CH3, CH4(S), M1, M2 [225]. In Fig. 6.10, the long and short chains are symmetric from left to right, their glycosylations, however, are generally not symmetric. Critical di-sulfide bonds are shown connecting between chains, each of the VH and CH regions typically have an internal disulfide bond as well. The lower portion of the antibody is water soluble and can be crystallized (denoted Fc). The upper portion of the antibody is the antigen binding part (denoted Fab).

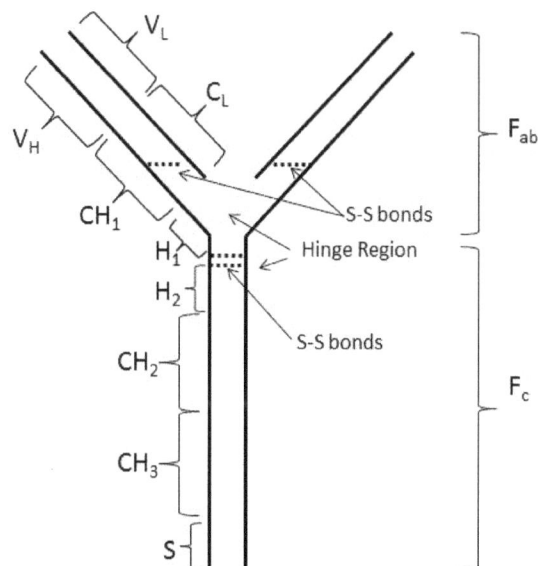

Fig. 6.10. The standard antibody schematic [225]. Standard notation is shown for the constant heavy chain sequence ('CH', 'H', and 'S' parts), variable heavy chain region ('VH' part), the variable light chain region ('VL' part), and constant light chain region ('CL' part). The full heavy chain sequence is derived from recombination of the VH part and {CH,H,S} parts (where the secretory region S is also called CH4). The long and short chains

are symmetric from left to right, their glycosylations, however, are generally not symmetric. Critical di-sulfide bonds are shown connecting between chains, each of the VH and CH regions typically have an internal disulfide bond as well. The lower portion of the antibody is water soluble and can be crystallized (denoted Fc). The upper portion of the antibody is the antigen binding part (denoted Fab).

Fig. 6.11 shows a typical antibody N-glycosylation (exact example for equine IGHD [275]). One possible N-glycosylation site is indicated in region CH2, and three possible N-glycosylation sites are indicated in region CH3. N-glycosylation consists of a covalent bond (glycosidic) between a biantennary N-glycan (in humans) and asparagine (amino acid 'N', thus N-glycan). The covalent glycosidic bond is enzymatically established in one of the most complex post translational modifications on protein in the cell's ER and Golgi organelles, and usually only occurs in regions with sequence "NX(S/T) – C-terminus" where X is 'anything but proline' and the sequence is oriented with the C-terminus as shown. Licensed therapeutic antibodies typically display 32 types of biantennary N-glycans, consisting of N-acetyl-glucosamine residues (GlcNAc, regions '1'); mannose residues (Man, regions '2'); galactose residues (Gal, regions '3'), and Sialic Acid Residues (NeuAc, regions '4'), as shown in Fig. 6.11. The N-glycans are classified according to their degree of sialylation and number of galactose residues: if disialylated (shown) have A2 class. If asymmetric and monosialylated have A1 class. If not sialylated then neutral (N class). If two galactose residues (shown) then G2 class, if one, then G1 class, if zero, then G0 class. If there is an extra GlcNAc residue bisecting between the two antennae +Bi class (–Bi shown). If a core fucose is present (location near GlcNAc at base), then +F (–F shown). So the class shown is G2-A2. The breakdown on the 32 types is as follows: 4 G2-A2; 8 G2-A1; 4 G1-A1; 4 G2-A0; 7 G1-A0; 4 G0-A0 [226]. The N-glycans with significant acidity (A2 and A1) are 16 of the 32, so roughly half of the N-glycans enhance acidity. The other main glycosylation, involving O-glycans, occurs at serine or threonine (S/T). The main non-enzymatic glycations occur spontaneously at lysines ('K') in proteins in the blood stream upon exposure to glucose via the reversible Maillard reaction to form a Schiff Base (cross-linking and further reactions, however, are irreversible and associated with the aging process).

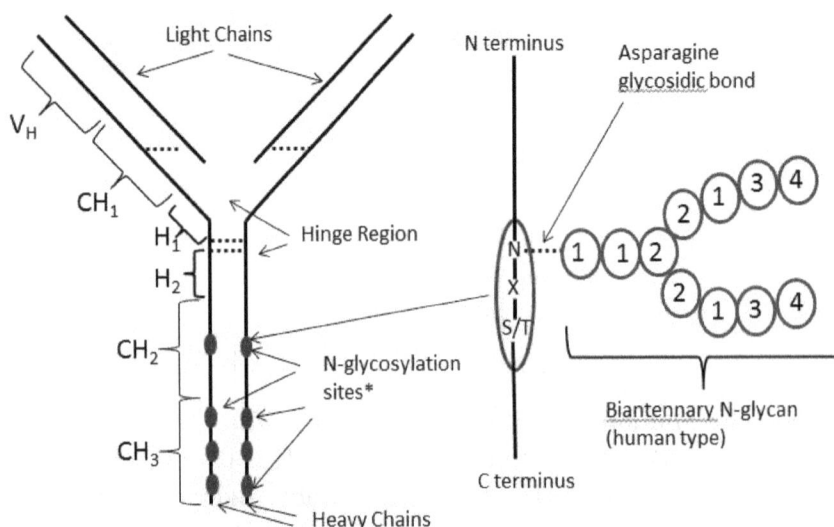

Fig. 6.11. Typical antibody N-glycosylation [225]. A schematic for typical antibody N-glycosylation is shown (drawn from results on the equine IGHD gene), where one possible

N-glycosylation site is indicated in region CH2, and three possible N-glycosylation sites are indicated in region CH3. N-glycosylation consists of a covalent bond (glycosidic) between a biantennary N-glycan (in humans) and asparagine (amino acid 'N', thus N-glycan). The covalent glycosidic bond is enzymatically established in one of the most complex post translational modifications on protein in the cell's ER and Golgi organelles, and usually only occurs in regions with sequence "NX(S/T) – C-terminus" where X is anything but proline and the sequence is oriented with the C-terminus as shown. Licensed therapeutic antibodies typically display 32 types of biantennary N-glycans, consisting of N-acetyl-glucosamine residues (GlcNAc, regions '1'); mannose residues (Man, regions '2'); galactose residues (Gal, regions '3'), and Sialic Acid Residues (NeuAc, regions '4'). The N-glycans are classified according to their degree of sialylation and number of galactose residues: if disialylated (shown) have A2 class. If asymmetric and monosialylated have A1 class. If not sialylated then neutral (N class). If two galactose residues (shown) then G2 class, if one, then G1 class, if zero, then G0 class. If there is an extra GlcNAc residue bisecting between the two antennae +Bi class (–Bi shown). If a core fucose is present (location near GlcNAc at base), then +F (–F shown). So the class shown is G2-A2. The breakdown on the 32 types is as follows: 4 G2-A2; 8 G2-A1; 4 G1-A1; 4 G2-A0; 7 G1-A0; 4 G0-A0. The N-glycans with significant acidity (A2 and A1) are 16 of the 32, so roughly half of the N-glycans enhance acidity. The other main glycosylation, involving O-glycans, occurs at serine or threonine (S/T). The main non-enzymatic glycations occur spontaneously at lysines ('K') in proteins in the blood stream upon exposure to glucose via the reversible Maillard reaction to form a Schiff Base (cross-linking and further reactions can be irreversible).

The base of the antibody plays the key role in modulating immune cell activity. The base is called the Fc region for 'fragment, crystallizable', which is the case, and to differentiate it from the Fab region for 'fragment, antigen-binding' that is found in each of the arms of the Y-shaped antibody molecule (see Fig. 6.10). The Fc region triggers an appropriate immune response for a given antigen (bound by the Fab region). The Fab region gives the antibody its antigen specificity; the Fc region gives the antibody its class effect. IgG and IgA Fc regions can bind to receptors on neutrophils and macrophages to connect antigen with phagocyte, known as opsonization (opsonins attach antigens to phagocytes). This key detail may explain the modulatory antibody interaction with the nanopore channel. IgG, IgA, and IgM can also activate complement pathways whereby C3b and C4b can act as the desired opsonins. The C-termini and Fc glycosylations of an antibody's heavy chain, especially for IgG, is thus a highly selected construct that appears to be what is recognized by immune receptors, and is evidently what is recognized as distinct channel modulator signals in the case of the NTD. Using NTD we can co-opt the opsonization receptor-binding role of the Fc glycosylations (and mAB glycations and glycosylations in general), and C-terminus region, to be a channel modulating role. This may also permit a new manner of study of the critical opsonization role of certain classes of antibodies (and possibly differentiate the classes in more refined ways) by use of the nanopore detector platform. The channel may provide a means to directly measure and characterize antibody Fc glycosylations, a critical quality control needed in antibody therapeutics to have correct human-type glycosylation profiles in order to not (prematurely) evoke an immunogenic response.

6.6.1 Small target Antibody-based detection (linked modulator)

IgG antibodies may vary in net charge but are nowhere near as negatively charged as the DNA hairpin molecules examined in [30]. Differences in channel interaction are often attributed to its net charge and its electrophoretic mobility. To improve the antibody's affinity for the channel and to aid in signal classification, a complex of antibody and DNA hairpin is sometimes used. The result is the increase in channel affinity and a significant reduction in capture class configurations (see Fig.s 6.12-6.14 for further details), while still retaining binding detection sensitivity. The *small*-antigen biosensing results (described here) complements those for *large*-antigen biosensing (presented in the next section). The large antigen study done here is also notable in that it involves direct use of the antibody as a bifunctional reporter molecule. This leads to complications with capture, and uniqueness in the orientation of that capture, but may offer a more sensitive detection approach since there is not a linkage separating the bound/unbound complex from the channel flow environment.

A DNA hairpin with EDC linkage to an antibody is shown in Fig.6.12, and examined in [30]. When the DNA portion of this linked complex inserts itself into the alpha hemolysin channel it creates a definable toggle signal that serves as reliable "carrier signal" for monitoring any changes of molecular state (such as binding). In our first study of DNA-hairpin linked antibody complexes [30], we used an anti-biotin-antibody (Stressgen) as our binding element linked to our DNA hairpin. (Note, as one of many control tests, we see that the blockade toggle signal is relatively unchanged after addition of excess biotin.)

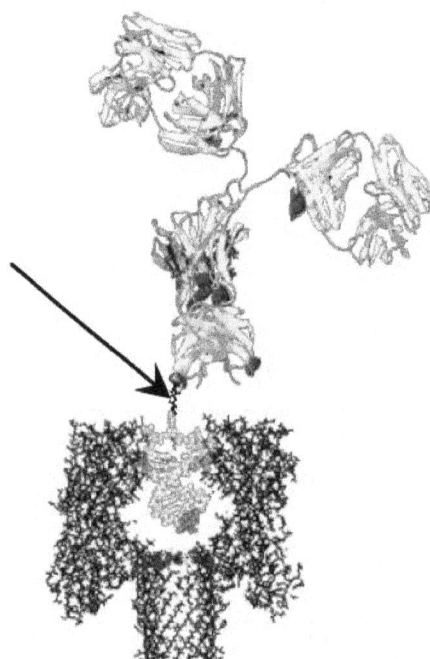

Fig. 6.12. DNA hairpin bound to Antibody via an EDC-linker. Approximately shown to scale. Arrow points to the Internal Amino Thymine Modification with Primary Amine on a six carbon spacer arm. Primary amine can be crosslinked using 1-Ethyl-3-(3-dimethylaminopropyl) carbodiimide hydrochloride (EDC) to the peptide carboxyl terminus of the antibody heavy chain. This crosslinkage results in a covalent bond between the primary amine and the carboxyl.

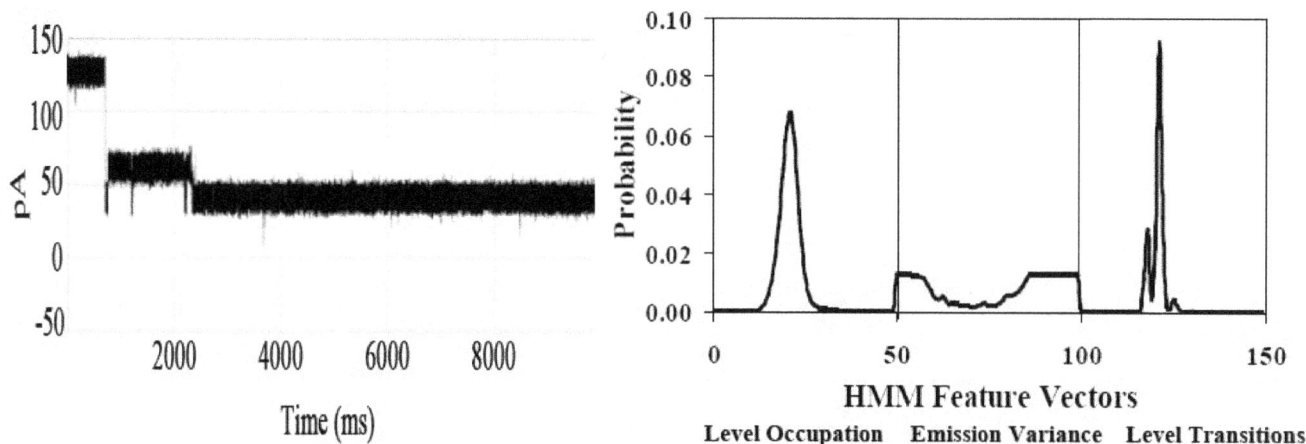

Fig. 6.13. Antibody linked to DNA-Hairpin Blockade signal and HMM Profile.

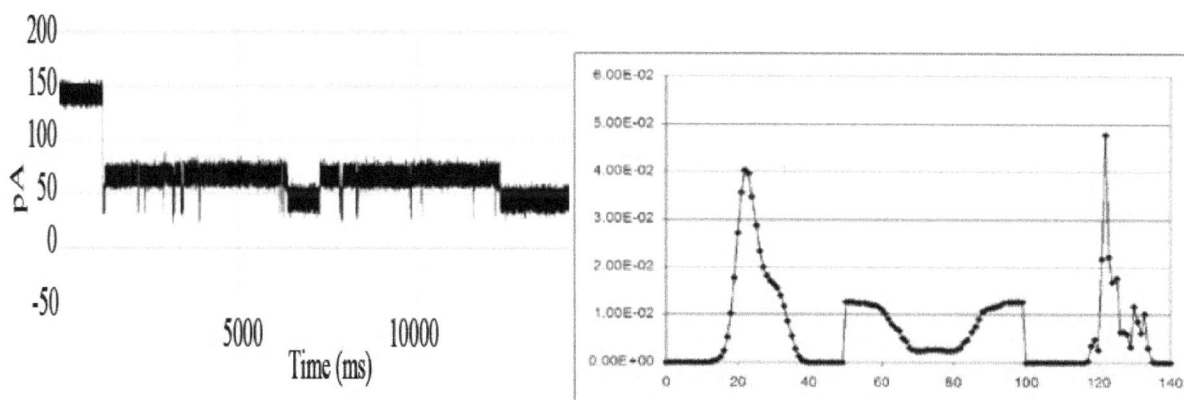

Fig. 6.14. Antibody linked to DNA-Hairpin, now bound to its target antigen (biotin) – new blockade signal, and associated HMM profile. Antigen binding to an EDC-linked Antibody/DNA-Hairpin, where stem of the hairpin is captured in the Nanopore Detector.

The clarity of the current blockade signal for Ab-antigen binding and Ab-pore interaction was examined by varying the composition of working buffer in presence of urea and $MgCl_2$. In one series of experiments, mentioned above, we used free antibody molecule interacting with the nanopore detector, where the antibody (anti-biotin) molecule is introduced to our nanopore device to produce the characteristic two-state telegraph signal (Fig. 6.13, 6.14). The blockade signal for the antigen is practically unaltered by excess antigen: even 100 fold excess of biotin does not change the blockade signal considerably (Fig. 6.15). The signal changes greatly in presence of urea, however, in a relatively small concentration. Here the duration of any event to occupy upper state level becomes shorter and the total probability value of upper level decreases with urea concentration rise.

129

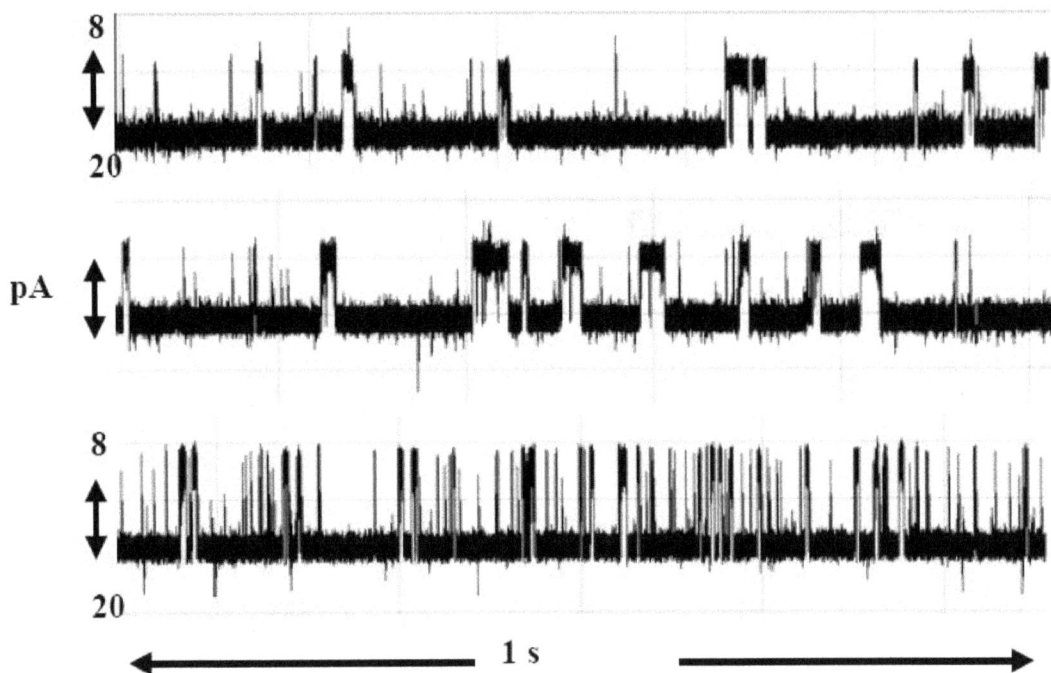

Fig. 6.15. DNA-hairpin signals. Top, No biotin concentration. Middle, low-to-high biotin concentration (1000-fold excess). Bottom. low urea concentration.

6.6.2 Large target Antibody-based detection (with direct antibody modulation)

For large-antigen antigen-binding studies, different versions of copolymer (Y,E)-A—K ('large' targets) were originally prepared to allow study of the effect of antigenic mass and valency of binding upon the observations in [30]. In this and other studies involving direct antibody interactions with the channel, however, we found that antibodies themselves typically produce a variety of 'long-lived' blockades at the channel themselves, sometimes modulatory, even possibly producing clear 'toggle' signals as shown in the study cases in Fig. 6.16.

It is found that the antibody blockade signal alters shortly after introduction of antigen, as Fig. 6.17 shows upon addition of a moderately high concentration (100 μg/ml) of 200kD multivalent synthetic polypeptide (Y,E)-A—K. Presumably, these changes are the result of antibody binding to antigen. The time before the blockade signal is altered is also interesting; it ranges from seconds to minutes (not shown). This presumably is a reflection of antibody affinity.

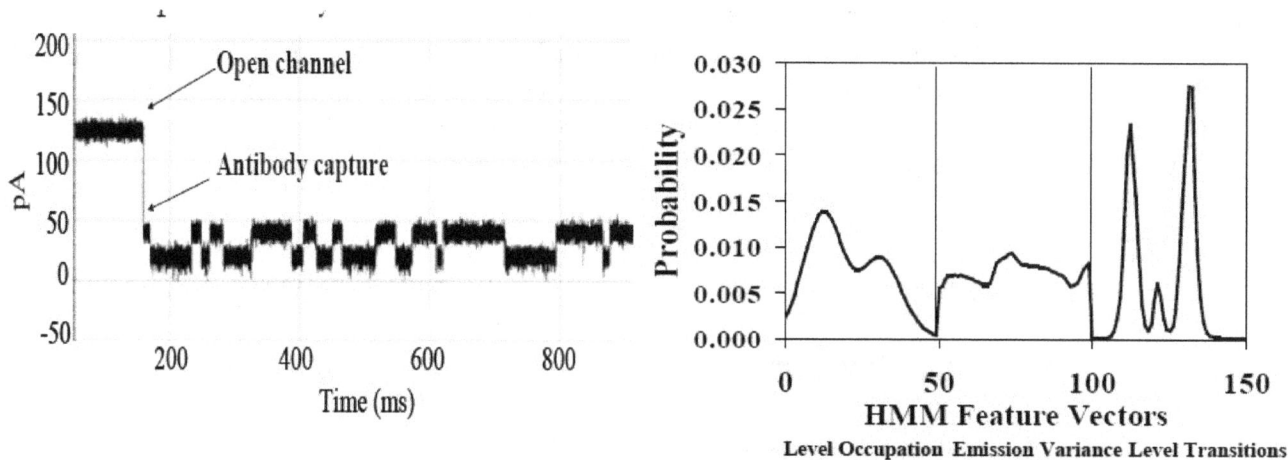

Fig. 6.16. Example that provides a very clear, stable, blockade direct by an Ab. Left: A toggle signal is generated as a channel-captured region of the molecule (IgG) wiggles above the limiting aperture of the alpha-hemolysin channel varying the ionic current between two transient states. **Right:** Antibody Toggle HMM Signal Profile. The 150 feature vectors obtained from the 50-state HMM-EM/Viterbi implementation in [1] are: the 50 dwell percentage in the different blockade levels (from the Viterbi trace-back states), the 50 variances of the emission probability distributions associated with the different states, and the 50 merged transition probabilities from the primary and secondary blockade occupation levels (fits to two-state dominant modulatory blockade signals).

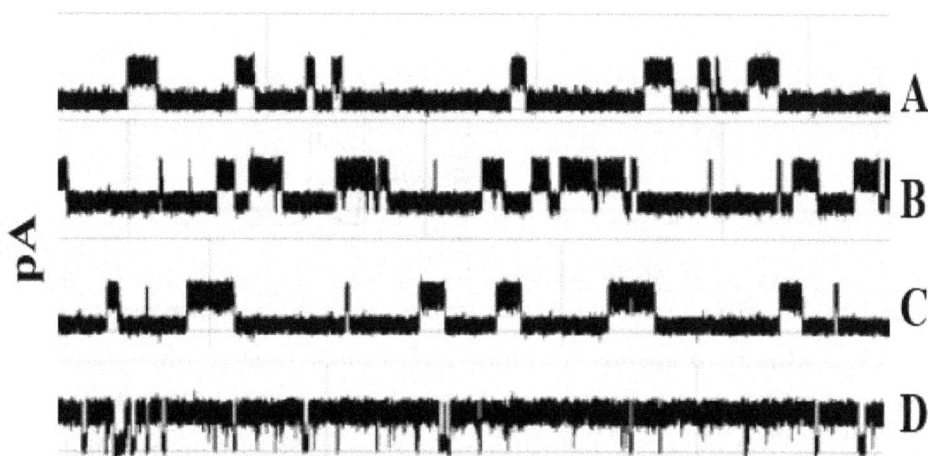

Fig. 6.17. Antibody-Antigen binding – clear example from specific capture orientation. Each trace shows the first 750 ms of a three minute recording, beginning with the blockade signal by an antibody molecule that has inserted (some portion) into the Alpha-hemolysin channel to produce a toggle signal (A). Antigen is introduced at the beginning of frame (A). Changes to the toggle signal are discernible in frame D, indicating the binding event between the antibody and antigen has taken place.

Direct antibody nanopore blockades are examined further in Fig. 6.18, Left, where the different capture signals provided by a single antibody species provide a 'nanopore epitope' mapping or assay of the antibody's surface features, including glycations and nitrosilations, as described in the following section. Typical captures seen after introduction of antigen are shown for the same system in Fig. 6.18, Right. Fig 6.19 shows a possible indication of a multivalent binding signal (the Ab being bivalent).

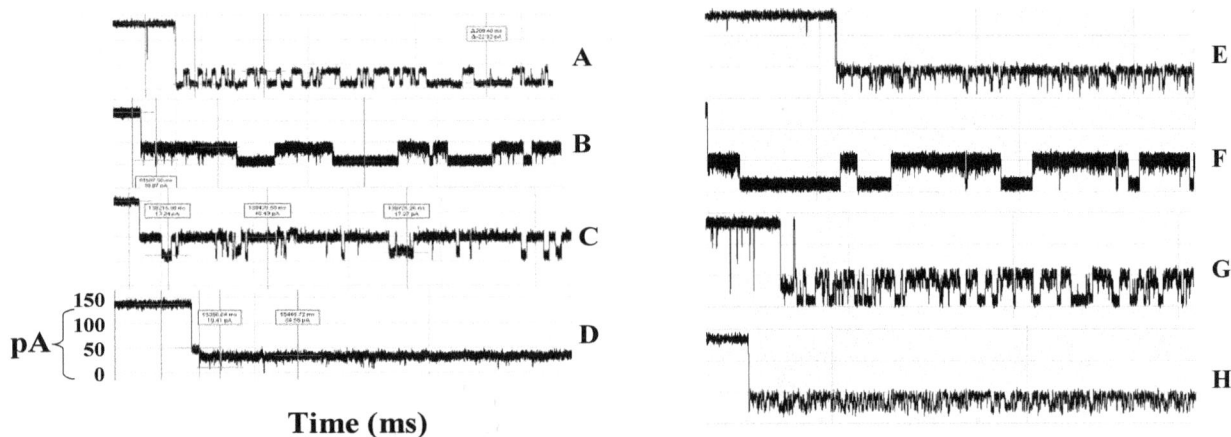

Fig. 6.18. Antibody Signal Classes and Ab-Antigen Signal Classes. A-D: various IgG region captures and their associated toggle signals (1 second traces). E-F: various IgG+Antigen region captures and their associated toggle signals (1 second traces). Each blockade signal was identified visually and represents a commonly observed signal class. Note the changes in dwell times for the upper and lower current levels in each signal class. We find a higher current level bias in the level occupancy as a result of binding with the antigen molecule.

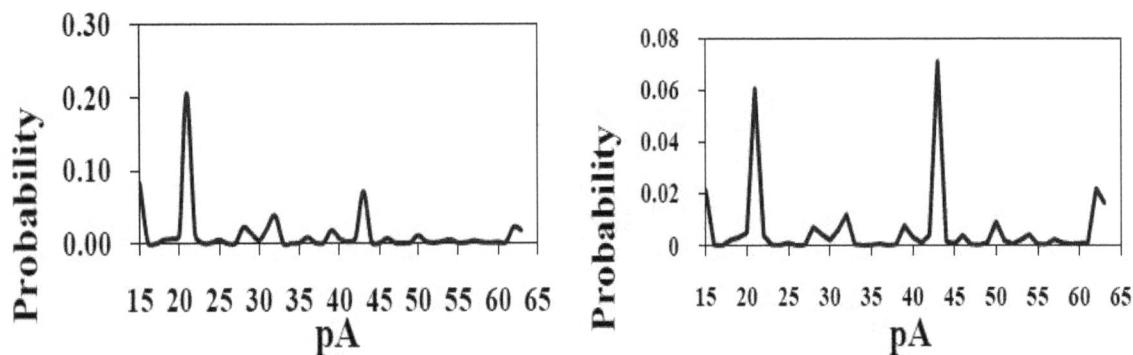

Fig. 6.19. Multivalent antigen binding. Left Panel: First Antibody Antigen Binding – 1st 50 feature components extracted from the HMM. **Right Panel:** Shifts in the values of these 1st 50 HMM feature components indicate a possible second Antibody Antigen Binding (same molecular capture). The first 50 components of the 150 feature vectors obtained from the 50-state HMM-EM/Viterbi implementation are the dwell percentages in the different blockade levels from the Viterbi trace-back states (approximately the Histogram in that range).

Chapter 7

NTD Assaying Methods

7.1 DNA enzyme analysis: Integrase
7.2 Single-molecule serial assaying
    7.2.1 DNA-Protein complex Assaying: Aptamer-TBP
    7.2.2 Glycoprotein assayer
    7.2.3 Antibody Assayer
7.3 Molecular capture via antibody, aptamer, capture-matrix, TERISA
7.4 NTD-Gel
7.5 Nanopore-based assays of cytosolic antigen delivery complexes
7.6 Transcriptome and Transcription-Factor based Drug Discovery
7.7 DNA Sequencing
    7.7.1 Single-molecule, processive
    7.7.2 NTD/Sanger DNA Sequencing

Using a NTD platform, a single bound/unbound system can be held, observed, and its history of bound/unbound states can be tracked. The *single* molecule state-tracking with lengthy time averages allows measurement of molecular characteristics that are obscured in ensemble-based measurements. The ensemble averages, that underlie most approaches, lose information about the true diversity of behavior of individual molecules. For complex biomolecules there is likely to be a tremendous diversity in behavior, and in many cases this diversity may be the basis for their function.

Molecular (protein) diversity via post-translational modifications can be examined as well, such as with heterogeneous mixtures of protein glycoforms that typically occur in living organisms (e.g., for TSH and hemoglobin proteins in blood serum and red blood cells, respectively). The hemoglobin 'A1c' glycoprotein is a disease diagnostic (diabetes) [227-230], and for TSH, glycation is critical component in the TSH-based regulation of the endocrine axis. Multi-component regulatory systems and their variations (often sources of disease) could also be studied much more directly using the NTD approach, as could multi-component (or multi-cofactor) enzyme systems. In what follows, NTD assaying applications will be described for enzyme studies and for nanopore-epitope sampling on (transient) protein complexes and configurations.

The NTD approach also provides an excellent method for examining enzymes, and other complex biomolecules, particularly their activity in the presence of different co-factors. There are two ways that these studies can be performed: (i) the enzyme is linked to a channel transducer, such that the enzyme's binding and conformational change activity may be directly observed and tracked or, (ii) the enzyme's substrate may be linked to the channel transducer and observation of enzyme activity on that substrate may then be examined. Case (i) provides a means to perform DNA sequencing if the enzyme is a nuclease, such as lambda exonuclease (discussed in Sec. 5.3). Case (ii) provides a means to do screening, for example, against HIV integrase activity (for drug discovery on HIV integrase inhibitors).

An example of a transient interaction that has been examined involves interaction of HIV integrase with its consensus DNA binding terminus [27,30]. One use of the nanoscope is as drug-discovery assayer in settings where measurements are made of transient interactions, such as HIV integrase interactions with DNA in the presence of interference agents or competitive inhibition molecules (decoy aptamers, for example).

HIV integrase binding to viral-DNA appears to favor the high flexibility of a CA/TG dinucleotide positioned precisely two base-pairs from the blunt terminus of the duplex viral DNA (and experimentally verified with the nanoscope in the conformational analysis shown in [35]). The CA/TG dinucleotide presence is a universal characteristic of retroviral genomes. Deletion of these base pairs impedes the integration process and it is believed that the unusual flexibility imparted by this base-pair on the terminus geometry is necessary for the binding to integrase. Once bound to integrase the viral DNA molecule is modified by removal of the two residues at the 3'-end together with subsequent insertion into the host genome. Further description of the HIV integrase studies are shown in the next section (Sec. 7.1). In Sec.s 7.2-7.8 other serial assays, also on a single molecules basis, are explored.

7.1 DNA enzyme analysis: Integrase

DNA termini are of critical importance for certain retroviral integrases and other biological processes – being able to study them, even comparatively, offers new avenues for understanding and drug selection (HIV integrase blockers). Information on the DNA molecules' variation in structure and flexibility is important to understanding the dynamically enhanced (naturally selected) DNA complex formations that are found with strong affinities to other, specific, DNA and protein molecules. An important example of this is the HIV attack on cells. The DNA terminus properties of retroviral DNA molecules are found to exhibit greater flexibility than similar sequences, often marked by an increase in the number of blockade states, such as in the upper-level fine structure for the molecule terminating with GACG-3' [41].

One of the most critical stages in HIV's attack is the binding between viral and human DNA. The DNA molecule studied in this instance consists of the HIV consensus terminus at the end of the Y-aptamer arm in Fig. 7.1 – where it is exposed for binding to integrase. Since this molecule presents another blunt-ended dsDNA for capture, it is no surprise that such events occur. The signal analysis must separate between two classes of signal associated with these two dominant forms of capture -- associated with capture of the two blunt-ended DNA regions (at the base of the Y and at the end of the integrase-binding arm). With appropriate capture of the molecule at the base of the Y, this permits direct examination of protein binding to the terminal DNA region.

The NTD approach may provide the best means for examining other enzymes, and other complex biomolecules, particularly their activity in the presence of different co-factors. There are two ways that these studies can be performed: (i) the enzyme is linked to the channel transducer, such that the enzyme's binding and conformational change activity may be directly observed and tracked or, (ii) the enzyme's substrate may be linked to the channel transducer and observation of enzyme activity on that substrate may then be examined. Case (i) provides a means to perform DNA sequencing if the enzyme is a nuclease, such as lambda exonuclease. Case (ii) provides a means to do screening, for example, against HIV integrase activity (for drug discovery on HIV integrase inhibitors).

A variation of the Y-aptamer used previously is used to observe interaction events between that terminus and HIV DNA integrase. Preliminary binding observations are shown in Fig. 7.1. More detailed signal profiles are shown in Fig. 7.2 (Left), where the three most common signal classes are shown for the HIV Y-aptamer (left side), with right side images zoomed in to a time-scale more than 100 times shorter. Similarly, a more detailed figure is shown in Fig. 7.2 (Right), where the a signal class is shown that is not seen when HIV Y-aptamer is introduced without addition of integrase. In that figure, a possible binding event might be shown at the change in signal pattern from fixed level that ends in the yellow box, (with actual end transition shown in the pink box).

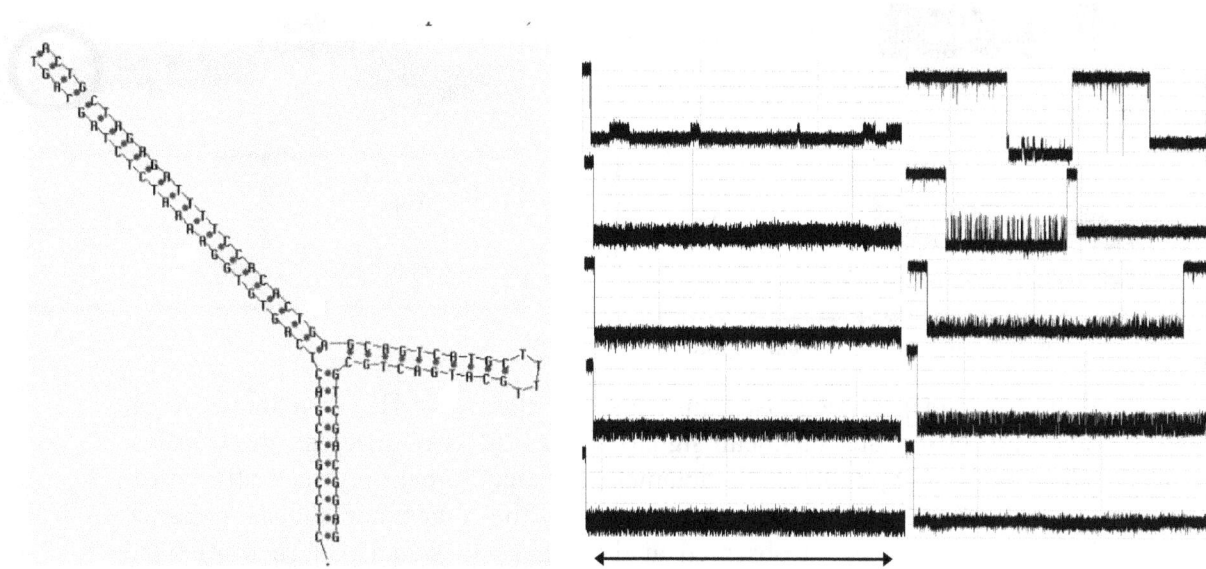

Fig. 7.1. Left: the mfold secondary structure map of the Y-aptamer used in the integrase binding study. Integrase will bind to the blunt-ended arm shown in the yellow circle, where the HIV DNA Terminus consensus sequence has been place. **Right:** Blockade signals produced before (left) and after (right) introduction and possible binding of HIV Integrase to the HIV-corresponding terminus of one arm of a channel-captured y-shaped aptamer. The time elapsed during each frame is approximately three seconds.

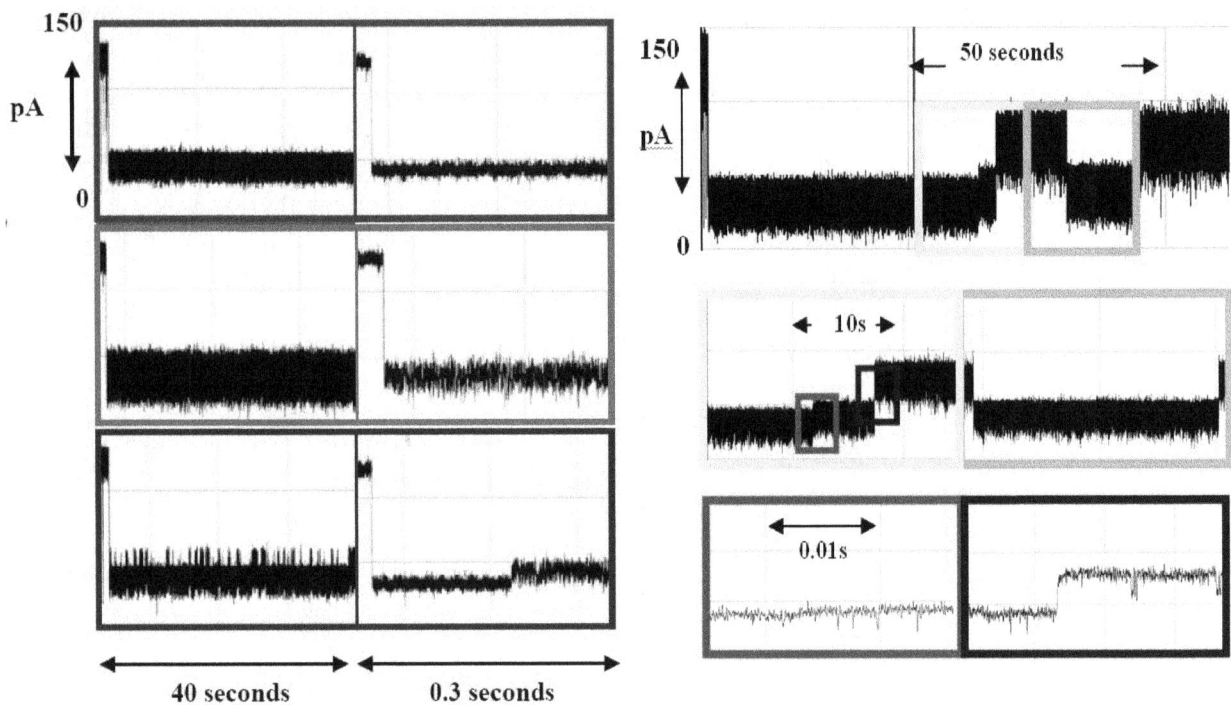

Fig. 7.2. (Left). Three most common signal classes for HIV Y-aptamer. Right and left boxes are identical signals shown at two different time scales. **(Right).** New type of HIV Y-aptamer Blockade Signal only seen after introduction of HIV Integrase to detector (with the Y-aptamer already present). Possible binding event observed at the change in signal pattern from fixed level that ends in the yellow box, (with actual end transition shown in the pink box).

7.2 Single molecule serial assaying
7.2.1 DNA-Protein complex Assaying: Aptamer-TBP
The TY10T1-GC aptamer was applied through refluxing to this environment and began to engage the alpha-hemolysin channel. Upon capture of a single TY10T1-GC aptamer at the channel there is an immediate and overall current reduction. Thereafter, the steady flow of ions through the channel was alternately blockaded at levels corresponding to approximately 40% and 60% of baseline, hypothesized to correspond with the binding/unbinding of the aptamer's blunt-ended terminus to the surrounding vestibule walls. These fluctuations in ionic current were measured and recorded as a blockade pattern. The two-level dominant blockade signal is shown in Fig. 7.3 for T-Y10T1-GC.

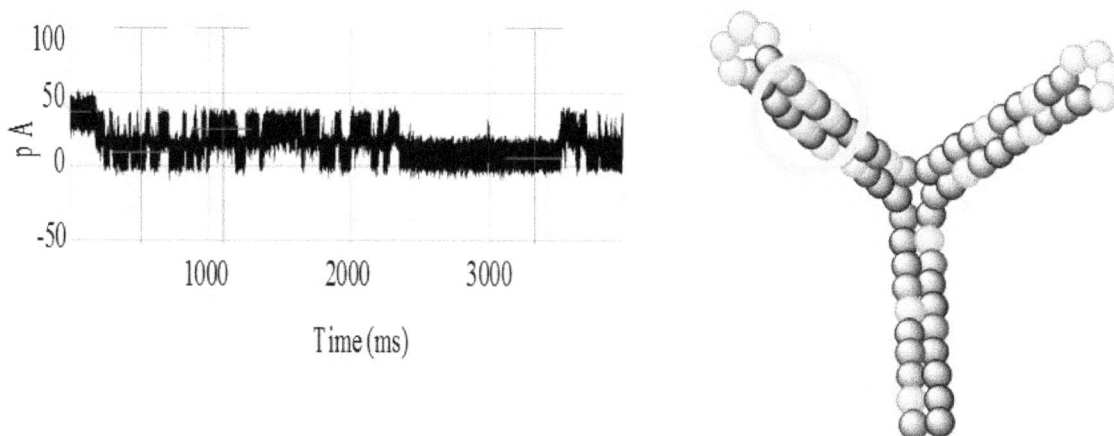

Fig. 7.3. The TY10T1-GC NTD-aptamer, with signal sample.

In an attempt to demonstrate the nanopore detector's capacity for describing the transcription factor/transcription factor binding site interaction, we examined the TBP/TATA box complex following the nanopore protocol. TBP, a subunit of transcription factor TFIID, was selected for its broad commercial availability and nominal price. TFIID is the first protein to bind to DNA during the formation of the pre-initiation transcription complex of RNA polymerase II (RNA Pol II). The TATA box, located in the promoter region of most eukaryotic genes, assists in directing RNA Pol II to the transcription initiation site downstream on DNA. For our transduction molecular system, the TATA box is located on a 4dT-loop terminating arm of our Y-aptamer, which was prepared in the lab by annealing to two DNA hairpin molecules. The base stem of our bifunctional Y-aptamer is designed to target and bind the area around the limiting aperture of the alpha-hemolysin channel, while the arm containing the TATA box binds the TBP.

We find that some of the blockade signals are only seen after introduction of TBP, which is hypothesized to be the sought after indication of TBP/TATA Box complex formation. The automated signal analysis profiles for T-Y10T1-GC w/wo TBP are shown in Fig. 7.4. The experiment is also repeated (not shown), with the receptor arm elongated several base pairs for more distal receptor placement from the channel environment, in order to ensure accommodation for the TATA binding protein (TBP), with similar indication of binding in our experiments.

137

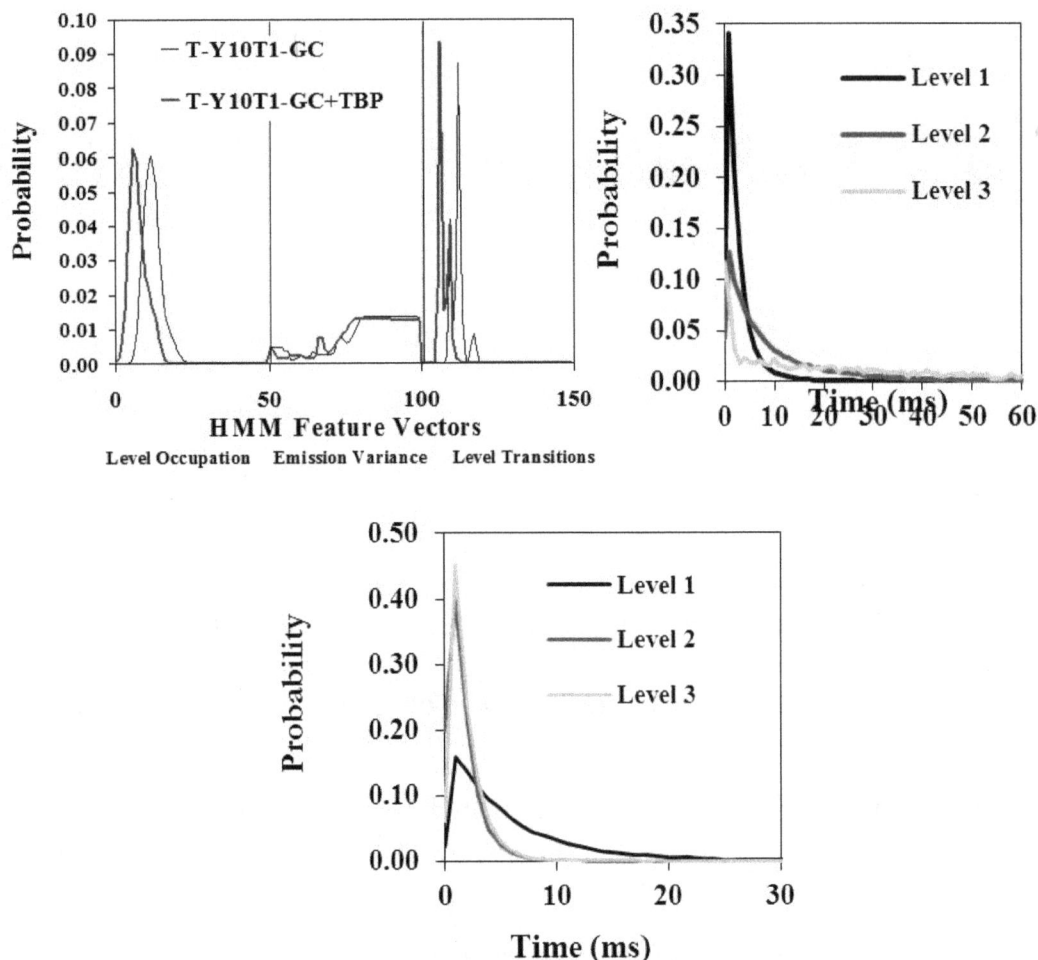

Fig. 7.4. Left. Standard 150-component HMM-based feature extraction for collections of T-Y10T1-GC blockade signals, w/wo TBP. After the EM iterations, 150 parameters are extracted from the HMM. The 150 feature vectors obtained from the 50-state HMM-EM/Viterbi implementation in [27] are: the 50 dwell percentage in the different blockade levels (from the Viterbi traceback states), the 50 variances of the emission probability distributions associated with the different states, and the 50 merged transition probabilities from the primary and secondary blockade occupation levels (fits to two-state dominant modulatory blockade signals). **Right.** Dwell Time at Each Level for T-Y10T1-GC (see Fig. 7.3 to visually identify the three levels – with two dominating). **Bottom Center.** Dwell Time at Each Level for T-Y10T1-GC + TBP (sample signal blockade not shown).

When TBP binds to the TATA box, it creates a nearly ninety degree bend in the DNA. This strong conformational change allows for strand separation. This is possible since the binding region of DNA is rich with the weaker two-hydrogen bond interactions of adenine and thymine. Once the strand separation occurs, RNA Pol II gains entry and begins transcription of the gene. The conformational deformity precipitated by the binding of TBP to DNA may be largely responsible for the alteration of the blockade signal originating after the introduction of TBP.

7.2.2 Glycoprotein assayer

NTD can operate as an HbA1c glycoform assayer (see next section for initial observations involving antibodies) to improve the knowledge of hemoglobin biochemistry (and that of heterogeneous, transient, glycoproteins in general). This could have significant medical relevance as a gap exists between what is known about hemoglobin biochemistry and how HbA1c information is used in the management of diabetic patients. The definition of 'HbA1c' is complex as HbA1c is a heterogeneous mixture of non-enzymatically modified hemoglobin molecules (whose concentration in blood is in part genetically determined). In clinical applications, HbA1c is used as if it were single complex with glucose whose concentration is solely influenced by glucose concentration. It may be possible, using an NTD platform, to improve diabetes management by introducing a new assaying capability to directly close the gap between the basic and clinical knowledge of HbA1c. It may be possible, perhaps optimal, to apply NTD in direct nanopore detector-to-target assays in combination with indirect NTD-to-target assays, for purposes of characterizing post-translational protein modifications (glycations, glycosylations, nitrosilations, etc.), see Fig. 7.5.

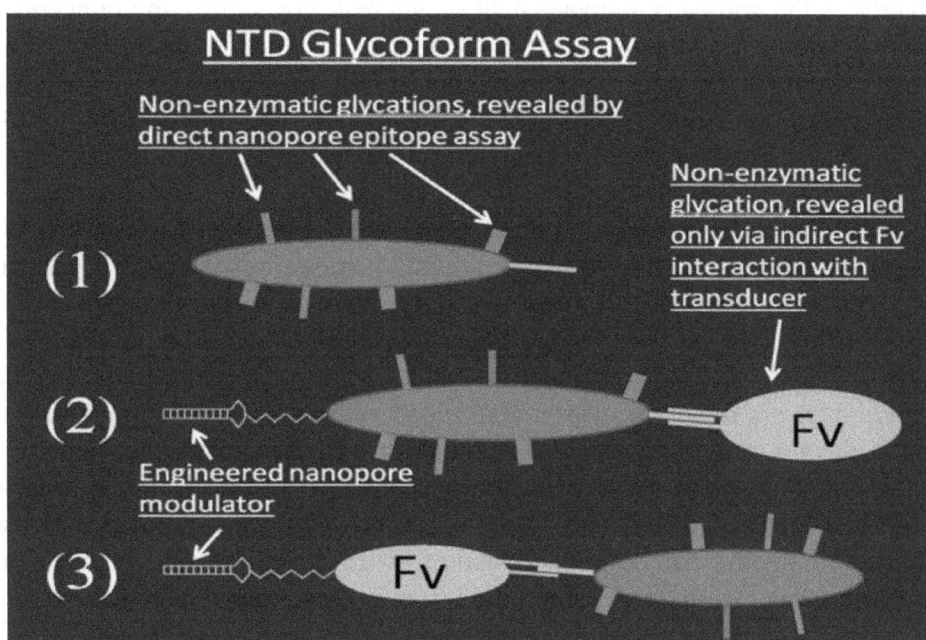

Fig. 7.5. NTD-based glycoform assays. Three NTD Glycoform assays are shown. Assay method (1) shows a protein with its post-translational modifications in orange (e.g., non-enzymatics glycations, glycosylizations, advanced glycation end products, and other modifications). Assay method (2) shows a protein of interest linked to a channel modulator. Direct channel interactions (blockades) with the protein modifications are still possible in this instance, but are expected to be dominated by the preferential capture of the more greatly charged modulator capture. Changes in that modulator signal upon antibody Fv interactions with targeted surface features provide an indirect measure of those surface feature. Assay method (3) shows an antibody Fv that is linked to modulator, where, again, a binding event is engineered to be transduced into a change of modulator signal.

The endocrine axis, thyroid stimulating hormone (TSH) in particular, is regulated via a heterogenous mixture of TSH molecules with different amounts of glycation (and other modifications). The extent of TSH glycation is a critical regulatory feedback mechanism. Tracking the heterogenous populations of critical proteins is critical to furthering our understanding and diagnostic capabilities for a vast number of diseases. Hemoglobin molecules provide a specific, on-the-market, example -- here extensive glycation is more often associated with disease, where the A1c hemoglobin glycation test is typically what is performed in many over-the-counter blood monitors. The NTD testing of surface features of the protein can be done before or after digestion or other modification of the test molecule as a means to further improve signal contrast on the identity and number of possible protein modifications, as well as other surface features.

Part of the complexity of glycoforms, and other modifications, of proteins such as hemoglobin and TSH, is that these glycoforms are present as a heterogeneous mixture, and it is the relative populations of the different glycoforms that may relate to clinical diagnosis or identification of disease (such as prion exposure [231]). To this end, a protein's heterogeneous mixture of glycations and other modified forms can be directly observed with a NT-detector, and this constitutes the clinically relevant data of interest, not simply the concentration of some particular glycoform. Furthermore, it is the transient, dynamic, changes of the glycoform profile that is often the data of interest, such that a 'real-time' profile of glycoform populations may be of clinical relevance, and obtaining such real-time profiling of modified forms (glycoforms, etc.) would be another area of natural advantage for the NTD approach.

The protein modification assays have indirect relevance for public health and biodefense. This is because the degree of glycation of a patients hemoglobin is an early indication of their disease state (if any, or simply 'glycation' age otherwise). This is because the *hemoglobin that is actively used in transporting oxygen throughout the body is analogous to a 'canary- in-the-coalmine' in that it provides an early warning about insipient complications or past chemical or nerve agent exposures.* Red blood cells (that carry hemoglobin) typically live for 120 days – providing a 120-day window into past exposures and a 120-day average on the regulatory load induced by those exposures. In the future, if a mysterious gulf-war syndrome is encountered, and there is concern about a low-level exposure to a nerve agent, examining the hemoglobin glycation profiles, and similar profiles on other blood serum constituents, would provide a rapid (30 min.) assessment of biodefense status.

NTD detection and assaying provides a new technology for characterization of transient complexes, with a critical dependence on 'real-time' cyberinfrastrucure that is integrated into the nanopore detection method using machine learning methods for pattern recognition and their implementation on a distributed network of computers for real-time experimental feedback and sampling control.

Thyroid stimulating hormone (TSH) is present as a heterogeneous mixture of TSH molecules with different amounts of glycation (and other modifications). The extent of TSH glycation is a critical regulatory feedback mechanism. Tracking the heterogeneous populations of regulatory proteins is required to further our understanding and diagnostic capabilities for a vast number of diseases. Hemoglobin molecules are an example where specific, on-the-market, glycation diagnostics are in use, where the A1c hemoglobin glycation test is typically what is performed in many over-the-counter blood monitors.

A nanopore-based glycoform assay could be performed on modified forms of the proteins of interest, i.e., not just native, but deglycosylated, active-site 'capped', and other forms of the protein of interest, to enable a careful functional mapping of all surface modifications. Pursuant to this, the methodology could also be re-applied with digests of the protein of interest, to further isolate the locations of post-translational modifications when used in conjunction with other biochemistry methods.

In conjunction with protein digests and HPLC, nanopore detection of glycation may provide a powerful new means to assay the post-translational modifications present for a given protein (in whole or via its digests), including their changing molecular complexations. This has profound significance for the understanding and treatment of a variety of diseases, including diabetes, where post-translational modifications to hemoglobin are an important biomarker for disease diagnosis and treatment.

7.2.3   Antibody Assayer: A new window into understanding antibody function

Upon binding to antigen, a series of events are initiated by the interaction of the antibody carboxy-terminal region with serum proteins and cellular receptors. Biological effects resulting from the carboxy-terminal interactions include activation of the complement cascade, binding of immune complexes by carboxy-terminal receptors on various cells, and the induction of inflammation. Nanopore Detection provides a new way to study the binding/conformational histories of individual antibodies. Many critical questions regarding antibody function are still unresolved, questions that can be approached in a new way with the nanopore detector. The different antibody binding strengths to target antigen, for example, can be ranked according to the observed lifetimes of their bound states. Questions of great interest include: are allosteric changes transmitted through the molecule upon antigen binding? Can effector function activation be observed and used to accelerate drug discovery efforts?

Thus, real-time analysis of antibody IgG binding affinity might be possible using a nanopore detector to better understand antibody-antigen binding affinities and the conformational changes that initiate signal pathways. Although some surface features clearly elicit blockade signals that are modulatory, not all surface features of interest will exhibit blockade signals when drawn to the channel and in these instances antibody or aptamer based targeting of those features could be used, where the antibody or aptamer is linked to a channel modulator that then reports on the presence of the targeted surface feature indirectly, e.g., the NT-biosensing setup.

A nanopore-based glycoform assay could be performed on modified forms of the proteins of interest, i.e., not just native, but deglycosylated, active-site 'capped', and other forms of the protein of interest, to enable a careful functional mapping of all surface modifications. Pursuant to this, the methodology could also be re-applied with digests of the protein of interest, to further isolate the locations of post-translational modifications when used in conjunction with other biochemistry methods.

7.3 Molecular capture via antibody, aptamer, capture-matrix, TERISA

It is possible to couple NTD methods with antibody capture systems, or any specific-binding capture system (e.g., MIP-capture or aptamer-based capture systems could be used as well, for example) to report on the presence of the target molecules via indirect observation of transduction molecule signals corresponding to UV cleaved NTD 'substrate' molecules (that are freed from the capture matrix). Commercially produced systems are available with matrices pre-loaded with immobilized Fc-binding antibodies, the secondary antibody can then be introduced, and bound by the Fc-binding Ab's, to establish the desired, immobilized, specific-binding matrix (analogous to sandwich-ELISA). If solution with target molecule is now repeatedly washed across the immunosorbant surface, an immobilized concentration of that target molecule can be obtained. We can now introduce our primary antibody that targets the immobilized antigen ('sandwiching' it). If the primary antibody can be attached to an NTD Biomarker as shown in Fig. 7.6, where the antibody-DNAhp linkage can be broken upon exposure to UV.

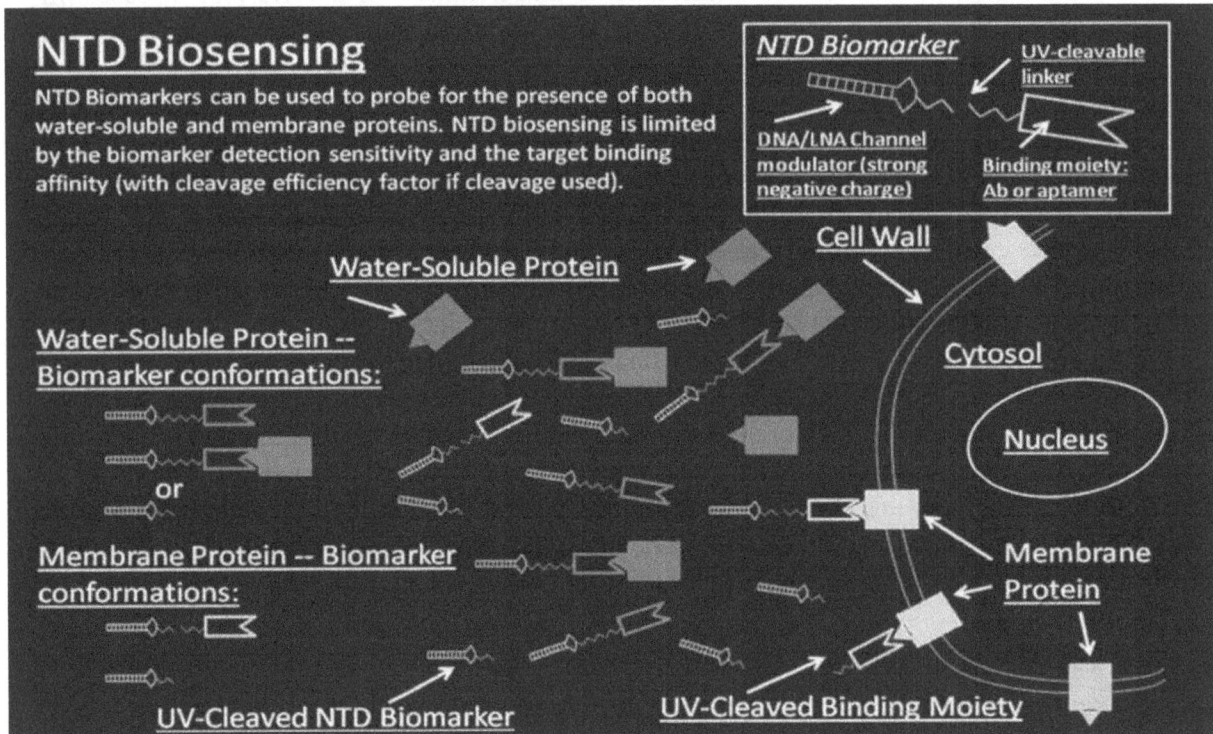

Fig. 7.6. The NTD biosensing approach facilitated by use of immuno-absorbant (or membrane immobilized) assays, such that a novel ELISA/nanopore platform results. The immune-absorbance, followed by a UV-release & nanopore detection process provides a significant boost in sensitivity.

The further novel aspect of this setup is to now have the primary antibody linked to an enzyme that acts on a NTD transducer substrate (analogous to a fluorescent substrate in ELISA). By taking some of the methodology from the ELISA approach (enzyme-linked immunosorbent assay), and merging it with unique aspects of our nanopore detection approach, we have the 'Transducer Enzyme-Release with ImmunoAbsorbent Assay' [88], where "Sandwich TERISA" assumed to typically be the case since specific immobilization is desired. This situation is shown in Fig. 7.7. Also shown in Fig. 7.7 is an example of an electrophoretic contrast (E-phi contrast) substrate. The idea being to have electro-neutral substrate and upon enzyme cleavage, to leave a highly negatively charged DNA hairpin to be electrophoretically driven ('report') to channel.

Anologous to real-time PCR, where a qualitiative PCR result is self-calibrated according to is real-time values to obtain a quantitative PCR results, we can do the same with the TERISA and TARISA biosensing methods outlined here. In other words, for all three methods with real-time observation (RT-TARISA, RT-TERISA, E-phi Contrast RT-TERISA), we can shift to a more quantitative footing (as with RT-PCR or RT-ELISA), but in our case this is trivially achieved since the data-acquisition and signal processing is already in use and operating in 'real-time'. This real-time tracking information helps to stabilize the method and complements the biosensing capability with a quantitative assaying capability (where highly accurate resolution of mixtures of DNA hairpin molecules was shown to be possible).

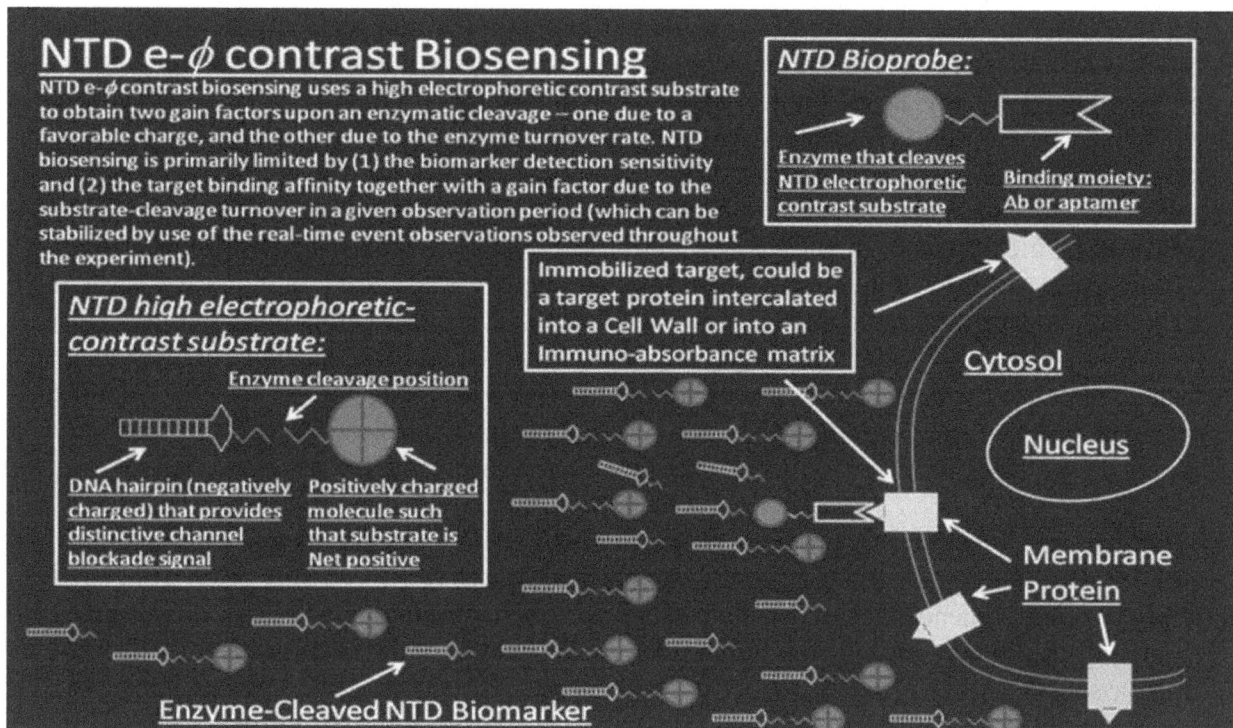

Fig. 7.7. The Detection events involved in the 'indirect' NTD biosensing approaches: TERISA and E-phi Contrast TERISA.

7.4 NTD-Gel

Nanopore detectors may offer the separation/identification information of gels, machine-learning based pattern recognition capabilities, and nanopore-based electrophoresis methods that can be used to discern clusters (like the bands or dots in a gel) in a higher dimensional feature space, for greatly improved 'gel band' resolution (such that isomers might be resolvable, etc.). For a nanopore to offer information equivalent to a gel, however, it must also sample a great number of molecules quickly, this requires active sampling control to optimize – i.e., once the sample molecule is identified it is ejected. To this end, pattern recognition informed sampling has been developed and used to boost the sampling rate on a desired species by two magnitudes over that obtainable with a passive recording. This lays the foundation for nanopore-based molecular analysis. The separation-based methods still have more infomation than the separation/grouping of molecules into clusters, however, since they also provide an *order* of separation, according to mobility, or according to isoelectric point, etc. For the nanopore-based methods to recover this critical ordering information on the observed data clusters something else must be considered. One possibility is the introduction of a mobility reducing agent, such as PEG, into the buffer. The change in average arrival time of the different species after introduction of PEG (using voltage reversal to clear a 'near-zone'), referred to as the 'PEG shift', can then be the basis for an ordering – the least PEG shifted molecules are those, it is hypothesized, with greater mobility and charge (where this is done by comparison of acquisition rates after introduction of PEG and use of voltage control). Just as with gels, all sorts of functionalized PEG, or other functionalized buffer media, can be introduced for different sieving results, and that provides numerous related functionalizations to the nanopore-gel approach.

7.5   Nanopore-based assays of cytosolic antigen delivery complexes

NTD Systems may provide an effective method for identifying good cytosolic antigen delivery complexes for use in evoking cytotoxic T lymphocyte (CTL) responses in organisms challenged by cytosolic virulence factors. In all such screening efforts, pore-forming toxins are used that allow introduction of peptides and other molecules into the cytosol of their host cell. The experiment here entails establishing a single pore-forming toxin in a membrane followed by measurement of peptide (antigen) transmembrane transport via channel current measurements in NTD conditions (where modulations are introduced or induced such that stationary statistics information flows are established). The experiments make use of established procedures for attaching target antigen to the recognition sequence of virulence factors associated with the pore-forming toxin in the natural setting. Via this transmembrane transport mechanism, a number of antigen/recognition molecules can be assayed, for NTD enhanced profiling, for effective use with the chosen pore-forming toxin.

Medicines and vaccines that provide resistance to cytosolic virulence factors can be rapidly assayed with this approach. Pore-forming toxins and viruses (e.g., HIV) are often associated with cytosolic virulence factors. Medicines and vaccines can provide treatment or resistance to such cytosolic virulence factors if an effective mechanism is devised for delivery of virulence-associated antigen to the cytosol of the host cell. Effective delivery of antigen to the cytosol is the first step in obtaining an effective agent for evoking a CTL response against virulence factors of the invading microorganism.

7.6 Transcriptome and Transcription-Factor based Drug Discovery

The examination of transcription factor binding to target transcription factor binding site (TF/TFBS interactions) affords the possibility to understand, quantitatively, much of the Transcriptome. This same information, coupled with new interaction information upon introduction of synthetic TFs (possible medicines), provides a very powerful, directed, approach to drug discovery.

Transcription factors (TFs) are proteins that regulate gene expression by binding to the promoter elements upstream of genes to either facilitate or inhibit transcription. They are composed of two essential functional regions: a DNA-binding domain and an activator domain. The DNA-binding domain consists of amino acids that recognize specific DNA bases near the start of transcription. TFs are typically classified according to the structure of their DNA-binding domain, which are of one of the following types: zinc fingers, helix-turn-helix, leucine zipper, helix-loop-helix, and high mobility groups. The activator domains of TFs interact with the components of the transcriptional apparatus and with other regulatory proteins, thereby affecting the efficiency of DNA binding. A cluster of TFs, for example, is used in the preinitiation complex (PIC) that recruits and activates RNA polymerase. Conversely, repressor TFs inhibit transcription by blocking the attachment of activator proteins.

Synthetic transcription factors (STFs) promise to offer a powerful new therapeutic against Cancer, AIDS, and genetic disease. STFs that can appropriately target (and release) their transcription factor binding sites (TFBS) on native genomic DNA provide a means to directly influence cellular mRNA production (to induce death or dormancy for Cancer and AIDs cells, or restore proper cellular function in the case of genetic disease). In synthetic TF drug discovery an effective mechanism for screening amongst TF candidates would itself be highly valued. Such may be possible with novel observation and analysis methods involving channel current observations of single molecule interactions/blockades.

Drug therapies that target the DNA binding site of individual TFs are currently impeded by insufficient sequence specificity for the appropriate binding site. While progress is being made to produce derivatives with greater specificity, another technique for altering transcription via single-stranded oligonucleotides has been implemented in such fields of medicine as cancer treatment. This approach employs a single-stranded oligonucleotide to bind a specific sequence in double-stranded DNA and form a triple helix which is not recognizable to a protein, such as a TF, that would usually bind to that site in double helical DNA. The creation of triple helixes was used, for example, to repress the transcription of the tumor necrosis factor gene, which in turn inhibited the growth of TNF-dependent tumor cells. Despite certain successes achieved with single-stranded oligonucleotides, synthetic TFs still provide a more directed, powerful means to influence the mechanism of transcription. Nanopore technology can be play an essential role in the further development of therapeutic agents by identifying and improving structural information on the binding of these drugs to DNA. Principally, the nanopore detector gives us the ability to observe the binding event in real time with a single molecule view. Furthermore, multi-component processes not discernible by standard cell-based analysis may also be resolvable using the nanopore detector.

7.7 DNA Sequencing
7.7.1 Single-molecule, processive
Nanopore transduced DNA-enzymatic activity has the potential to be an inexpensive and versatile platform for DNA sequencing. In the proposed DNA sequencing scenario, the transducer molecule (NTD probe) captured in the nanopore channel is engineered to modulate the channel current with four discernably different signals as a linked lambda exonuclease processively excises the four different types of nucleotides from a strand of bound duplex DNA.

An NTD experiment has been designed (see Fig. 7.8) to discriminate between the four nucleotides that are excised by lambda exonuclease as it enzymatically and progressively excises the 3' strand of bound duplex DNA. Other exonucleases are of interest as well but lambda exonuclease is known to work in a broad range of buffer conditions, including the standard buffer conditions used in the NTD platform, with magnesium added as co-factor. DNA sequencing occurs by observing the different back-reaction events (possibly conformational-change mediated) that are observed with an enzyme-coupled NTD probe-- according to whether an 'a', 'c', 'g', or 't' is excised. Additionally, the NTD probe can be engineered such that a coincidence detection event is enabled via the associated translocation disturbance associated with the excised nucleotide as it passes through the nanopore channel. We believe that the translocation event alone (see Fig. 7.9) will not supply enough information to discriminate between the four nucleotides.

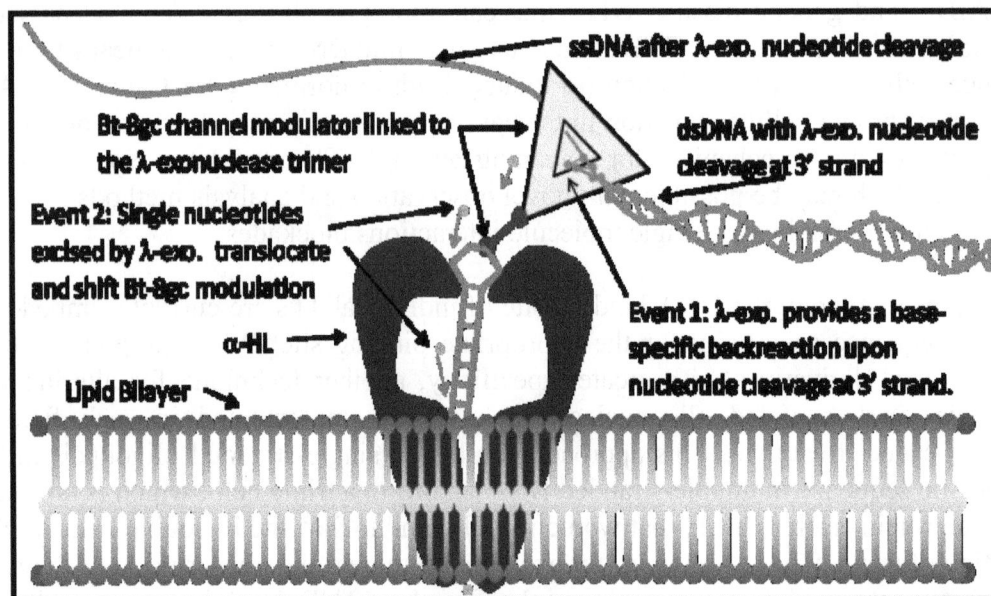

Fig. 7.8. Schematic diagram of the nanopore with DNA-enzyme event transduction as a means to perform DNA sequencing. A Bt-8gc DNA hairpin captured in the channel's cis-vestibule, with lambda nuclease linked to the Bt-8gc modulator molecule as it enzymatically processes the duplex DNA molecule shown.

DNA-hairpin modulators linked to processive DNA enzymes can report on the binding to DNA substrate and possible enzyme activity with introduction of cofactors such as magnesium. The enzymes listed below are all known to work in buffers compatible with the buffer requirements of the alpha-hemolysin channel heptamer. Items (i)-(iii) to follow are a non-exhaustive listing of possible DNA enzymes to use in the proposed method.

(i) DNA sequencing may be possible via examination of the Klenow fragment (KF) of E. coli DNA polymerase I, which processively grows a dsDNA strand from a dsDNA/ssDNA primer, via terniary complexation with the appropriate matching 'a', 'c', 'g', or 't' from an dNTP substrate that has been introduce (along with magnesium). To the extent that the magnesium acts as an on/off switch for the enzyme, rate control is best established via concentration control on the dNTPs present. This provides a substrate concentration variable-speed control mechanism.

(ii) DNA sequencing may be possible via examination of the base excision process as source of signal, via use of lambda exonuclease. Now the only cofactor needed is magnesium.

(iii) DNA sequencing may be possible via examination of the base excision process as source of signal, via use of Exo.

If the enzyme is a DNA exonuclease, the excised molecular bases can themselves interact with the channel modulator to produce a synchronization or coincidence detection enhancement to the detection, or be the main detection event for DNA sequencing itself, in some engineered scenarios. Linkage to any enzyme, thus, permits potential direct assays of that enzymes activity in the presence of cofactors. This has direct application in assays to identify molecules that can block HIV integrase activity, among other things.

146

It is possible to develop computational/experimental architectures and machine-learning (ML) based pattern recognition software to perform real-time channel blockade classifications that operates at the *single*-molecule level. The importance of this can be understood in the context of the single-molecule selection 'demon' posited by Maxwell. With such a demon, and some operational idealizations, Maxwell showed how to defy the equilibration of the second law of thermodynamics, and thereby lay the foundation for a perpetual motion device. Here, using artificial intelligence & machine learning methods we are able to establish a single-molecule selection demon such that the channel appears to always be open (in a non-blocking sampling mode), which happens to be critical in high concentration probe experiments (pushing the biosensing limits). The importance of this selection-activity 'demon' capability in the context of the above is that a coincidence coherence/synchronization demon may be critical to having the signal-to-noise for the aforementioned, coincidence synchronized, DNA sequencing. The problem with the weaker signal-to-noise may, initially, be due to loss of 'framing' information that delineates the different phases of blockade signal. To address this problem, in the case of lambda exonuclease, we can set up signal modeling and signal processing that accounts for two streams of 'coincidence' information. The problem is that the 'coincidence event', of excision/addition back-reaction accompanied by nucleotide translocation, may not exist for all nanopore detector settings. It may be that the 'coherence' of the timing between the two event series (one back-reaction phase changes, the other nucleotide traversal phase changes) may require active feedback by the nanopore detector setup. Fortunately, we have fully enabled the signal processing requirements for the feedback timescales involved, as demonstrated in the PRI Results, so establishing coherence stabilization appears to be possible. Control molecules, carrier references, can be introduced as well, to further inform the signal processing, and enable the coherence stabilization that may be needed.

Four-phase resolution may not be possible once the enzyme turnover (processive) rate is increased. In such an instance two-phase resolution might be attempted, for different DNA modifications/buffers/channels so as to recover four-state sequence info from a set of two-state sequencings.

Some processive DNA enzymes may have much more distinctive conformational change than others, according to base polymerization, allowing single-molecule sequencing at the processive rate of the enzyme at that temperature (which typically doubles for every added 10 C above the standard operating temperature of 23 C). By adjusting magnesium concentration and temperature the processive rate could be quite fast, with thousands per second easily possible. Thus, the success of the NTD-based DNA sequencing approach would present a radically new form of DNA sequencing.

Observation of DNA monomer translocations in presence of dsDNA modulator
Part of information in an exonuclease based DNA sequencing method: single-base translocations with a transduction modulator molecule present are shown in Fig. 7.9.

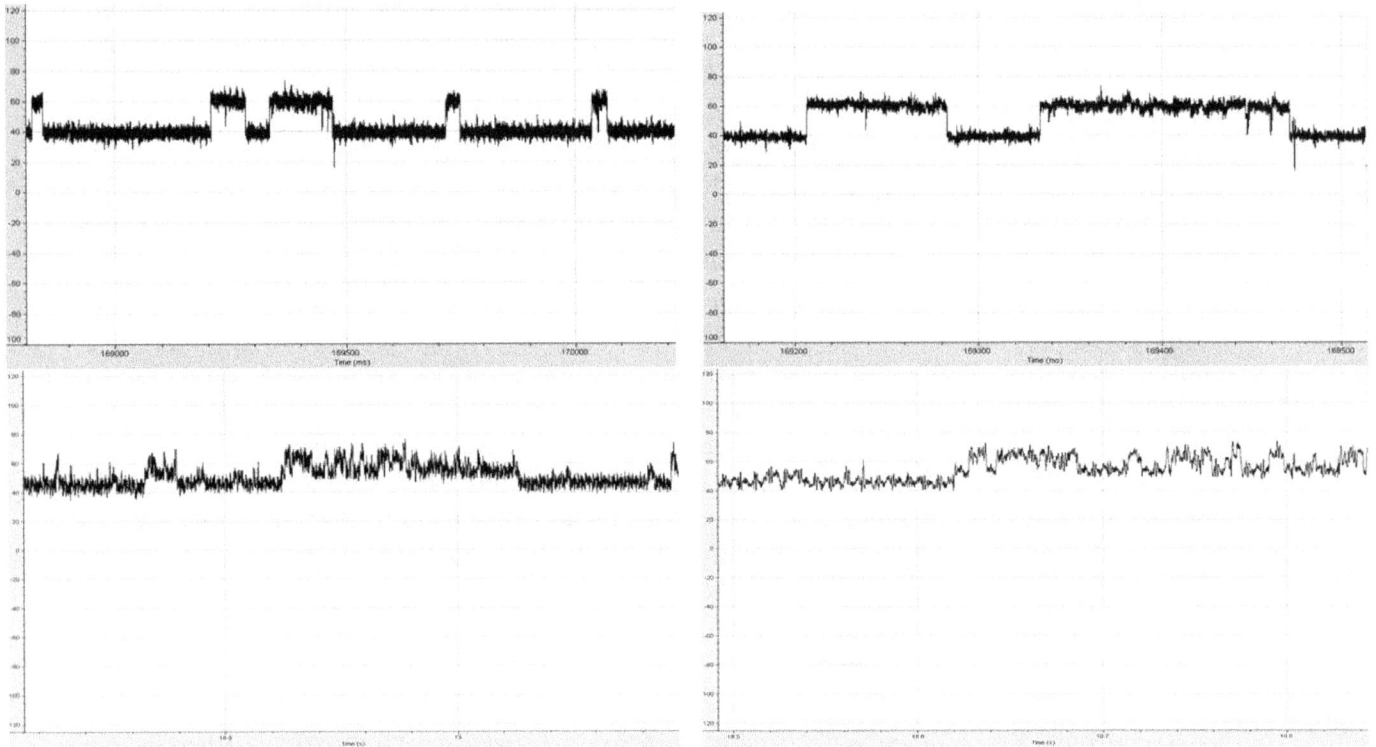

Fig. 7.9 TOP. Left: 9gC hairpin toggling; 500ms per column. Top Right 9gC hairpin toggling; 100ms per column. BOTTOM Left: 9gC hairpin toggling w/ thymine nucleotides added; 500ms per column. BOTTOM Right: 9gC hairpin toggling w/ thymine nucleotides added; 100ms per column.

7.7.2 NTD/Sanger DNA Sequencing

There is a NTD/Sanger sequencing scenario where sequencing is on a Sanger-sequencing type mixture, where copy terminations are designed to be blunt-ended dsDNA rather than DNA with a dye attachment or other expensive linkage. The blunt-ended DNA is then identified by its (blunt-ended) terminal base-pair and by its length, as with Sanger, to arrive at information usable, if complete, to determine the parent sequence (see Fig. 7.10). The terminal base pair is classified according to the distinctive blockade signals that captured dsDNA ends can provide (laser, or other, modulations may be needed to excite the captured blunt end to force it to exhibit its blockade toggle signal – this latter technique already done in a proof-of-concept experiment). The strand length is classified according to channel blockade signal under a variety of nanopore detector modulations (applied potential, laser (electric) pulsing, electromagnetic field modulations, to list a few methods for externally driven modulations).

The key aspect of the success of the length discrimination method lies in the fact that the physical mechanism (producing the discriminatory signal found to be useful) need not be understood. Rather, a model-independent machine-learning approach to the signal analysis can latch onto discriminatory aspects of the information. SVM are well-suited for that purpose here, together with feature extraction performed by a HMM.

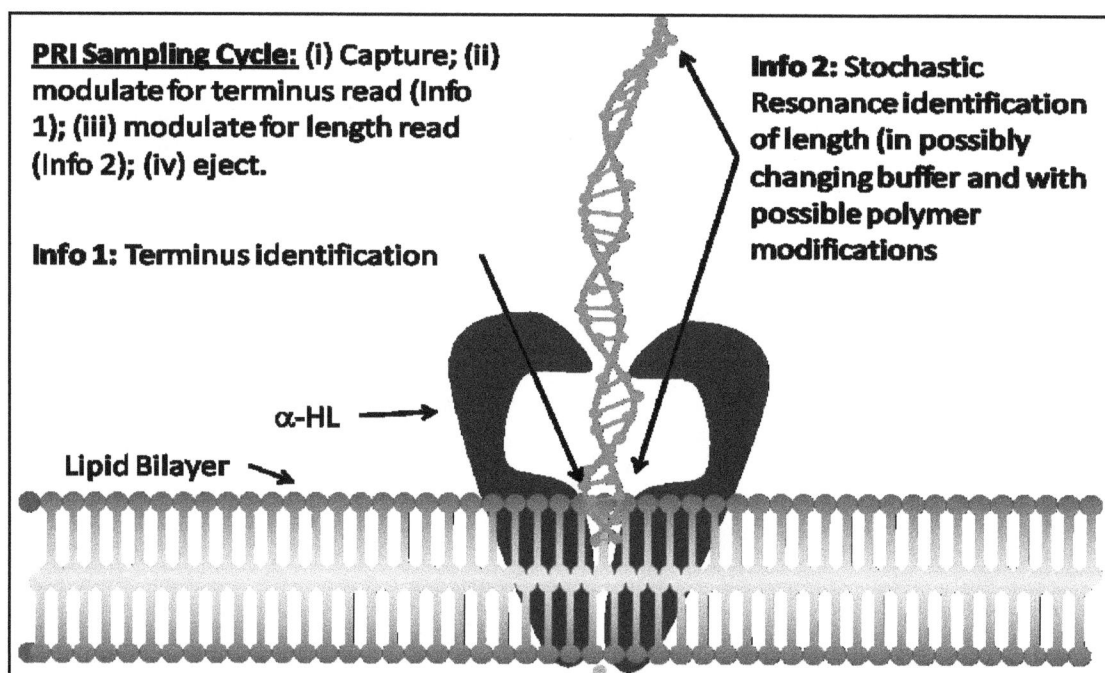

Fig. 7.10. A blunt-ended dsDNA molecule captured in the channel's *cis*-vestibule.

The idea is to expose the channel to a mixture of PCR amplified DNA sequence with random termination (or other mixture of DNA), that is in a dsDNA annealed form with channel size such that the channel blockades correspond to single, non-translocating, dsDNA blockades ('captures') of one end of the dsDNA molecule, while extracting from the blockade channel current signal, a set of one or more pattern features to establish over a period of time either a blockade channel current signal pattern or a change in the blockade channel current signal pattern, with each sampling of the mixture.

Modulation responses may enable the PCR analytes (or any analytes for that matter) to be discerned with better resolution (such as for discerning the length of the captured dsDNA molecules in Fig. 7.10). Modulations serve to sweep through a range of excitations, with response possibly allowing classification of lengths given pre-calibrated (trained on known length) test cases, response also used to establish identity of captured end (terminal base-pair identification, for example).

Also note that very small reagent usage is necessary in NTD/Sanger due to the possible nano-scale reduction in operating analyte chamber volume, competive with established methods (standard Sanger sequencing) where larger analyte volumes are needed, and more expensive reagents such as dyes (and associated suite of lasers) are required.

Chapter 8
NTD Signal Stabilization and Interference Handling

Operation of an alpha-hemolysin nanopore transduction detector is found to be surprisingly robust over a critical range of pH (6-9), including physiological pH=7.4 and PCR pH=8.4, and extreme chaotrope concentration, including 5M urea. The engineered transducer molecule that is captured in the standard alpha-hemolysin nanopore detector, to transform it into a transduction detector, appears to play a central role in this stabilization process by stabilizing the channel against gating during its capture. This enables the nanopore transduction detector to operate as a single molecule 'nanoscope' in a wide range of conditions, where tracking on molecular state is possible in a variety of different environmental conditions. In the case of streptavidin biosensing, results are shown for detector operation when in the presence of extreme (5M) urea concentration. Complications involving degenerate states are encountered at higher chaotrope concentrations, but since the degeneracy is only of order two, this is easily absorbed into the classification task as in prior work. This allows useful detector operation over a wide range of conditions relevant to biochemistry, biomedical engineering, and biotechnology.

Extensive results are shown to validate the NTD Nanoscope results using standard methods from isoelectric focusing (IEF) gels and capillary electrophoresis (CE). Further results for the Streptavidin-Biotin biosensor are shown in Sec. 8.1, confirming the NTD Nanoscope binding of Bt-8gc to streptavidin in urea with concentrations up to 5M, where IEF Gel and

CE validation results are found to be in agreement. In Sec. 8.2 results on antibody binding are provided, along with validation results, building off the preliminary work already mentioned. The validation results show antibody binding with biotin as antigen in urea concentrations up to 2M, with validation by IEF Gels.

8.1. Biotin-Streptavidin Binding Experiments
8.1.1. The BT-8gc transducer viability in urea up to 5M concentrations

In some instances, chaotropes (such as urea) are used to weaken the binding affinity, or DNA-DNA annealing affinity, of molecules studied with the nanoscope, such that binding tests can be performed with numerous on/off transitions in the lifetime of the experiment. In the case of DNA-DNA annealing, the collective binding that occurs can remain sufficiently strong in the presence of chaotrope such that it provides a clear contrast with non-collective binding interactions and can greatly improve signal quality. For this reason, and others, understanding the response of the channel and transducers in the presence of chaotropes is useful. The NTD approach will benefit most where the transducers provide little change, or have just a few states, when in channel blockade with change in chaotrope concentration. From high voltage capture strain prior studies it was found that the Bt-8gc blockades exhibit two different capture blockade signals. This is hypothesized to be due to two states of the transducer itself, probably due to two accessible loop 'twists' conformations, one not normally accessible without capture-strain. Two transducer states that are degenerate (being simply due to hypothesized racemization on molecules with different loop conformations) is a manageable complication with the automated pattern recognition, but clearly reveals how at the single DNA-hairpin level of resolution we can see changes in molecular conformation (and terminus regions, as shown previously). Thus, a racemization over capture states with two loop "twists" was hypothesized to occur upon introduction of chaotropic agents (urea 2.0 M – 5.0 M), and this result is confirmed in Fig. 8.1. (A crude schematic for the twists is envisaged, from a top-down view of the hairpin loop, to look like the yin-yang symbol boundary that bows in to the left at the top, then to the right at the bottom, and the reverse for the other twist conformation.)

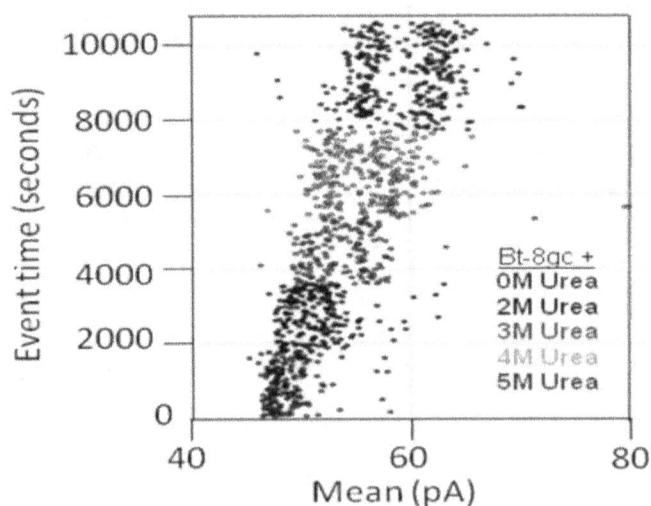

Figure 8.1. Sufficiently strong Urea concentration (5M) results in racemization of the two loop capture-variants, while weaker urea (<2M) does not. The results show Bt-8gc measurements at 30 minute intervals (1800 s on vertical axis) with urea concentration 0, 2, and 3M, 45 minutes at 4M, and 60 minutes at 5 M, with signal blockade mean on the x-axis, with results consistent with the two-state loop hypothesis, and consistent with the observation of such in

[14], but not due to zero or weak urea content but due to high strain due to mass and charge effects upon binding to the large streptavidin molecule.

The α-hemolysin channel demonstrates a high tolerance to high salt concentration and the presence of chaotropic agents, which is important to establish a platform for the study of binding between other molecules under such conditions. By varying the composition of running buffer it is possible to control the interaction of analyzed molecules with the nanopore or with each other. In what follows, tests are shown of the impact on binding affinity between streptavidin, or mAb, and biotin upon introduction of chaotropic agents. This demonstrates new capabilites in nanopore detector applications.

8.1.2. Observations of Biotin-Streptavidin Binding on the NTD Nanoscope

Preliminary results on streptavidin biosensing were shown in Sec. 8.4 for urea concentration up to 3.5M. The resolution of the bound/unbound Bt-8gc is greater than 99.99% accurate in less than 100 ms, with greater accuracy if longer observation time is used. The analysis uses the signal processing pipeline described in Sec. X on channel current cheminformatics where a 150-component feature extraction is done on each blockade signal. Using just two 'human-friendly' features based on each signal's maximum and minimum blockade values in that 100 ms observation window, a surprisingly clear separation of the molecular classes is easily discerned, as well as the role of urea in weakening interactions where bound states are reduced in observation frequency and unbound states increased. Initially, at 0M urea, two clusters are easily discerned by eye. One corresponds to the Bt-8gc blockades, the other corresponds to the (Streptavidin) – (Bt-8gc) complex. Upon introduction of urea, signals for Bt-8gc unbound start to shift the Bt-8gc cluster, where direct quantification of the cluster results is directly accessible from the cheminformatics analysis.

8.1.3. Bt-8gc -- Streptavidin Binding validation using IEF Gels w/wo chaotropes

Complex formation between the biotinylated DNA hairpin (Bt-8gc) and streptavidin is shown on the NTD Nanoscope -- this result is validated via electrophoretic mobility shift analysis with isoelectric focusing in Fig. 8.2. The standard gel analysis can't resolve presence of different isoforms in a single 'band' of gel, but Nanopore augmentation of gel electrophoretic methods, may offer a means to resolve components within the bands.

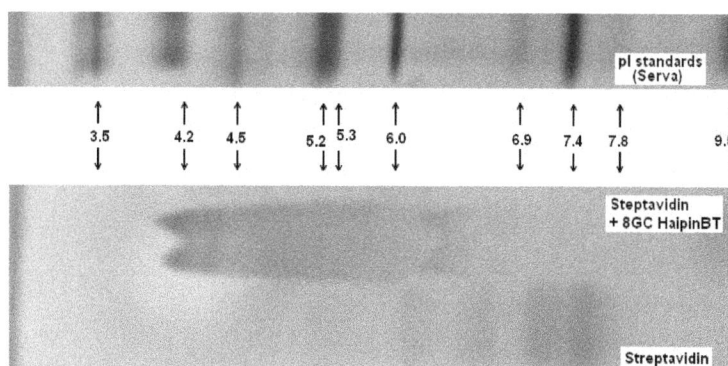

Figure 8.2. Biotinylated DNA hairpin (Bt-8gc) and streptavidin complex verification.

Electrophoretic methods provide a means to study the process of complex formation. Depending on the affinity (thermodynamic constant value) and the kinetics of the reaction, different electrophoretic techniques can be used. For highly stable complexes, the isoelectric focusing technique can be applied [232,233]. This electrophoresis technique has the advantage of extremely high resolution that allows maximally complete detection of existing heterogeneity in complex population, due to both multi-valent interaction and initial heterogeneity of interacting species.

In Fig. 8.3 we show the gel IEF results describing the interaction of streptavidin and biotinylated hairpins. Due to very strong interaction between the streptavidin and biotin the complex is extremely stable: it does not break apart for hours and IEF detects practically no presence of free streptavidin. In Fig. 8.3 the IEF spectra of streptavidin and streptavidin incubated with an excess of the biotinylated hairpins Bt-8gc and Bt-9gc, are shown. For the streptavidin, the two major components are visible (with their pIs at 7.1 and 7.5, approximately). After targeting with hairpin those streptavidin isoforms convert to two new bands (pI 4.2 and pI 4.35). We hypothesize that there exists a one to one correspondence between the two above pairs of major components (before and after the complexation takes place). According to our theoretical calculation such a high pI shift can be achieved when all four binding sites of the streptavidin molecule are targeted. Here we used the technique allowing for predicting the electric charge vs. pH relationship for a protein molecule based on the amino acid composition, or more generally, any biopolymer with known content of so-called ionogenic groups. The approach has limitations connected with the dissociation scheme selected for the model and the exact values of the dissociation, but typically serves as a reasonably good approximation for isoelectric point calculation or protein titration curve behavior. The latter are often used as tool for optimizing various electrophoretic of chromatographic separations of intact or labelled proteins (with covalent or non-covalent interaction). With an excess of hapten, heterogeneity does not become more pronounced (although, by shorter incubation time or deficit or hairpin, some reaction products are detectable in the middle acidic range – pH 5-6.5).

One should expect that during the electrophoretic experiment, the reacting mixture becomes quickly divided to single components, so the complex is subjected to decay. The decay above, still, occurs rather slowly, as it may be seen from the Fig 8.3. When the interaction is not as strong, the IEF method may not detect the complex formation. In particular, we did not detect any product that may correspond to a complex for anti-GFP Mab and its binding partner, GFP (data not shown).

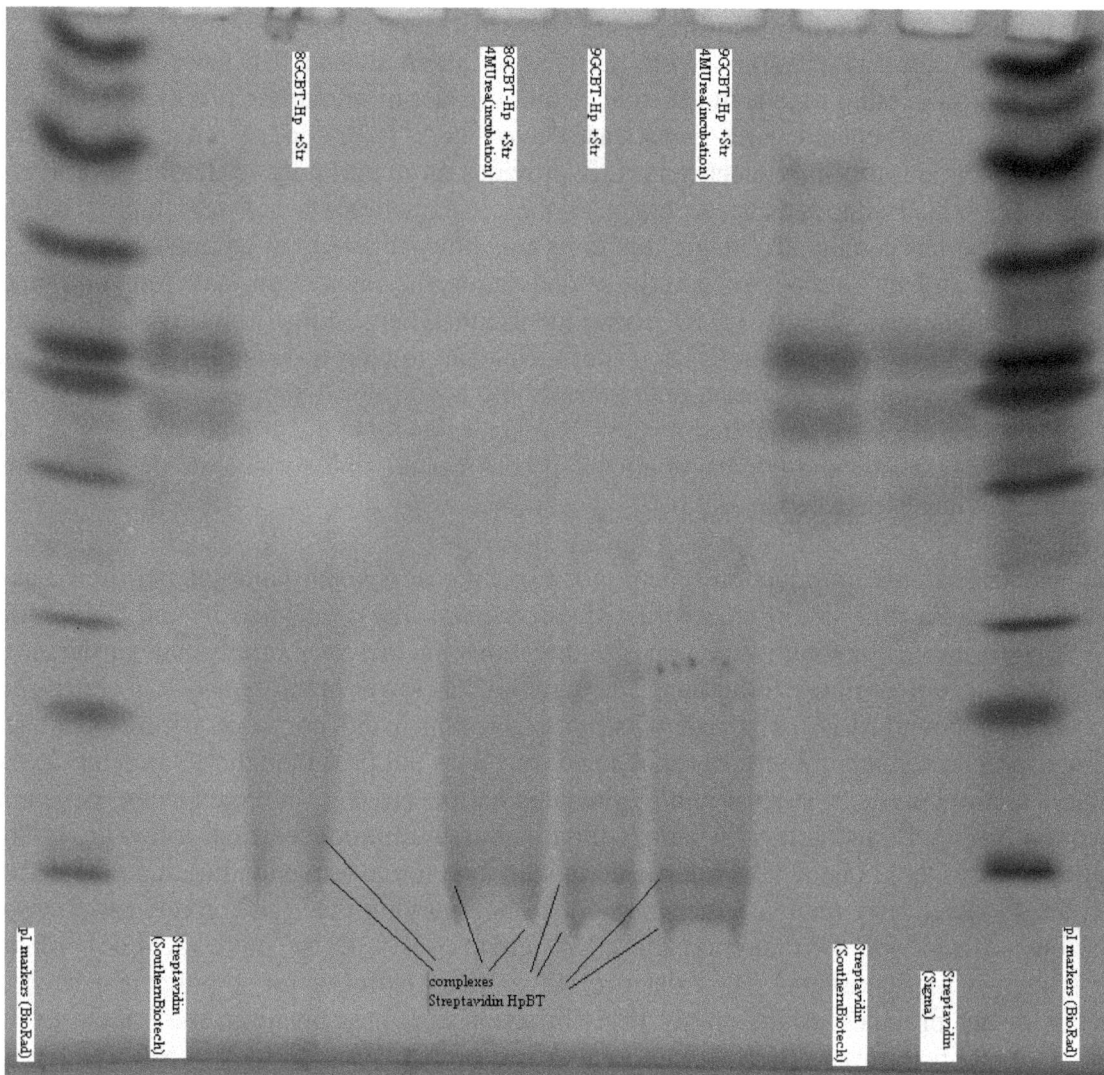

Figure 8.3. Complexes examined for streptavidin and biotinylated hairpins (Bt-8gc and Bt-9gc). Isoelectric focusing 3-10 pH range is implemented using a vertical system. Incubation time was 40min. Urea concentration in the sample buffer was 4M (with no urea in the gel). Outer lanes: pI markers (Bio-Rad). Inner lanes from left: (1) Streptavidin (Southern biotech); (2) Biotinylated 8GC hairpin + Streptavidin; (3) Biotinylated 8GC hairpin + Streptavidin + 4M urea incubation; (4) Biotinylated 9GC hairpin + Streptavidin; (5) Biotinylated 9GC hairpin + Streptavidin + 4M urea incubation; (6) Streptavidin (Southern biotech); (7) Streptavidin (Sigma). The notation at the bottom of inner lanes (2)-(5) marks the complexes of streptavidin and the biotinylated DNA hairpins. (The well pronounced pI –shift of the protein-hairpin complex is due to the presence of strong acidic moiety of DNA.)

155

8.1.4. Bt-8gc -- Streptavidin Binding using CE w/wo chaotropes

We also used capillary electrophoresis (CE), as an alternative to IEF gel electrophoresis, since the CE processing time is much shorter, on the order of minutes. CE may be employed for analysis of fast chemical reactions (fast decay, etc.) [234-236]. Similar to chromatographic separation, capillary electrophoresis provides an opportunity to determine reaction kinetics [234-237] although the accuracy of these calculations is not very high. The CE technique also has certain advantages due to its suitability for study of complex formation at different pH and in presence of additives modulating the interaction (salt ions and other charged compounds). Results of CE experiments on (streptavidin)--(biotinylated hairpin) complexation have been obtained. The experiments aim to confirm complex formation, and its relative concentration decrease, under chaotropic conditions. Complex formation (streptavidin-Bt-8gc) is clearly exhibited as new peak appearance on electropherogram when the mixture of streptavidin and DNA is analyzed. It becomes possible to separate the same components, previously detected by gel IEF.

The standard sample introduction scheme for two interactions substances is performed as a test: Streptavidin plug is introduced first (hydrodynamically), followed by the DNA plug. The two substances moving in the opposite directions interact very briefly, but sufficient to see side effects of complex formation. The complex has lower mobility and it is eluted second, after unbound DNA. The part of streptavidin which did not react with biotinylated DNA continues its moving towards anode and thus does not pass though the detector. In capillary electrophoresis of streptavidin/biotinylated hairpin (Bt-8gc) complex using sequential injection (Fig. 8.4), a streptavidin sample plug is pressure introduced first, following by the second one of DNA. The DNA plug passes though the protein (streptavidin) and the interaction time is 2sec. By reducing the sample load and varying the DNA/protein ratio it was possible to separate two streptavidin-DNA complexes. (The more acidic complex pI=4.3approx. is eluted first). The existence of two major isoforms for complex is in accordance with our previous results on gel IEF. (Injection time/pressure from the bottom to the top: 5s/0.5psi(prot)-0.5psa(DNA); 5s/0.5psi(prot)-0.3psa(DNA); 5s/0.2psi(prot)-0.1psa(DNA). Run at 250Kv/cm.

By adding urea in the running buffer, even in the absence of urea in the sample buffer, one changes the electropherograms, beginning with indications of population shift, i.e., different proportion between the complex and unbound hairpin. Further increase in urea concentration decreases the ratio between the complex and free Bt-8gc (here the concentration of urea in running buffer does not have a significant impact). Finally, very high urea concentration results in essential changes: the streptavidin is apparently mostly in its denatured form, although some capability of binding biotin still remains. Urea concentration increase influences the elution time. Several different effects act simultaneously, in particular, dielectric constant and viscosity change. In addition, there is a possibility of electroosmotic flow modulation. The most pronounced effect, apparently, is the conformation changes induced by urea; this explains considerable reduction in migration times both for denatured protein and DNA.

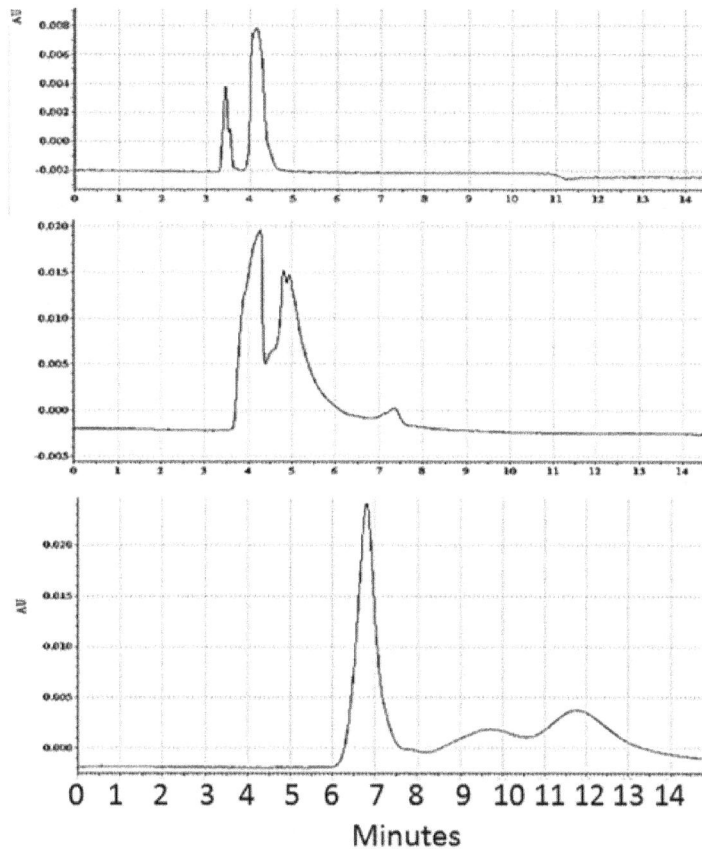

Figure 8.4. Capillary electrophoresis of equilibrium mixture, streptavidin/biotinylated hairpin (Bt-8gc) in presence of urea. Urea concentration increase suppresses the complex formation. Upper panel: 2.5M urea running buffer. Sample – equilibrium mixture, no urea. Middle panel: 2.5M urea running buffer. Sample – equilibrium mixture, 4M urea. Left and right peaks on the two upper panels represent DNA and streptavidin-DNA complex, accordingly. The concentration of complex decreases with chaotrope concentration. In the case of 8M urea concentration (lower panel) no complex formation is observed. The markings on the x-axes are in minutes.

8.2. Biotin-mAb Binding Experiments
8.2.1. Observations of Biotin-mAb Binding on the NTD Nanoscope

In one series of experiments, mentioned previously, we used free antibody molecule interacting with the nanopore detector, where the antibody (anti-biotin) molecule is introduced to our nanopore device to produce the characteristic two-state telegraph signal (Fig. 8.5). The blockade signal for the antigen is practically unaltered by excess antigen: even 100 fold excess of biotin does not change the blockade signal considerably (Fig. 8.5). The signal changes greatly in presence of urea, however, in a relatively small concentration. Here the duration of any event to occupy upper state level becomes shorter and the total probability value of upper level decreases with urea concentration rise.

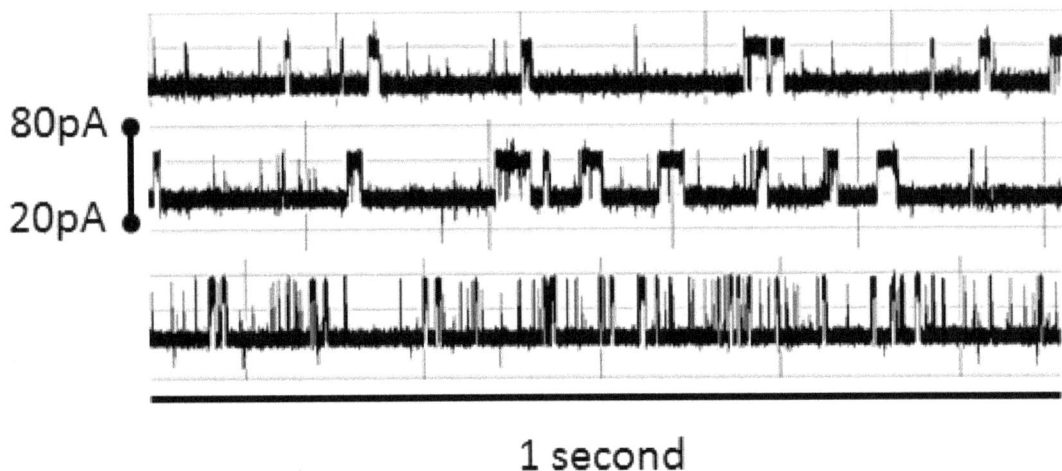

1 second

Fig. 8.5. Robust NTD mAb-Bt-8gc (the mAb has biotin as antigen) binding signal under 100-fold biotin excess shows minimal interference effect (signals top middle show before after), while introduction of small amounts of chaotrope (<1M urea) change the blockade signal significantly (bottom signal). Each signal is shown in a pA range from 20 to 80. The window of observation time is 1s.

8.2.2. Bt-8gc -- mAb Binding validation using IEF Gels w/wo chaotropes

Complex formation between the biotinylated DNA hairpin (Bt-8gc) and mAb is shown on the NTD Nanoscope and validated via electrophoretic mobility shift analysis with isoelectric focusing in Fig. 8.6.

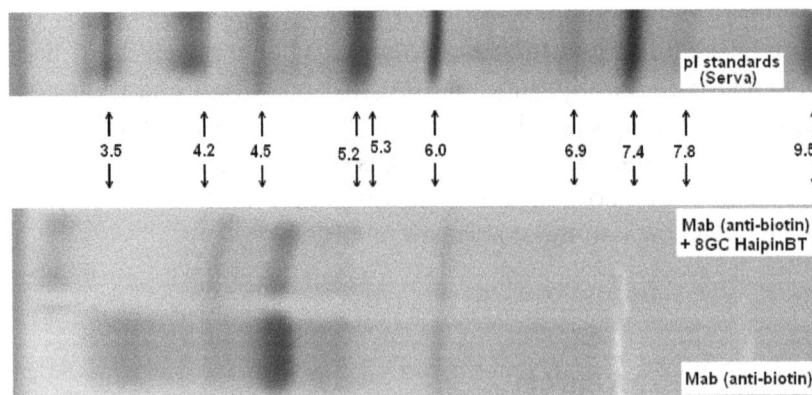

Figure 8.6. Biotinylated DNA hairpin (Bt-8gc) and mAb complex verification.

With addition of chaotropic agents, Bt-8gc – mAb interactions are weakened, which results in a significant decrease in the relative concentration of the complex. We observe this in IEF experiments where the complexation in the system is seen between anti-biotin Mab and biotinylated hairpin in Fig. 8.6. With progressive increase of urea the presence of complex becomes completely undetectable, as shown in Fig. 8.7, where complex is no longer discerned at a urea concentration above 2M.

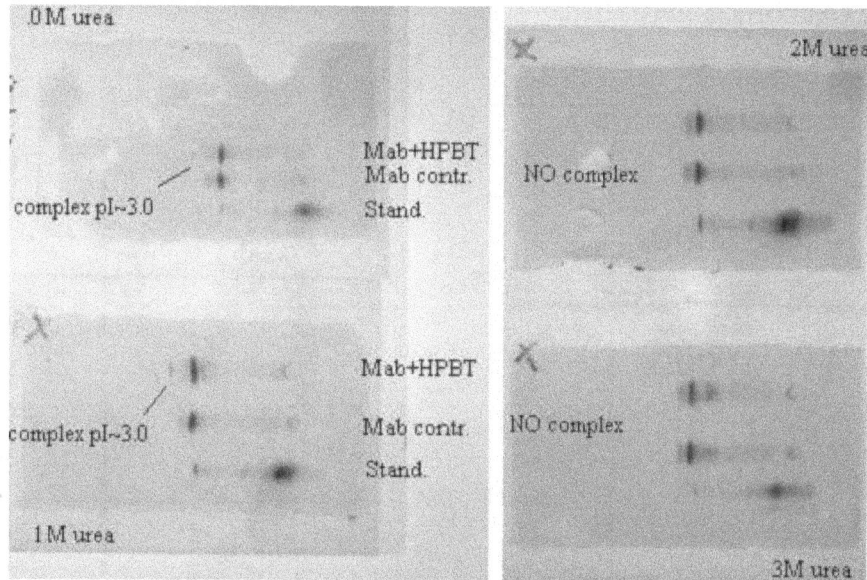

Figure 8.7. Urea suppresses complex formation between the Mab (anti-biotin) and biotinilated hairpin. Mab IEF spectra in presence of HP-BT (Bt-8gc in newer notation)) are shown at 0,1,2 and 3M urea. 1st and 2nd line on each panel – Mab and Mab +HP-BT, correspondingly. Anode is on the left. Isoelectric focusing in 3-10 pH gradient (horizontal system).

8.3. Alpha-Hemolysin Nanoscope Operational pH range

Since the nanopore detector we described was implemented using an alpha-hemolysin protein channel, the operational range of the detector is partly governed by the pH range over which the channel geometry remains relatively unchanged (see Fig. 8.8). At pH 8.0 the channel is very stable with infrequent gating even when using higher voltages than 120 mV, or sampling frequencies above 10 Hz (with polarity switching on voltage). At pH 9.0 some gating does occur (a rare example is shown in Fig. 8.8), but a surprisingly large range of pH appears to be accessible if the occurrences of channel gating can be ignored, or analyzed separately, or alleviated by introduction of the transducer molecule. Such is easily managed with the automation software, thereby allowing us to operate in a wide range of pH values, particularly those involving enzyme activity and protein-protein interactions. *One beneficial characteristic is that channel gating and other complications appear to be further reduced when a transducer is captured.* Evidently the captured, nearly channel-filling molecule, helps to stabilize the channel in its main conformation. In practice, minor partial gating in the channel under such conditions can be entirely absorbed into the pattern recognition task and be automatically handled with the pattern recognition pipeline.

Once the channel is established there exists the possibility of variation in composition of the upper electrode reservoir (according to the design we employed). Those changes allow for the possibility of regulating the protein-ligand interaction, as described earlier. It has to be mentioned that with changes in pH, viscosity, dielectric permeability etc., one can influence not only the current trough the nanopore to some extent, but also the transport of the analyte of interest to the channel (or through the channel). While the letter effect mostly depends on electrophoretic phenomena, the effect of electroosmotic transport also has to be taken into account [138-140]. Additionally, during prolonged experiments, some effects of buffer electrolysis could potentially start playing an effect [241,242]. The latter may influence the local pH value inside the nanopore, especially when an experimental setup with different buffers in electrode chambers is considered.

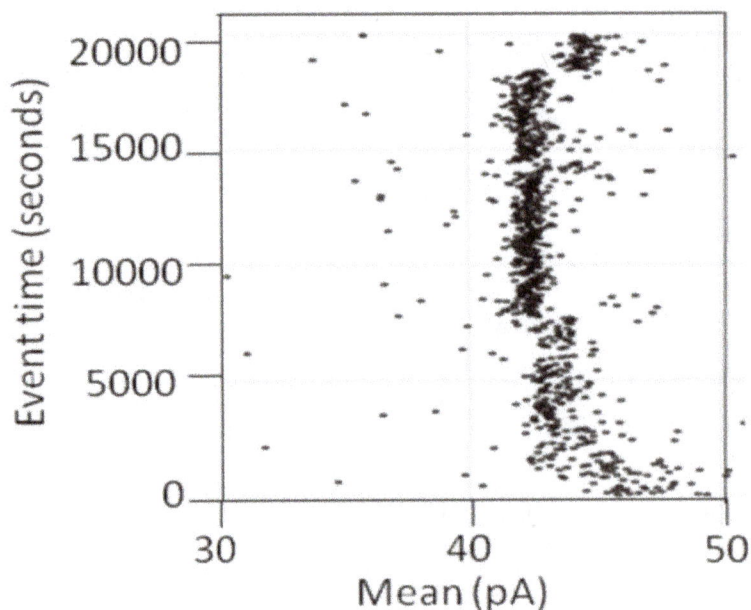

Figure 8.8. Part of the time-trace is shown for Bt-8gc events observed in an experiment that ran for two days. For T=0 to T=3000 the device is not at standard temperature and other conditions. At T=3,000 to 7,500 the channel operates in its normal, exposed chamber, evaporative mode, which leads to a concentration in the cis-chamber, including a concentration of salt (from 1.0 M KCl). At T=3,000, the standard detector is established, aside from having its operational pH set at 9.0 instead of 8.0. At T=7,500 the channel changes configuration, and the hairpin signals obtained are now notably different in just the one attribute (mean). A brief return to a normal channel conductance occurs around T=14,000, with a return to gated configuration at T=14,500. A final return to normal channel conductance occurs around T=18,000.

8.4 Experimental Methods
8.4.1 Bilayer & Channel Setup
Each experiment is conducted using one alpha-hemolysin channel inserted into a diphyt-anoyl-phosphatidylcholine/hexadecane bilayer across a, typically, 20-micron-diameter horizontal Teflon aperture. The alpha-hemolysin pore has a 2.0 nm width allowing a dsDNA molecule to be captured while a ssDNA molecule translocates. The effective diameter of the bilayer ranges mainly between 5-25 μm (1 μm is the smallest examined). This value has some fluctuation depending on the condition of the aperture, which station is used (each na-nopore station, there are four, has its own multiple aperture selections), and the bilayer applied on a day to day basis. Seventy microliter chambers on either side of the bilayer contain 1.0 M KCl buffered at pH 8.0 (10 mM HEPES/KOH) except in the case of buffer experiments where the salt concentration, pH, or identity may be varied. Voltage is applied across the bilayer between Ag-AgCl electrodes. DNA control probes are added to the *cis* chamber at 10-20 nM final concentration. All experiments are maintained at room tempera-ture (23 ± 0.1 °C), using a Peltier device.

8.4.2. Control probe design

Since the five DNA hairpins studied in the prototype experiment have been carefully characterized, they are used in the antibody (and other) experiments as highly sensitive controls. The nine base-pair hairpin molecules examined in the prototype experiment share an eight base-pair hairpin core sequence, with addition of one of the four permutations of Watson-Crick base-pairs that may exist at the blunt end terminus, i.e., 5'-G|C-3', 5'-C|G-3', 5'-T|A-3', and 5'-A|T-3'. Denoted 9GC, 9CG, 9TA, and 9AT, respectively. The full sequence for the 9CG hairpin is 5' CTTCGAACGTTTTCGTTCGAAG 3', where the base-pairing region is underlined. The eight base-pair DNA hairpin is identical to the core nine base-pair subsequence, except the terminal base-pair is 5'-G|C-3'. The prediction that each hairpin would adopt one base-paired structure was tested and confirmed using the DNA mfold server (http://bioinfo.math.rpi.edu/~mfold/dna/form1.cgi).

8.4.3. NTD-Aptamer Design

The Y-shaped NTD-aptamer molecule design we are currently using has a three-way DNA nexus geometry: 5'-CTCCGTCGAC GAGTTTATAGAC TTTT GTCTATAAACTC GCAGTCATGC TTTT GCATGACTGC GTCGACGGAG-3'. Two of the junctions' arms terminate in a 4T-loop and the remaining arm, of length 10 base-pairs, is usually designed to be blunt ended (sometimes shorter with an overhang). The blunt ended arm has been designed such that when it is captured by the nanopore it produces a toggling blockade. One of the arms of the Y-shaped aptamer (Y-aptamer) has a TATA sequence, which is meant to be a binding target for TBP binding studies. In another, variant a DNA aptamer is placed at one arm, instead of a 4dT loop, and similarly with an HIV integrase consensus terminus sequence for use in studies of HIV integrase inhibitors. In general, any transcription factor binding site or DNA enzyme could be studied (or verified) in this manner.

8.4.4. Gel electrophoresis and image analysis

Gel electrophoreis was performed mostly in vertical (Invitrogen) or horizontal (Pharmacia) system. Alternatively, for IEF 11 cm IPG strips (Bio-Rad) were used. The slab gels were fixed in ethanol / acetic acid mixture (10% / 10%), stained with Comassie Blue or Sypro-Ruby dye and further scanned at 100 dpi resolution using a Bio-Rad Molecular Imager FX. The resulting images were analyzed using the PDQuest software (Bio-Rad, V7.1).

8.4.5. Capillary electrophoresis (CE)

CE was carried out with a P/ACE MDQ apparatus (Beckman Coulter) equipped with a UV detector. A 30-cm long, coated, low electro-osmotic flow, capillary with an inner diameter of 75um and outer diameter of 360um was used. The sample buffers and the electrophoresis run buffer were identical: 25 mM sodium tetraborate at pH. The capillary was rinsed with the run buffer for 5 min prior to each run. Electrophoresis was carried out for a total of 10 min by an electric field of 600 V/cm with a positive electrode at the injection end of the capillary. The temperature of the capillary was maintained at 10 ± 0.1 C. At the end of each run, the capillary was rinsed with the same buffer at 10psi for 2 min, followed by a rinse with deionized water for 5 min.

For sample injection, the inlet and outlet reservoirs are established with run buffer, and the capillary is prefilled with the run buffer. Normally, pressure injection 1-0.2 psi was used, or alternatively, sample injection was performed electro-kinetically.

8.4.6 Chemicals
Anti-biotin monoclonal antibodies obtained from Vector Laboratories (9100 (Hyb-8)) and from Stem Cell Technologies (#01405(C6D5.1.1)) were used for binding studies. The antibodies, stored as supplied, were brought to a final dilution 1-4 ug/mL in the electrode chamber. Ampholytes (pH 4-9), and CE buffers were purchased from Bio-Rad (Hercules, CA). GFP was obtained from Molecular Probes (Eugene, OR). Streptavidin was supplied by Sigma-Aldridge. Potassium chloride, HEPES and magnesium chloride were purchased from Sigma, St. Louis, MO. Other chemicals were from Fisher Scientific, Atlanta, GA.

8.5 Validation of NTD complexation with binding target
8.5.1 Validation of NTD complexation using standard methods
In this section we describe efforts to test the hypothesis that conventional electrophoretic methods (gel electrophoresis, IEF and SDS-electrophoresis), as well as capillary electrophoresis, can serve as excellent tools in guiding nanopore signal interpretation. With electrophoretic techniques it's become possible to detect the complex formation, the number of different states (for multivalent systems) and, sometimes, the microheterogeneity of interacting molecules. Electrophoretic techniques are also an excellent tool for experimental monitoring the population distribution between different states as the concentration of chaotropic agent varies in the system. Since some traditional electrophoretic techniques require the presence of ionic detergent (SDS), however, they have limited application in studying the process of complex formation. With the validation results shown here we see how NTD methods, in turn, offer a means to inform and validate conventional electrophoretic methods, as well as offer an SDS free method for analyte separation according to molecular weight. In other words, nanopore detectors can operate as specialty gels, where the "gel" representation of the information is recovered computationally, refered to as NTD *in-silico* gels in [1].

8.5.2 Sample Purity Tests
The protein species examined were subjected to a careful purity tests in order to determine the presence of contamination and existence of microheterogeneity (if any). The controls included electrophoretic techniques: IEF and SDS electrophoresis in gel and microchip (Agilent). The IEF analysis shows subtle differences in isoelectric point values, while no heterogeneity is revealed by SDS electrophoresis. We observed microheterogeneity for the monoclonal antibodies (Mabs) we analyzed. The different mAb's exhibit non-similar IEF-spectra and different levels of contamination. In addition, they differ by the degree of glycosylation (MB-9100 show much higher sugar content being stained Pro-Q-Emerald stain). This difference possibly explains particular features of IEF spectra for these Mabs and the IEF spectra changes in presence of urea.

To discriminate the contribution coming from the low MW impurities we tested a number of such substances: amphoteric dyes, polypeptides and neutral compounds (PEG), in order to recognize such contribution in the future when we are dealing with the signal processing. The electrophoretic mobility tests are consistent with the binding results observed using the NTD method.

We successfully employed the electrokinetic method for controlling the purity heterogeneity and complex formation analysis. We anticipate further refinements can be made with parallel use of mass-spectroscopy information [243,244].

8.6 NTD Capabilities and Limitations

The NTD idea is a noise-state transduction detection method, and can even be used in very-low-current nanopore transduction detection (NTD), where laser pulsing can be used to induce the coherency modulation, if not already present, in the observed channel current noise that is monitored for transducer state change. This is a generalized channel transduction detection setting insofar as the stationary statistical profiles are obtained from the stationary noise fluctuations induced via laser modulation of the channel's environment (not the channel's DC ionic current observations). In the general device-enhancing setting, any introduction of system modulations that results in stationary signals with stationary statistical profiles can be leveraged in a similar manner. One application of note is to live whole-cell studies, where large fertilized sea urchin egg cells, for example, provide a very accessible and well-studied model biological system for complex biosystem analysis, where a single sea urchin egg cell could be merged directly onto the operational NTD bilayer/aperture (with transducer in place), and non-destructive live-cell cytosol assaying might be possible.

The NTD idea also relates to single-molecule analysis and characterization using the nanopore transduction detection method and the stochastic carrier wave signal analysis method, whereby real-time assaying of transient molecular complexes, such as glycoprotein complexes, and intricate protein-protein interactions, such as STAT dimerization, can be done. Single-molecule based analysis, performed sequentially on captured analytes, may allow NTD-glycoassays to be performed on blood samples for a single test to provide detailed analysis of the individual's globin glycosylations. The NTD setting can also be used in gene-circuit analysis using a biosystem extra element theorem (EET) analysis method together with a method for non-destructive analysis of gene interaction networks using NTD with a weakly binding reporter molecule.

By use of pre-processing with simple capture matrices, or microarrays co-opted for that purpose (if nucleic acid involved), it also appears possible to perform detection on very low concentration analytes, with application in broad-based pathogen exposure assays, miRNA detection and haplotyping schemes, and SNP detection and haplotyping schemes. When capture matrices don't suffice, such as with membrane bound analytes of interest, the TERISA and TARISA methods can be used in pre-processing instead.

The engineered transducer molecule central to the transduction approach is shown to offer the added benefit of channel stabilization, and thus overall device stabilization, when working with buffer conditions involving extreme pH, chaotrope, or interference concentration. This enables the nanopore transduction detector to operate as a single molecule 'nanoscope' in a wide range of conditions, where what is seen is not the molecule in a visual sense as with the microscope, but molecular state. The tracking on molecular state is critical to a complete understanding of many allosteric proteins and enzymes. This allows useful device operation in a wide range of conditions relevant to biochemistry, biomedical engineering, and biotechnology.

Binding affinity results upon introduction of chaotropic agents (2.0-3.5 M urea) show agreement between nanopore transduction detection (NTD) and standard electrophoretic-separation methods, including: (i) isoelectric focusing; and (ii) capillary zone electrophoresis.

8.7 Interference Testing

The electrophoretic part of the NTD detector provides a huge advantage when dealing with possible contaminants. Electrophoresis drives strong negative charges to the nanopore detector during normal operation. Nucleic acids in particular will be separated and driven to the detector, along with certain proteins and other molecules that have a low pI. Most proteins with low pI are found to have very little interaction with the nanopore channel, as already mentioned, the main exception being antibodies. Consider the common level of interference agents used to demonstrate robust medical testing applications (see Table 8.1). Actual levels of interference agents seen in (healthy) human blood samples are far lower (see Table 8.2). Consider working with a 1uL sample (such as with a pinprick sample) that contains high levels of common interference agents from blood, or other biological sources, Table 8.3 shows the very high contaminant levels that have been tested on the NTD with very low concentrations of reporter molecule. Most interference agents pose little channel interaction and the occasional channel blockade that does occur is short and non-modulatory. As a group antibodies are the exception, where a single monoclonal antibody (mAb) is found to produce a variety of distinct channel modulation signals types.

Bilirubin:	10mg/dL = 0.10mg/mL
Cholesterol:	800 mg/dL = 8.00 mg/mL
Hemoglobin:	250mg/dL = 2.50 mg/mL
Triglyceride:	500mg/dL = 5 mg/mL

Table 8.1. Common level of interference agents used to demonstrate robust medical testing applications.

Bilirubin	5mg/L (10uM)
Cholesterol (healthy)	< 2mg/mL (5mM)
Hemoglobin in plasma	2mg/dL = 0.02mg/mL (300nM)
Hemoglobin in whole blood	150mg/mL(2.5mM)
Triglyceride	1g/L (1mM)
Serum DNA (no cell ruptures)	1-200ng/ml
Albumin	35-50 g/L (600uM)
Immunoglobulin G (IgG)	15mg/mL (at 160kDa → 93.75nmol/mL)
Urea	15 mg/dL (3mM)
Glucose (fasting)	100 mg/dL (5mM)

Table 8.2. Actual levels of interference agents seen in (healthy) human blood samples.

Cholesterol (healthy)	8mg/mL > 2mg/mL
Hemoglobin	4mg/mL > 2.5mg/mL
Immunoglobulin G (IgG)	30mg/mL > 15mg/mL
Urea	> 5M >> 3mM
Glucose	>> 50mM > 5mM

Table 8.3. Contaminant levels that have been tested where reporter molecules are easily discerned.

In studies with interference on the control 9GC molecule it is found that 1uL of 0.7nM 9GC can easily be seen in the detector (that has 70uL wells) in presence of 1uL of 1uM 7GC (approximately a 1:1000 ratio of 9GC to 7GC but easily discerned due to the distinctive channel modulation of the 9GC molecule). If analyzing the trace amounts of DNA present in blood serum (such as for early fungal pathogen identification), suppose 10ng/mL of total DNA is present of which 1/1000 is due to fungal pathogen. If the fungal pathogen is 'reported' by a modified form of the 9GC molecule (or a Y-transducer) then it is necessary to 'see' 1/1000 of 10ng/ml 9GC at the detector. Since 10ng/mL concentration of 9GC is 1.5nM, and we can see even less, 0.7nM, when the rest of the serum DNA is interference (from accidental cell ruptures, etc.), then it is clear that we can detect on trace DNA targets. Interference from other biomolecules that have higher pI is handled much more easily: 1 uL of 0.7nM 9GC in the presence of 4mg/mL hemoglobin (Hb) is easily resolved. Hb has a pI = 6.87 (normal, sickle cell pI=7.09), so in the standard pH=8 buffer it is expected that some Hb should be delivered to the channel, but even when this occasionally occurs, it has no apparent interaction. This is in agreement with albumin interference results, where concentration = 8mg/mL, and with a pI of 4.7, it is expected that many of the albumin molecules should be delivered to the channel, but no significant channel blockade events or even brief 'noise-spike' blockades are seen (this is thought to be because albumin is not glycosylated). In practice, an albumin capture matrix could be used to prevent the normally high levels of blood albumin (the main protein in blood plasma) from entering the nanopore detector. This would not be to prevent interference with the channel detection per se, but to prevent bilayer interactions. Having entered the nanopore detector albumin can still potentially be blocked from bilayer interference by having a surface scaffolding on the bilayer from PEG linked albumin.

Cholesterol acts similarly to albumin, where high concentrations are not found to have observable channel blockade effect. This is not to say that albumin and cholesterol have no effect whatsoever, they appear to have a beneficial effect at physiological or lower concentration via stabilizing the bi-layer against rupture and to overall reduced current leakage (membrane permeability), and result in a lower RMS noise to the overall single channel current (no cholesterol, typical channel current RMS noise is 1.32 pA; with cholesterol it drops to 1.02 pA). The suspected role of albumin in channel nucleation is also revealed in these studies as late channel additions (bad news for single channel experiments) are observed to occur with introduction of albumin. Bilirubin has similar isoelectric point to albumin and similar non-reactivity with the channel.

8.8 Polyethylene glycol (PEG) for size exclusion chromatography and filtering
Introducing PEG into the buffer reveals strong size-exclusion chromatography fractionation effects, allowing species to be computationally grouped according to their PEG shift measurements then presented as an ordered 'computational gel-separated' list of species (affording gel-separation and blot-identification entirely on the NTD apparatus). In the results shown in Fig.s 8.9 & 8.10 we see representative channel blockades for two types of DNA hairpins (Fig. 8.9), each with 4dT loops capping one end, one with seven base-pair stem (7CG molecule in Methods), and one with a twelve base-pair stem (12CG molecule in the Methods). Fig. 8.10 shows observations on mixtures of 7CG and 12CG before and after addition of PEG. The PEG-shift in this instance should see a shift in channel events to favoring more channel events with the larger nucleic acid, 12CG over 7CG in these experiments. Before addition of PEG hundreds of 7CG and 12CG events were observed with the ratio of 12CG to 7CG events: 0.82. After addition of PEG the ratio favors 12CG: 1.33. There are also more counts overall. So have the overall appearance of greater concentration of 12CG (roughly twice), when it should be halved by the removal of volume to accommodate the

dilute PEG solution addition. In other words, an effective ionic concentration increase due to the volume excluding effect of PEG on charged analytes, with increased volume exclusion effect on larger charged molecules like 12CG vs. 7CG.

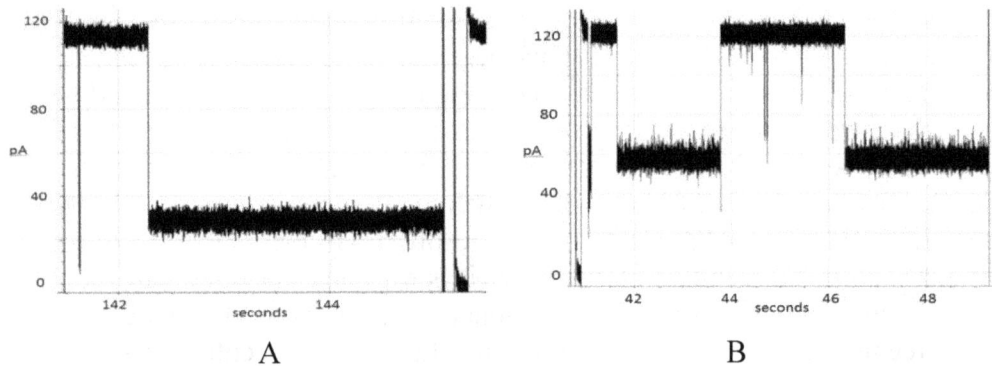

Figure 8.9. DNA hairpin blockade signals before addition of PEG. (A) 12CG blockade; (B)   7CG blockades.

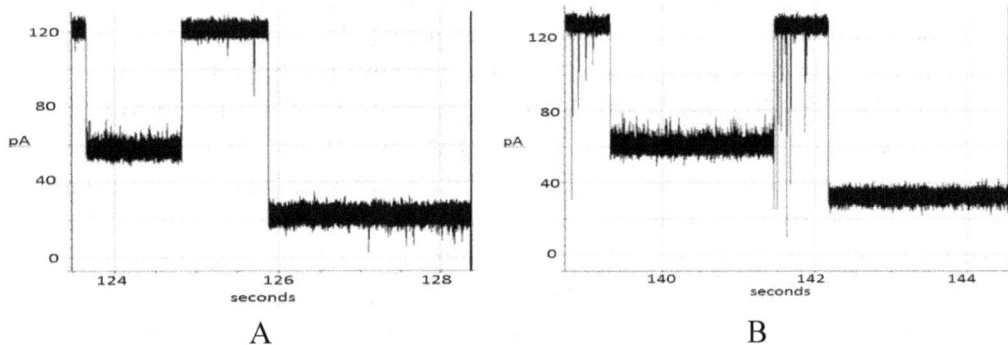

Figure 8.10. 7CG and 12CG DNA hairpin mixture blockade signals before and after addition of PEG. (A) Before. **(B)** After.

The idea here is that nanopore detectors may offer the separation/identification utility of gels, but under physiological buffer conditions and using non-destructive pattern recognition on blockade events to do the clustering "*in-silico*". Using the PEG-shift approach, for example, the nanopore-based methods may be able to match the information content of drift-separation methods, such as mobility shift gels, but this won't resolve the topology mapping of the isoelectric *focusing* methods. Although a nanopore can be easily coupled to capillary electrophoresis geometries, for hybrid separation/clustering using capillary/nanopore, there is still no simple way for a *single* nanopore detector to 'read' the focusing clusters without new plumbing being introduced, so will not be discussed further here.

Chapter 9

Nanopore Transducer Engineering and Design

Biomolecules, such as DNA, RNA, protein, and glycoprotein, typically provide channel blockades at a fixed level. If their blockades can be induced into telegraph-like signals via introduction of laser modulations, then the critical modulatory signal aspect of the transducer can be made ubiquitous, allowing close inspection of any molecule, via its states, when coupled to such a modulator function while interacting with the nanopore.

9.1 Transduction channel-modulator capability via laser modulation
Biomolecules are in size-ranges that are well-sized for interaction with the alpha-hemolysin based nanopore detector shown in Fig. 5.3. Duplex DNA can't translocate the channel, for example, being captured at one end instead, but ssDNA can translocate. It is discussed in [25,26] that the end of the DNA molecule can be read for nine base-pair DNA molecules with very high accuracy based on the telegraph-like modulatory signals directly elicited during their channel interactions. DNA hairpins with lengths greater than roughly twelve base-pairs no longer elicit channel modulations, residing at a fixed-level blockade. If the high accuracy of the DNA terminus read can be extended to DNA hairpins at longer lengths, then highly efficient Sanger-style DNA sequencing might be possible on the Nanopore platform. In [16-19], a 20 base-pair hairpin with a magnetic bead attachment was studied with this in mind. The 20 base-pair hairpin (bphp) with magnetic bead produced a fixed level blockade that was similar to the blockade of the 20 bphp with no bead attachment (see Fig. 9.1). In the presence of appropriate laser modulations with a chopped beam, channel blockade modulations resulted (Fig. 9.1 Right Panel). It was found that the modulatory signals were distinctive in this 're-awakened' configuration. Regarded in a different sense, the captured 20 bphp provides a terminus-dependent transform on the injected laser modulation that allows the terminus to be identified as in the 9bphp analysis, presumably with similar high accuracy given sufficient observation time. Thus Sanger sequencing on the NTD platform appears possible with use of laser modulations (but without dyes). Perhaps what's more interesting, however, is simply that a molecule producing a fixed level blockade upon capture was successfully induced into a unique telegraph-like blockade signal by use of laser modulations.

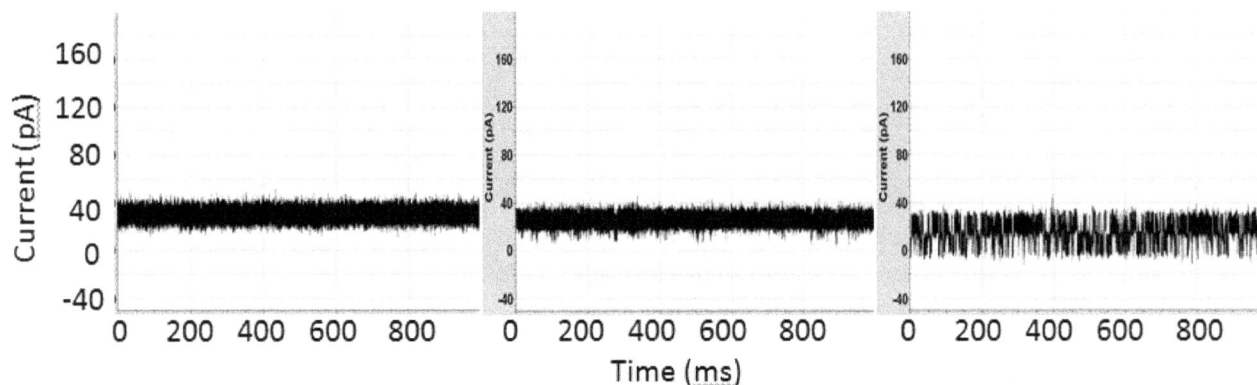

Figure 9.1. A (Left) Channel current blockade signal where the blockade is produced by 9GC DNA hairpin with 20 bp stem [36]. (Center) Channel current blockade signal where the blockade is produced by 9GC 20 bp stem with magnetic bead attached. (Right) Channel current blockade signal where the blockade is produced by c9GC 20 bp stem with magnetic bead attached and driven by a laser beam chopped at 4 Hz. Each graph shows the level of current in picoamps over time in milliseconds.

9.2 NTD Transducer Design

The *bound* state of the transducer/reporter molecule is sometimes found to not transduce to a different toggling ionic current flow blockade, but to a fixed-level blockade (i.e., the transducer provides distinctive channel modulation when unbound, but not so distinctive fixed-level channel blockades when bound). It is important for *both* the bound and unbound transducers to have distinctive channel modulations in order to have automated high-precision state identification and tracking (and allow for multiplex assaying). In this instance, the switch to a fixed-level blockade was thought to be an effect of the large electrophoretically held complex forcing the channel-captured end to reside in one blockade state. This was previously explored in experiments where a streptavidin-coated magnetic bead was attached to biotinylated DNA hairpins known to be good modulators or poor channel modulators [36]. Once a streptavidin coated magnetic bead was attached to the biotinylated hairpins, it was found that gently pulsing the nanopore channel environment with a chopped laser beam (a laser-tweezer tugging) allowed a distinctive channel modulation to result (see Fig. 9.1). It was found more recently that the induced blockade modulations occur in two types, for early laser-tweezer induced results. Further laser tweezer results showing the different, overlapping, modes will be given in later sections, where the experiments are performed with a DNA-hairpin transducer as in previous studies. In terms of the convenient Y-transducer, however, the same could be done by simply making use of the unused arm, as shown in Fig. 9.2.

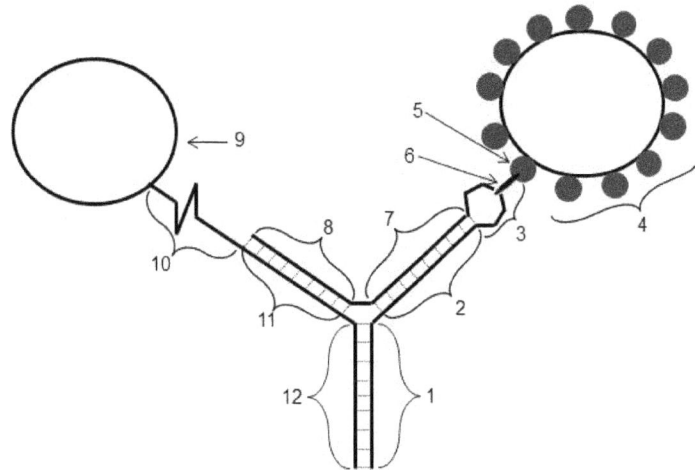

Figure 9.2. Y-laser transducer for high-specificity binding detection or individual protein binding & conformational change study. The Y-transducer is meant to have a study molecule, region 9, attached by a single stranded nucleic acid linker, region 10, that is possibly abasic (non-base-pairing), that is linked to a single stranded nucleic acid region, region 11 & 12, that is meant to anneal to a second nucleic acid to create the Y-shaped nucleic acid construct shown.

In Fig. 9.2 the annealed Y-transducer is comprised of two, possibly LNA/RNA/DNA chimeric, nucleic acids, where the first single stranded nucleic acid is indicated by regions 1-3 and 7-8 and the second nucleic acid is indicated by regions 10-12. The paired regions {1,12}, {2,7}, and {8,11} are meant to be complements of one another (with standard Watson-Crick base-pairing), and designed such that the annealed Y-transducer molecule is meant to be dominated by one folding conformation (as shown). Region 3 is a biotin-modified thymidine loop, typically 4-5 dT in size (here 5dT shown with 2 dT, a biotinylated dT, then another 2 dTs), that is designed to be too large for entry and capture in the alpha-hemolysin channel, such that the annealed Y-transducer only has one orientation of capture in the nanopore detector (without bead, region 4, attached). Region 4 is a streptavidin coated magnetic bead (that is susceptible to laser-tweezer impulses). The base region, comprising regions {1,9}, is designed to form a duplex nucleic acid that produces a toggling blockade when captured in a nanopore detector. The typical length of the base-paired regions is usually 8, 9 or 10 base-pairs. The study molecule (region 9), an antibody for example, has linkage to single stranded nucleic acid via a commoditized process due to the immuno-PCR industry so is an inexpensive well-established manufacturing approach for the molecular construction. The Y-transducer on the left will not form if the 'immuno-PCR tagged' antibody is not present, which provides an additional level of event detection validation. If region 9 is a DNA enzyme that is processively acting on a DNA substrate this may provide a new means for nucleic acid sequencing.

NTD transducers are typically constructed by covalently linking a binding moiety of interest to a nanopore current modulator, where the modulator is designed to be electrophoretically drawn to the channel and partly captured, with its captured end distinctively modulating the flow of ions through the channel. Using inexpensive (commoditized) biomolecular components, such as DNA hairpins, this allows for a very versatile platform for biosensing, and given the high specificity high affinity binding possible, this also allows a very versatile platform for assaying at the single molecule level, even down to the single isoform level, e.g., molecular substructure profiling, such as glycosylation profiling. (Glycosylation profiling can also be done directly for some molecules that directly produce toggling blockades, antibodies in particular [13]. Glycosylation profiling is of critical importance in the development of the most effective antibody treatments [31,245-249].) Two complications with the transducer design, however, are (1) the convenient DNA-based modulators are often short-lived; and (2) the overall transducer's bound state often doesn't modulate. The first is shown to be solved using locked nucleic acid (LNA) nucleosides, the second is solved by introducing a third functionality for receiving laser-tweezer impulses by means of a covalently attached magnetic bead (another commoditized component). A description of the detector's robust performance in the presence of numerous interference agents with very low analyte concentration was also needed, and this is now much more clearly affirmed. LNA Y-transducers with magnetic bead attachment and laser pulsing gives rise to a generic modulator arrangement (see Fig. 9.2), that modulates even when bound, to allow NTD probing over long timescales on biological system components. An inexpensive commoditized pathway for constructing nanopore transducers is thereby obtained.

Previously it was shown that the presence of a specific five base length nucleic acid could be ascertained using NTD [28], and that an eight base sequence of DNA could be ascertained with very high specificity which could be further enhanced with the introduction of urea as a chaotrope [17]. In [17] is shown the binding results at the population-level where numerous single-molecule events are sampled and identified and fall into clear classes. Without chaotrope, the classes aren't as distinctive.

Eight and nine base-pair DNA hairpins have been used as channel modulators [17], where the modulator has had a covalently attached binding moiety (biotin or linked antibody) that was tracked as to its binding state according to the channel modulation exhibited by their channel-captured DNA hairpin ends. Further developments along these lines without using a linker arrangement, where a more commoditized immuno-PCR tagging methodology is used, has led to the DNA 'Y-transducer' platform (for aptamer-based biosensing and for monoclonal antibody biosensing [11,27], see Sec. 9.3,). The Y-transducer was used in experiments showing DNA-DNA annealing on 5-9 base nucleic acids, and in transducing DNA-protein binding events: HIV integrase and TATA-binding protein (TBP) [13,29]. A limitation in all of these efforts was that the critical length of duplex nucleic acid needed for modulation, even in an unbound state, ranged from 8 to 10 base-pairs for the alpha-hemolysin nanopore platform that was being used. (Reasons for the alpha-hemolysin platform being highly preferred can be found in [1], and won't be discussed further here.) The short duplex lengths meant that the reporter molecule could only be observed for seconds or minutes before melting, forcing the NTD to operate in a rapid-sampling 'ensemble' detection mode on the transducer/reporter molecules [20], and less in the single-molecule event-tracking mode that might otherwise be optimal for some applications.

Two problems with the NTD approach are, thus, revealed in the Bt-8gc transducer study: (1) the aforementioned fixed-level blockade by the *bound* transducer; and (2) isomer splitting on the transducer itself under high strain conditions (such as high chaotrope). In this Section we show how to eliminate both of these problems using transducer designs.

Transducer instability: short lifetime and isomer splitting under strain

Two twist conformations, due to different configurations in the hairpin loop and stem duplex conformation (such as B, B*, or A/B conformation duplex DNA), have been suspected from results on the DNA hairpins under other strain conditions, such as high voltage settings [1]. Thus, it is consistent that two types of DNA hairpin channel blockade modes appear in the laser-tweezer experiments. The two modes are thought to be rigid-body configuration changing, or 'toggling', and internal DNA hairpin configuration changing, or 'twisting'. Although the resulting toggle/twist mode signal analysis is more complicated when working with channel modulators, especially if induced by laser-tweezer, this is actually a highly favorable result since additional modulatory modes beyond the physically associated toggling and twisting are not seen. A bound transducer can now generically provide a modulatory state by use of a bead attachment with laser excitation, with high-strain modal proliferation apparently limited to two types, which is a very manageable situation given sufficient observation time. Thus, the stochastic carrier wave analysis [1] can proceed as before, only with more training data needed to 'learn' the more complicated background 'carrier wave' signal's characteristics. The transducer problem, thus, remains tractable with laser-tweezer generalized (ubiquitous) transducer design. Furthermore, there is the ability to turn the twist mode type of internal signaling to our advantage in specialized transducer designs, as will be discussed in the section that follows.

The problem with the DNA transducers with short lifetimes (in the electrophoretically-driven capture strain environment), and the internal mode transmission (excessive twist mode) transducers, is they have too much internal freedom. If it was possible to 'lock-up' some of the internal twist motion, then a stronger hairpin might result, and one less likely to have twist modulations on top of toggle modulations. Such nucleic acid variants exist and are known as locked nucleic acid nucleosides (LNAs). They are a nucleic acid analogue where the ribose ring is locked into a highly favorable configuration for Watson-Crick base-pairing. The locking is accomplished by forming a methylene bridge from the 2'-O atom to the 4'-C atom of the ribose ring. LNA oligonucleotides can be synthesized using standard phosphoamidite chemistry (e.g., is compatible with standard enzymatic processes) and can be incorporated into chimeras with RNA and DNA [250]. The high affinity of LNA for complementary DNA or RNA provides improved specificity and stability, and is resistant to exo- and endonucleases for use in both *in vivo* and *in vitro* settings. The increased affinity leads to much more stable LNA hairpin and other LNA duplex configurations. This has special significance in the NTD setting where specially designed DNA hairpin and Y-transducer molecules have already been identified for use as event transduction molecules, and minor alterations on these transducers for the LNA form (see Methods) are obtainable that retain the transduction properties, but now with the long-lived and improved specificity and affinity attributes of LNAs [11]. LNA versions of the biotinylated hairpins studied in [17] are explored in [10,11], where streptavidin binding occurs where one twist appears to dominate (so only toggle modes are non-trivial), and the lifetimes of the LNA/DNA chimeric transducer molecules in the high-strain capture environment of the nanopore is now on the order of hours instead of minutes.

The generic Y-transducer for annealing-based detection (no laser-tweezer needed) could have a form like in Fig. 9.3, where the regions with high LNA content are shown in dashed boxes, protecting the molecule in those regions from terminus fraying, loop opening, or nexus opening.

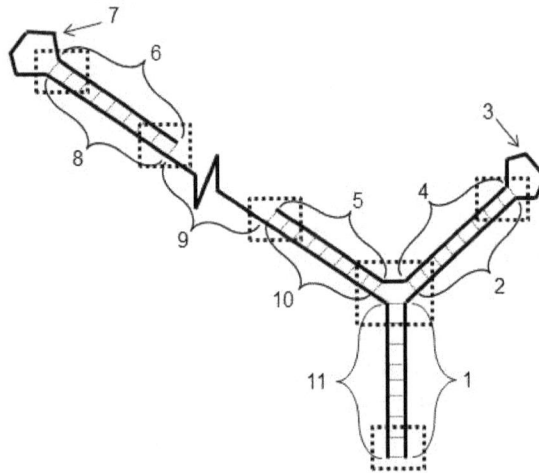

Fig. 9.3. Y-transducer for annealing-detection for presence of specified viral digests. The boxed regions indicate favorable areas for LNA substitution to protect the molecule in those regions from base-pair fraying at the terminus, loop-opening, or nexus-branchings.

9.3 Y transducers: Aptamer and Antibody based

Aptamers are nucleic acids with high specificity and high affinity for a target molecule, the same properties found to be so useful in monoclonal antibody (mAb) diagnostics and bio-sensing applications. Aptamer selection is done by a rapid artificial evolutionary process known as SELEX. Nanopore-directed (NADIR) SELEX offers a means to accelerate the SELEX process and arrive at improved outcome, where the standard aptamer sequence library has the constraint that a portion of the sequence self-assemble (anneal) such that it provides an interface with a nanopore detector to provide a modulatory blockade and thereby introduce a 'stochastic carrier wave' (SCW) allowing a NADIR design/detection process [1,16]. Subject to the SCW constraint the bifunctional aptamer construct already satisfies the criteria to be a nanopore transduction reporter or event 'transducer'. If the transducer has a magnetic bead attachment 'arm', then we are now talking about a trifunctional molecule, thus the Y-shaped DNA molecule in the discussions that follow. Aptamer design can be quite complicated in some settings, however, such as when the binding target of interest involves large molecular features (for some air or water pollutants), large cell-surface features, heavy metal chelation binding, or because the aptamer transducer is inherently more complex with multiple binding moieties or functionalities. (See Disc. for examples of linked double-aptamer constructs and dual aptamer/antibody binding moieties. For the tissue-targeted antibody/aptamer quadfunctional transducer arrangements a 4-way, Holliday-junction, type of DNA molecule could be used, or a linkage via more complicated EDC linker technology [36].) The NADIR augmented SELEX procedure is even more advantageous in the more complicated aptamer design settings.

In Fig. 9.4, the Center and Right Y-transducers are comprised of two, possibly RNA/DNA chimeric, nucleic acids, where the first single stranded nucleic acid is indicated by regions 1-5 (Center) or regions 1-6 (Right) and the second nucleic acid is indicated by regions 6-11 (Center) or 10-15 (Right). In the Left Figure, the paired regions {1,9}, {2,4}, and {5,8} are meant to be complements of one another (with standard Watson-Crick base-pairing), and designed such that the annealed Y-transducer molecule is meant to be dominated by one folding conformation (as shown). The region 3 is a loop, typically 4 dT in size, that is designed to be too large for entry and capture in the alpha-hemolysin channel, such that the annealed Y-transducer only has one orientation of capture in the nanopore detector. The base region, comprising regions {1,9}, is designed to form a duplex nucleic acid that produces a toggling blockade when captured in a nanopore detector. The typical length of the base-paired regions is 8-10 base-pairs.

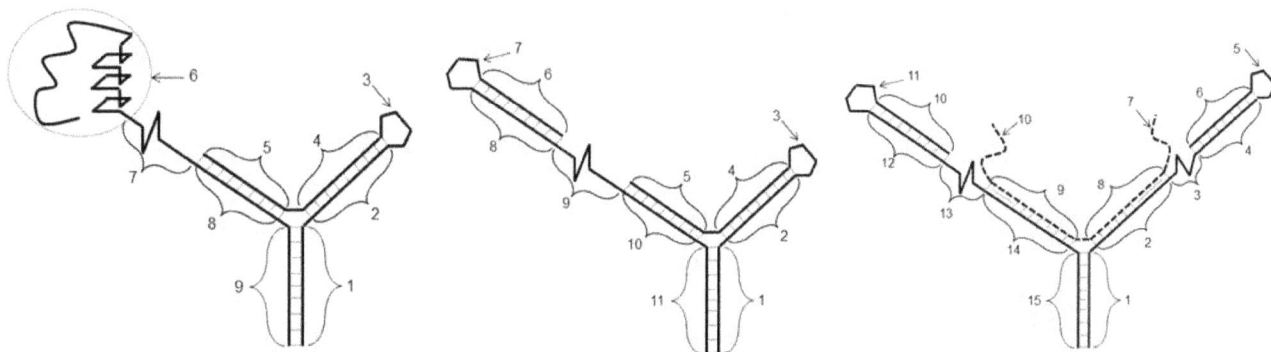

Figure 9.4. Left. Y-transducer for high-specificity aptamer binding detection. Center and Right: Y-transducers for testing hypothesized miRNA binding sites and/or miRNA interactions with a known miRNA binding site. The Y-transducer is meant to have a high-specificity aptamer attached by a single stranded, possibly abasic (non-base-pairing), nucleic acid linker, region 7, to an aptamer in region 6. The sketch of the aptamer in region 6 is meant to suggest the 3D conformational aspect of the aptamer, where stacking of G-quadruplexes is a common, but not necessary, feature of aptamers.

Fig. 9.5 [250] shows Y-transducer variations where the aptamer in Fig. 7 Left is replaced with an antibody, while Fig. 9.5 Right shows a variation in the Y-transducer where the base 'Y' conformation can be achieved with or without the binding target, and its nucleic acid attachment, needed to anneal to for the Y-transducer's stem region.

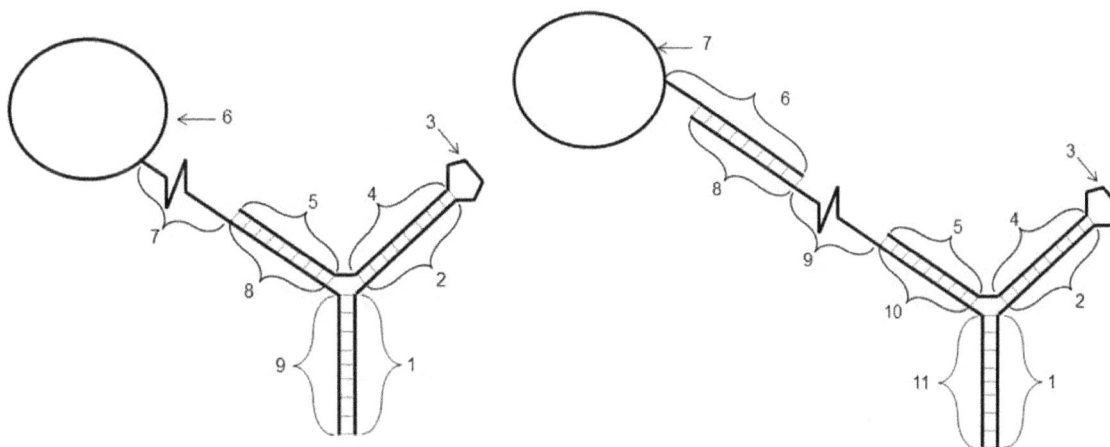

Figure 9.5. Y-transducer for high-specificity antibody binding detection.

9.4 Experimental Methods
NTD Y-transducer/Reporter probe
The Y-shaped NTD-transducer molecule design used in the SNP experiments [1,4], and referenced in the Discussion, has a three-way DNA nexus geometry: 5'-CTCCGTCGAC GAGTTTATAGAC TTTT GTCTATAAACTC GCAGTCATGC TTTT GCATGACTGC GTCGACGGAG-3'. Two of the junctions' arms terminate in a 4T-loop and the remaining arm, of length 10 base-pairs, is usually designed to be blunt ended. The blunt ended arm, or 'stem', has been designed such that when it is captured by the nanopore it produces a toggling blockade. The Y-transducer can be linked with aptamer-based therapeutics [251-253] as will be discussed in later sections.

Biotinylated DNA transducers (from IDT DNA, purification by PAGE)
8GC-BiodT: 5'- GTCGAACGTT/iBiodT/TTCGTTCGAC -3'
9GC-BiodT: 5'- GTTCGAACGTT/iBiodT/TTCGTTCGAAC -3'

Biotinylated LNA/DNA Chimeric transducers (from Exiqon, purification by HPLC)
8GC-BiodT: 5'- +G+TCGAA+C+GTT/iBiodT/TT+CGT+T+CG+AC -3'. The LNA version of 8GC-Bt has 8 LNA bases shown preceded by '+', 12 DNA bases, and 1 biotin dT base.
9GC-BiodT: 5'- +G+CTTGAA+C+GT/iBiodT/TT+CGTT+CAA+GC -3'. The LNA version of 9GC-Bt stem does not have the same sequence as the DNA-based 9GC, and has only a 3dT loop aside from the modified dT with biotin attachment, and has 7 LNA bases shown preceded by '+', 14 DNA bases, and 1 biotin dT base.

Laser Trapping transducers (from IDT DNA, purification by HPLC)
The 20bp hairpin with 4dT loop:
9GC-ext:
5'- GTTCGAACGGGTGAGGGCGCTT TTGCGCCCTCACCCGTTCGAAC -3'

The 20bp hairpin with 5dT loop , where the central loop dT was modified to have a linker to biotin:
9GC-BiodT-ext: 5'-
GTTCGAACGGGTGAGGGCGCTT/iBiodT/TTGCGCCCTCACCCGTTCGAAC -3'

Conjugation to Magnetic Beads
The streptavidin-coated magnetic bead diameters were approximately 1 micron and the mass about 1 pg. Some of the bead preparations involved use of bovine serum albumin (BSA) buffer, which required tolerance of BSA at the nanopore detector. This was separately confirmed for the concentrations of interest, up to the level of 8mg/mL BSA [11].

Laser Setup
Laser illumination provided by a Coherent Radius 635-25. Output power before fiber optic was 25mW at a wavelength of 635 nm. The beam was chopped at 4Hz. During laser excitation studies the Faraday cage was removed. Significant 60 Hz wall-power noise was not seen with cage removed when there was no laser illumination, but with cage removed and under laser illumination 60Hz line noise could clearly be seen. The 60 Hz line noise was, thus, picked up at the laser's power supply and transmitted via the laser excitation process into the detector environment as a separate modulatory source. After fiber optic, approximately 5-10mW illumination is focused into in an approximate 1mm illumination diameter produced at the nanopore detector's aperture.

9.5 Indications of Twist modes

Experiments are done with biotinylated DNA20bp hairpins (9GC-biodT-ext) that are linked to streptavidin-coated magnetic beads (forming (GC-ext-mag) in pH 8 buffer. The transducer DNA hairpin has stem length twenty base-pairs (20bp) and loop size 5 dT, with the central thymidine modified with a linker to biotin. The hairpin in this form is referred to as 9GC-biodT-ext because it is a 20bp extension of the biotinylated 9GC control molecule that has a 9bp stem. The hairpin is then mixed with a solution of magnetic beads that have a streptavidin coating, leading to complexes of magnetic beads attached to a DNA hairpin channel modulator (9GC-ext-mag) by way of a streptavidin-biotin linkage. The mass of the magnetic bead is substantially greater than the hairpin, such that upon capture the likelihood of twist mode being excited is even greater (an even greater angular momentum impulse would occur on capture), even though it is still relatively rare initially. As the experiment proceeds, however, the twist modulating captures increase in likelihood due to more beads becoming more bound with hairpin and thus more mass and charge, thus greater angular momentum impulse on capture (Fig. 9.6).

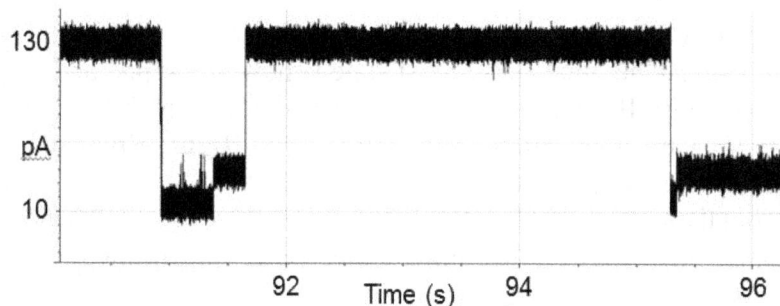

Fig. 9.6. A less common, short duration, full-length 9GC-ext-mag block-ade signal is shown (before diffusional escape) with the beginning of another at far right. The Faraday cage is in-place for this trace, and the 42pA level is seen as before as the upper level toggle (but is less noisy than before since the cage in place). Two clear levels of blockade can be seen, and are thought to correlate with two distinct molecule-channel blockade configurations/conformations as usual. The toggle signals are thought to describe a switching between molecular loop/stem 'twist' conformations, however, and not between two channel blockade configurations (where the molecule in the same internal conformation).

It was shown in [36] that laser-tweezer pulsing could induce a transition from a fixed-level to a toggling blockade on biotinylated 20bp DNA hairpins. It was not clear at that time however, that there was both spatial configuration switching and twist configuration switching, because the latter switching hadn't been seen before. The existence of two loop/stem twist configurations began to become apparent, however, as experiments began to explore a variety of strain conditions, such as high urea (such as 2-5M concentration of chaotrope) [14], higher than the 120 mV applied potential (such as 150-180 mV), higher pH (9 or greater), or in the presence of large bound charge/mass objects (e.g., streptavidin, streptavidin-bead, antibody, or large-antigen attachment).

In Fig. 9.7 we see a channel blockade due to 9GC-ext-mag in the presence of laser-tweezer pulsing (using a chopped laser beam focused with an off-target edge-illumination intensity gradient). The upper 'twist-level' is briefly seen (the 42pA level), followed by a switch to the lower-level twist blockade that has its own, laser-induced, toggle, before sticking at the lower twist state's lower blockade level at the end of the trace (the sticking could be due to the magnetic bead attaching other biotinylated hairpins with increase in charge and overall electrophoretic driving force). The image on the Right in Fig. 9.7 is an enlarged view of the lower twist state's laser induced toggle as it finally becomes 'stuck' at one level. Note the clear 60 Hz line noise evident in the enlarged view. This noise is not present in 9GC-ext-mag blockades without laser illumination (and without cage), so the 60Hz line noise is being transmitted in the laser beam not via the unshielded surroundings. The laser was found to induce the most notable switching in the lower-level twist state when chopped at 4Hz.

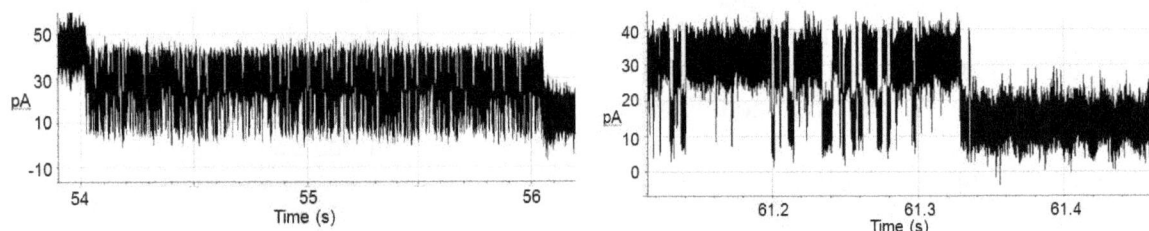

Fig. 9.7. Left. A 9GC-ext-mag blockade signal with Laser-Tweezer modulation (no cage). The upper-level twist state is now at ~45pA, and the lower-level twist state itself toggles between 30pA and 15pA (with fraying 'spikes' to 0pA as seen in other hairpin studies [45,48]). **Right. Enlarged view of lower-level twist toggling**, then sticking at its lower level (note the 60Hz noise is from the laser; without laser, but no cage, there is no 60Hz noise). The beam chopper frequency is 4Hz, but the 'awakened' stochastic modulation is non-periodic.

Clearly the twist toggle adds complication on top of the spatial configuration-toggle and this impacts the design of the transducers. Use of LNAs to lock the twist configuration is expected to eliminate the loop-stem twist toggle complication, but it's not as if the signal processing can't manage the two-toggle mode signal for most cases. So, the main purpose in tuning the LNA content in the LNA/DNA chimeras is to select the most effective transmission of binding event to the channel modulator, where most effective could be via twist mode transmission with large-mass long-tether via a long DNA arm (see. Fig.'s discussed in next section), while for low-mass short-arm tethering linkages the most effective transducer may require a very rigid LNA/DNA chimera (lots of LNA). Further results on twist modes and long-lifetime LNA/DNA chimeric transducers are given in [11].

The laser-tweezer excitation clarified the new twist modes, by effectively allowing them to be turned on and off. Returning to the spontaneous added twist modes occasionally seen with 9GC-Bt/streptavidin, we expect even more notable twist mode to be spontaneously induced with the larger 9GC-ext-mag molecule, especially upon initial capture, due to its larger mass attachment to the nucleic acid modulator portion. This is found to be the case and, in turn, allows new understanding of the complex blockades previously observed when the channel transducer was examined with magnetic bead attached, but not driven by laser (and fully shielded under Faraday cage). A large amount of twist toggling is associated with 9GC-ext-mag blockades: the extended 20bp DNA hairpins typically lasted ~50-60s, often several minutes, before diffusing away, but in some instances only lasted a few seconds (Fig. 9.8). In order to understand the signal, imagine that there are two dominant twist configurations for the molecule and two dominant channel-blockading configurations (toggles). When first captured the transducer molecule might reside in a twist configuration that has yet to reside in one of the dominant channel-configuration (toggle) modes. The twist blockade has two forms of blockade associated with the two dominant twists. We see this as two levels of blockade, each with its own noise properties, where one is referred to as the 'lower level' (LL), and the other the 'upper level' (UL). Once a capture has settled down, it is usually describable in terms of twist {LL,UL} blockades and toggle {LL,UL} blockades and their overlap occurrence. Consider Fig. 9.8 in this regard. It starts in a twist-LL capture and shifts (at about 49.5s) to a twist-LL/toggle-{LL,UL} switching blockade (according to the toggle {UL,LL} mode), which lasts from 49.5s to about 50.3s, when a transition to a twist-UL blockade occurs (non-toggling, from 50.30s to 50.45s). At about 50.45s the blockade returns to a twist-LL/toggle-{LL,UL} switching until 50.80 s when a transition to a twist-LL blockade occurs (non-toggling, from 50.80s to 51.00s). From 51s to the end of the blockade signal it then returns to a twist-LL/toggle-{LL,UL} switching.

Sometimes there is observed a lengthy twist-toggle event that eventually settles down to a fixed-level twist-LL blockade. One such signal is shown in its lengthy toggle portion (Fig. 9.9) that eventually trails off to the fixed level. An enlarged view of one of the twist-LL blockades from the middle of Fig. 9.9 Left is shown in Fig. 11. The twist-LL state clearly exhibits the familiar fraying terminus type of blockade signal observed in other DNA hairpin studies on the alpha-hemolysin nanopore detector.

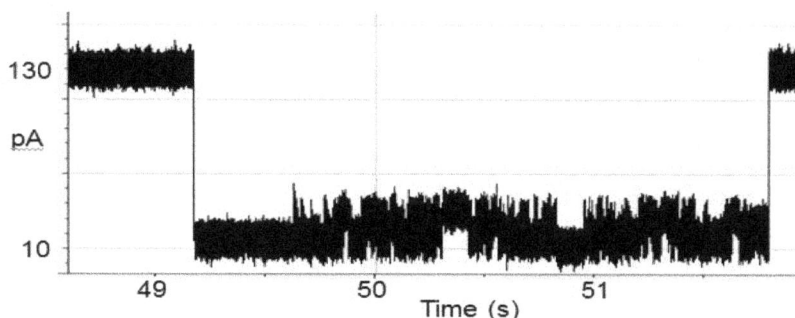

Figure 9.8. A 2.5s 9GCext_mag blockade with cage, starts at twist-LL then twist-LLtoggle, then twist-UL (which doesn't notably toggle), then twist-LL toggle, then twist-LL briefly stuck in its lower level, then twist-LL toggle.

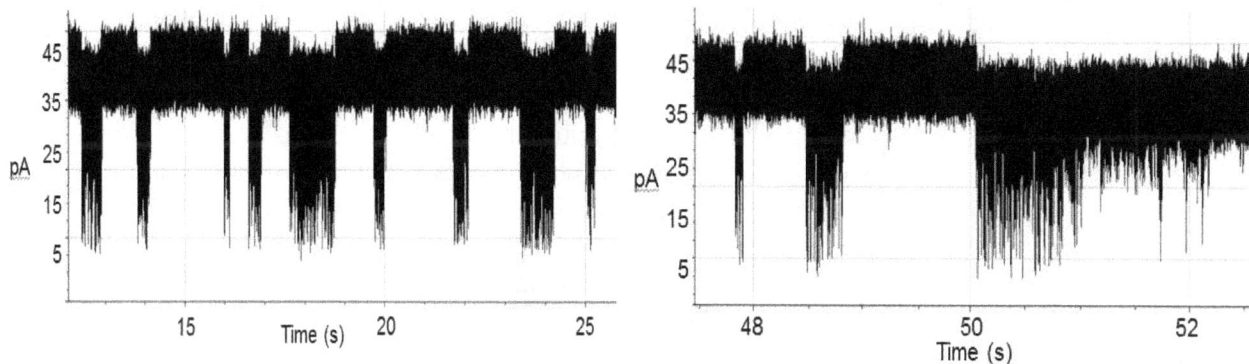

Figure 9.9. Left: A portion of the lengthy body of a 9GCext_mag blockade signal (with cage), about a tenth of the signal in this mode of toggling is shown. The molecule in the twist-LL state appears to be experiencing fraying-type 'spike' blockades (15sec trace). **Right: The end of the twist-toggling**

part of the 9GCext_mag blockade signal. The final transition is to the twist-LL state, for which the fraying falls off until entirely gone, the blockade then continues in the twist-LL fixed blockade (without 'fraying spikes) for several minutes before ending. As the twist-LLstate with frequent fraying events 'settles down' to the twist-LL upper blockade level. Eventually the downward spike fraying events stop entirely. This may be due to the twist-LL having an overall channel-orientation that is slightly pulled out of the channel due to the large magnetic bead attachment, or may simply be due to the molecular excitation damping out, with subsequent less frequent fraying events.

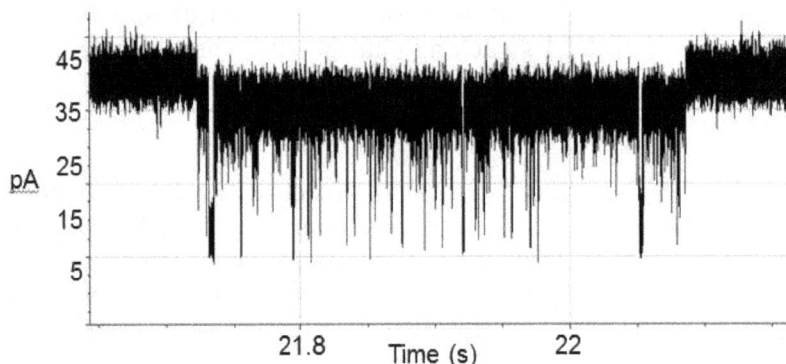

Figure 9.10. An enlarged view of the 9GCext_mag blockade shown in Fig. 9.9 Left for one of the twist-LL sections. The twist-LL state clearly exhibits the familiar fraying terminus type of blockade signal observed in other DNA hairpin studies on the alpha-hemolysin nanopore detector.

9.6 Transducer design: aptamer or antibody, and possible bead attachment

When working with longer stem-length DNA hairpin captures the duplex end is observed to mainly reside in one, fixed, blockade configuration [27], probably due to the electrophoretic force strongly drawing the larger charge nucleic acid into the channel. In early tests with annealing a fixed blockade states often weren't a problem with the Y-shaped transducers, especially for the unbound case, but became a complication when bound if the bound object was 'large' or a significant length of nucleic acid, or when working with a commoditized linkage by annealing to an immuno-PCR tagged antibody or other protein. When working with longer DNA hairpins with large mass/charge attachments sometimes the opposite occurred, the large bound extension appeared to induce occasional toggling where none was observed before (such as seen in streptavidin-coated bead binding to biotinylated 20bphp hairpins described in [11]). In order to have a controlled way to have a simple nucleic acid based transducer, it was then attempted to recover unique modulatory blockade signaling by linking to a magnetic bead where laser-tweezer 'tugging' could then be used for injection of kinetic energy at the single molecule level [36]. Initial results indicated the possibility for a simple inexpensive probe design, as described in what follows, but it was unclear if the stochastic carrier wave signal processing on the more complex transducer modulation signal would be possible given the proliferation in blockade modes observed under laser-tweezer modulation. In more recent studies with chaotropes, the isoforms of the DNA hairpin modu-

178

lators is better understood, however, indicating that there are two mode types of modulation in captured duplex nucleic acid: position/orientation and twist/stretch, where the new signal complication is due to the appearance of the twist/stretch modes. Since there is just the one new mode class for duplex nucleic acid channel blockade (twist modes), the mode proliferation is limited, and SCW signal processing can be performed.

So, nanopore-captured DNA hairpin modulators can exhibit not only spatial/orientation toggling but also torsional/twisting toggling when sufficiently excited. This effect becomes most notable when channel modulations are induced by laser-tweezer pulsing. The new understanding of the laser-tweezer induced modulations suggests a limit for the induced modulator's signal classes to those already seen and a manageable signal analysis platform can thereby be implemented. In practice a stochastic channel modulator that produces the simplest, non-fixed-level, stationary signal blockade is desired, such that the stochastic carrier wave (SCW) signal processing methods can be employed. The position and twist toggle modes in the modulator together pose a more complex SCW system, but can be managed with sufficient sample observations on modulator during its different states (such as linked to bound or unbound analyte). As mentioned earlier, a related complication with using DNA-based channel modulators has been their short lifetimes until melting. This problem was eliminated by use of locked nucleic acid nucleosides (LNAs), as shown in [11] in applications involving isomers, where LNAs serve to reduce twist modes by locking the nucleic acid and thereby restricting its internal degrees of freedom in term of twist/stretch. This can be a good thing in that it will simplify the SCW signal training mentioned above. A simpler SCW analysis is not critical, however, as long as the SCW learning can be done with a manageable amount of training data. So the main optimization to be accomplished by 'locking up' the modulator with increased LNA is effectively a tuning over molecular variants with greater or lesser twist mode event transmission. For annealing-based detection this could be critical since the properly annealed nucleic acid duplex will transmit twist mode excitations notably differently than improperly annealed DNA (if even present). For this reason some modulator arrangements with laser-tweezer pulsing may have their bead attachment on the same arm as the annealing binding site (see Fig.s 9.11 & 9.12), and have a low number of LNA bases in the LNA/DNA chimeras in the binding template (keeping blunt terminus and Y-nexus regions strongly LNA based to prevent melting as much as possible, but permitting twisting). In Fig. 9.11 a description is given for a Y-transducer for single molecule studies using twist mode modulations. A 'twist mode' specialized Y-transducer for single molecule *annealing* studies using twist mode modulations is shown in Fig. 9.12. Related discussions are in [11,250]. Fig. 9.13 shows a 4-way transducer (a.k.a, a Holliday Junction transducer), for dual aptamer/antibody tissue-targeting functional aptamer delivery studies [251-253], where a modulatory transducer is enabled by laser-tweezer coupling.

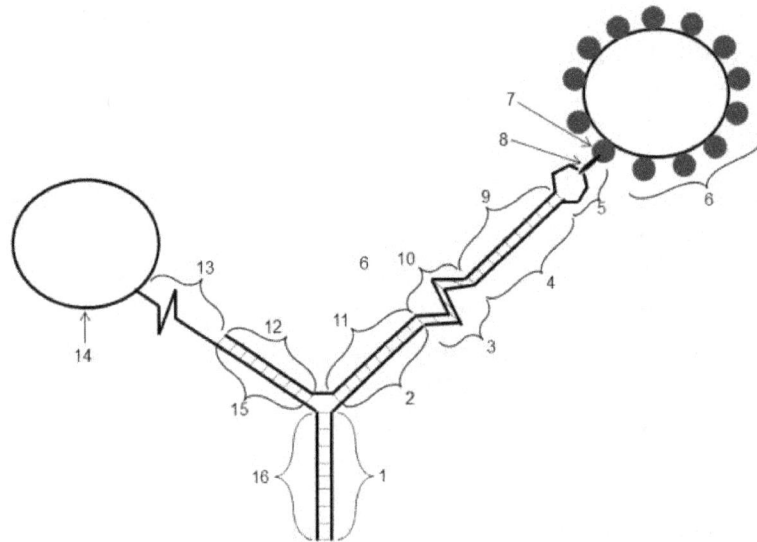

Fig. 9.11. Y-transducer for single molecule studies using twist mode modulations. Region 14 indicates the study molecule of interest (an antibody for antibody-detection or a protein for conformation/binding studies), where a magnetic bead is attached for laser-tweezer modulation (region 6), where the transducer is designed with sufficient LNA substitutions to allow laser-tweezer excitations to be transmitted as a twist-mode impulse (shown in regions numbered {3,10}) while maintaining discernibly different signals according to the study molecules state.

In Fig. 9.11, paired regions {1,16}, {2,11},{4,9},{12,15} are designed to anneal with the dominant conformation shown, and are typically a minimum of 8 or 9 base-pairs in length. The linker arm in Region 13 and Region {12,15}, is whatever is needed to provide sufficient steric clearance between region 14 and region 6 when the stem (region {1,16}) is captured and held at the nanopore.

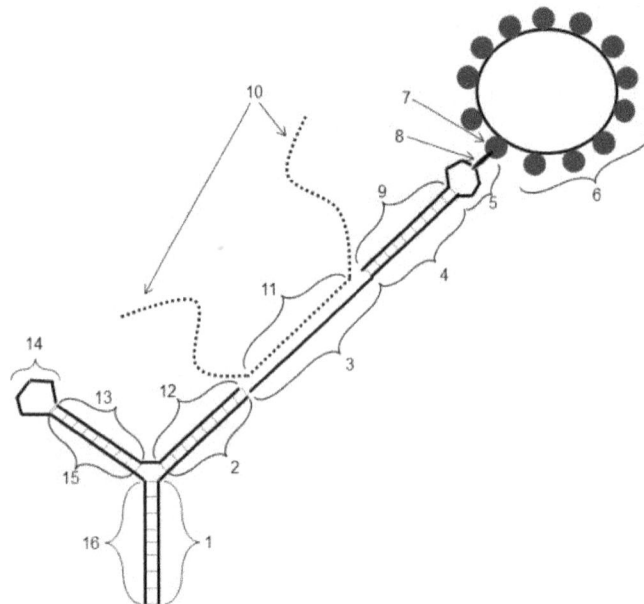

Fig. 9.12. Y-transducer for single molecule annealing studies using twist mode modulations. Region 10 indicates the single stranded nucleic acid

180

study molecule of interest: a nucleic acid whose region 10 section is annealed to region 3 of the transducer; where a magnetic bead is attached for laser-tweezer modulation (region 6), where the transducer is designed with sufficient LNA substitutions to allow laser-tweezer excitations to be transmitted as a twist-mode impulse through the annealed-target region. A twist mode will only transmit if the annealing-target is bound, giving rise to very different channel modulation signals.

In Fig. 9.12, paired regions {1,16}, {2,12},{4,9},{13,15} are designed to anneal with the dominant conformation shown, and are typically a minimum of 8 or 9 base-pairs in length. The loop in Region 14 is designed to not favor channel capture and strongly favor a single conformation for the Region 13-15 stem-loop region.

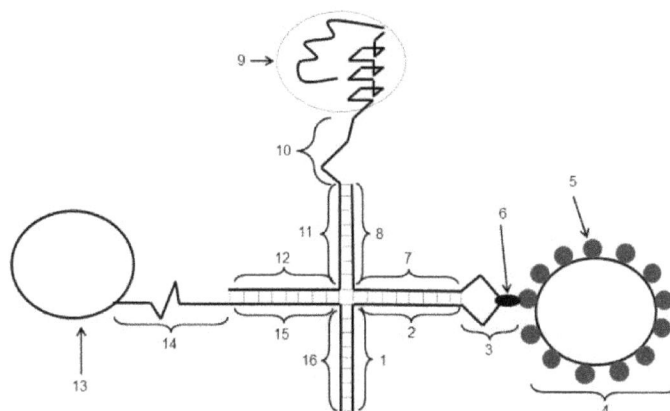

Fig. 9.13. A 4-way transducer (a.k.a, a Holliday Junction transducer). Used for dual aptamer/antibody tissue-targeting functional aptamer delivery studies, where a modulatory transducer is enabled by laser-tweezer coupling.

The Y-shaped DNA transduction molecule is also a versatile construct to test for as an intermediate annealed complex (see single nucleotide polymorphism SNP detection efforts [27] for further details). Highly accurate SNP detection with the Y-shaped DNA transduction molecule was possible in [27] by designing the Y-transducer to anneal to nucleic acid target sequence such that the SNP variant occurs in the Y-nexus region, giving rise to a clear difference in the annealed Y-transducer's channel modulation. The NTD method provided a means to perform SNP variant detection to very high accuracy that will likely be improved further when using the higher specificity LNA form of the transducers indicated by the LNA improvements seen in [11].

9.7 Chelator design, and targeted therapeutic aptamer delivery design

Functionalizing both Y-transducer arms for binding actually leads to a number of other options, such as double aptamer molecules (see Fig. 9.14 for possible use in heavy metal chelation), dual aptamer/antibody transducers (see Fig. 9.15, used in tumor-directed aptamer delivery), where two drugs are FDA approved of this type [251-253], and dual DNA annealing testing (see Fig. 9.15).

In Fig. 9.14, the Y-transducer is comprised of three, possibly RNA/DNA chimeric, nucleic acids, where the first single stranded nucleic acid is indicated by regions 1-4, the second single stranded nucleic acid is indicated by regions 5 and 6, and the third nucleic acid is indicated by regions 7-10. The paired regions {1,10}, {2,5}, and {6,9} are meant to be complements of one another (with standard Watson-Crick base-pairing), and designed such that the annealed Y-transducer molecule is meant to be dominated by one folding conformation (as shown). The base region, comprising regions {1,10}, is designed to form a duplex nucleic acid that produces a toggling blockade when captured in a nanopore detector. The typical length of the base-paired regions is usually 9 or 10 base-pairs. The same Y-transducer is shown on the right, but with a binding target, object 11, positioned for a chelation-type binding configuration, this in addition to any possible chelation binding on the part of the individual aptamers during their individual binding to object 11.

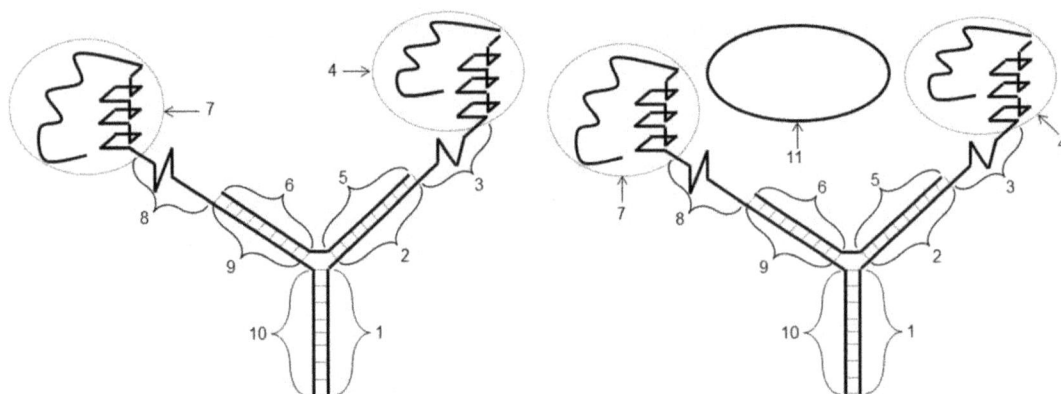

Fig. 9.14. Left and Right: Y-transducer for high-specificity dual-aptamer binding detection. The Y-transducer shown on the left is meant to have two high-specificity aptamers attached by single stranded, possibly abasic (non-base-pairing), nucleic acid linkers, with region 8 for the left arm linker to the left arm aptamer in region 7 and region 3 for the right arm linker to the right arm aptamer in region 4. The sketch of the aptamers in regions 4 and 7 is meant to suggest the 3D conformational aspect of the aptamer, where stacking of G-quadruplexes is a common, but not necessary, feature of aptamers.

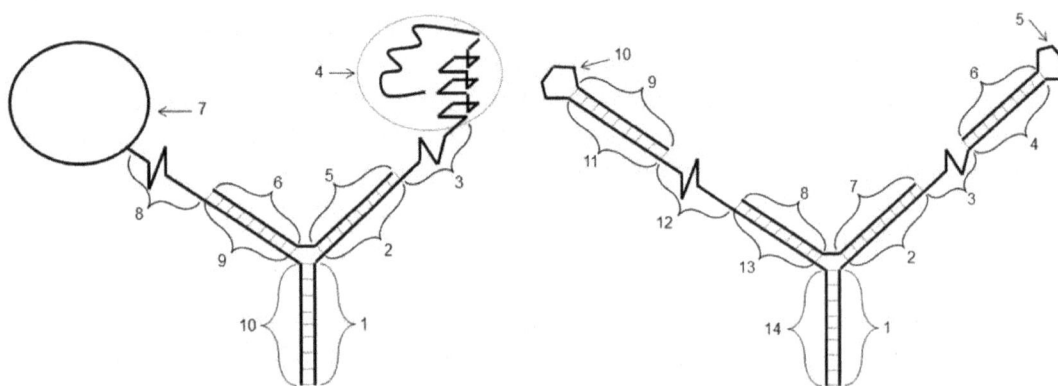

Fig. 9.15. Left and Right: Y-transducer for high-specificity dual aptamer/antibody binding detection; and Y-transducer for dual testing for presence of specified viral digests.

Two types of DNA hairpin channel blockade modes appear in the laser-tweezer experiments. The two modes are thought to be rigid-body configuration changing, or 'toggling', and internal DNA hairpin configuration changing, or 'twisting'. By use of LNA/DNA chimeras with varying amounts of LNA, the twist modes can be locked out. Melting times on the transducers can be directly designed similarly. High-strain bound transducers can thus be used with a modulatory state by use of a bead attachment with laser excitation, with high-strain modal proliferation limited to two types (in the results seen thus far). Using stochastic carrier wave analysis with sufficient training data to 'learn' the transducer's 'carrier wave' signal characteristics, a stable NTD biosensing platform can be then established. The transducer problem, thus, appears to be solvable with laser-tweezer generalized (ubiquitous) transducer design involving LNA/DNA-chimeric transducers.

9.8 Long-lived chimeric LNA/DNA transducers

Experiments are done with a biotinylated LNA/DNA chimeric 9bp hairpin (LNA 9GC-Bt) in pH 9 that has been linked to a streptavidin-coated magnetic bead. LNA 9GC-Bt with streptavidin bound shows a new mode of toggle (Fig. 9.16 & 9.17). Possible twist mode switching is found for the large mass binding case here as with high pH (>9 pH), high voltage (>200mV), and under laser-tweezer inducement to follow. In Fig. 9.17, the molecule appears captured in one twist/configuration, then shift to the other twist may briefly occur, from which a configuration toggling commences. The configuration toggle appears to involve blockade positions favored by neither of the twist conformations. The captured molecular excitations typically start, as it does here, in what is thought to be a DNA-hairpin twist-modulation mode (a direct consequence of conservation of angular momentum and the large mass streptavidin attachment), eventually this settles into a configuration-toggle mode -- where one configuration is sufficiently deep that DNA terminus fraying and extending can sometimes be observed, as in previous studies [1].

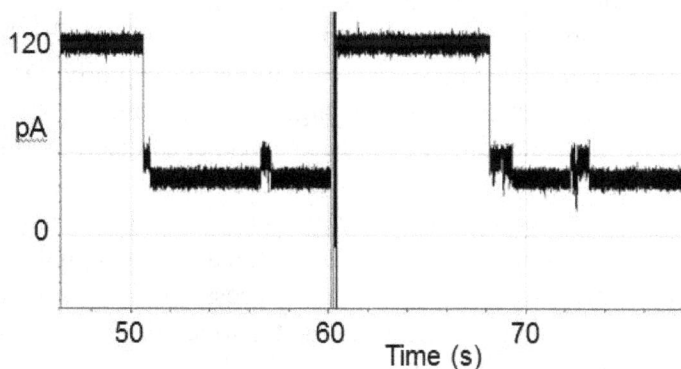

Fig. 9.16. LNA 9GC-Bt blockade signals at 500pM concentration in the detector well at pH9. Auto-eject time is set at 10s. The LNA 9GC-Bt blockades have a faster 'toggle' than LNA 9GC-Bt at pH8.

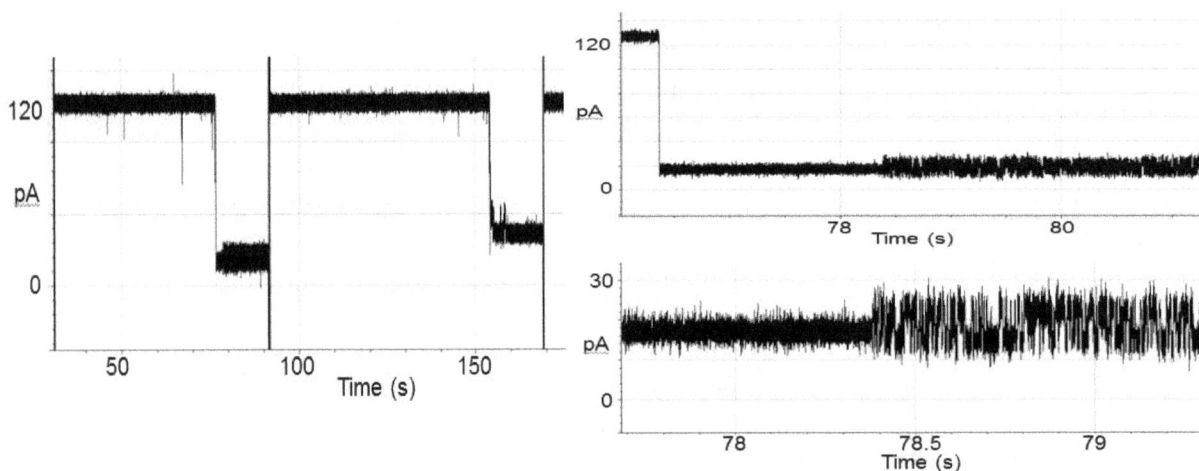

Figure 9.17. Left: LNA 9GC-Bt blockade signals at 250pM concentration in the detector well at pH9, with streptavidin added 1:1, taken in the first 10 min. Auto-eject time is set at 10s. Bound blockade signals are now seen (the one with the lower blockade level on the left). The bound LNA 9GC-Bt blockades now occasionally have a 'toggle' or switch to a toggle mode (the bound blockade on the left shows a transition to toggle 2s into the bound blockade). **Right Top: Enlarged view of bound LNA 9GC-BT blockade. Right Bottom: Further enlarged view of bound LNA 9GC-BT blockade.**

9.9 Two molecular modes – two molecular signal classes

It is shown that nanopore-captured DNA hairpin modulators can exhibit not only spatial/orientation toggling but also torsional/twisting toggling when sufficiently excited. This effect becomes most notable when channel modulations are induced by laser-tweezer pulsing, but has been observed in other high-strain conditions for captured DNA hairpin channel modulators, such as high chaotrope, high pH, high applied voltage, and high mass/charge capture events. The new understanding of the laser-tweezer induced modulations suggests a limit for the induced modulator's signal classes to those already seen and a manageable signal analysis platform can thereby be implemented. In practice a stochastic channel modulator that produces the simplest, non-fixed-level, stationary signal blockade is desired, such that the stochastic carrier wave (SCW) signal processing methods can be employed [1]. The position and twist toggle modes in the modulator together pose a more complex SCW system, but could be managed with sufficient sample observations on modulator during its different states (such as linked to bound or unbound analyte).

A related problem with the DNA-based channel modulators has been their short lifetimes until melting. This problem has been eliminated by use of LNAs, where LNAs also serve to reduce twist modes as needed as well, to simplify the SCW basis mentioned above. Since the simpler SCW analysis is not critical, however, the main optimization to be accomplished by 'locking up' the modulator with increased LNA is effectively a tuning over molecular variants with greater or lesser twist mode event transmission. A general method for transducer construction is thereby suggested with twist-mode dominated state tracking for large charge/mass biomolecular complexes with long duplex DNA tether constructions, and configuration-switching dominated state tracking for small charge/mass biomolecule complexes with short-linkage constructions.

9.10 Isomer Analysis

A method is needed for inexpensive assaying of isomer mixtures that provides not only the ability to specifically bind a particular isomer with high affinity, but a means to multiplex profile a mixture of isomers with high accuracy. A nanopore transduction detector could be used to address this problem, where an aptamer or monoclonal antibody selected for the specific isomer binding of interest is linked to a uniquely modulating NTD transducer for direct quantification via the relative abundances of the different isomers-transducers observed at the nanopore. The nanopore device itself is quite inexpensive (the cost of the key components, including patch clamp amplifier, comes to less than $20,000). So the main cost barrier to NTD applications has been an easy (non-expert) and inexpensive procedure for designing and obtaining transducer molecules for binding interactions of interest. It is shown in this paper, however, that inexpensive assaying can be done with commoditized ('mail-order') components, such as from DNA labeling procedures, biotinylation procedures, and streptavidin-coating procedures. The transducer order-on-demand design process is made possible due to two results shown in this paper: (1) the complexity of transducer blockade modes appears to be limited to the orientation-toggling and configuration-twisting semi-rigid-body motions of the dsDNA transducer molecule, which allows SCW signal processing to be performed in a manageable context. And, (2) laser-tweezer excitations can be used to drive transducers with magnetic bead attachments to arrive at high success-rate transducer designs. Regarding the latter, a general method for nanopore transduction detection transducer construction based on LNA/DNA chimeras that have twist-mode dominated state tracking for large charge/mass biomolecular complexes with long duplex DNA tether constructions is proposed, as is configuration-switching dominated state tracking for small charge/mass biomolecule complexes with short-linkage constructions.

The general-use nanopore transduction detector system offers the prospect for high-specificity molecular, molecular feature, and particulate testing (whether air-quality and water-quality testing), not only in the lab setting, but also the field setting (portable nanopore-based DNA sequencing devices are already commercially available, e.g., Oxford Nanopore Technologies). High-specificity detection is possible by incorporating the high binding specificity of aptamers and monoclonal antibodies for their binding targets into a nanopore binding-event transduction system. Once a binding event is transduced to an electrical ionic current flow measurement, novel channel current cheminformatics and machine learning methods are introduced for event classification. A quantification of the amount of bound versus unbound reporter molecule detected at the nanopore transduction detector then allows the concentration of the target molecule or particulate to be determined. A general procedure thus results for inexpensive LNA/DNA transducers to be obtained, where inexpensive low-power laser tweezer excitations can be used to retain the modulatory role of the transducer in its various states.

Exploring protein conformation-binding relationships and antibody glyco-profiles using a nanopore transduction detector

The relation between protein structure and function is well known and minor changes in protein folding or isoform variants, or surface modifications such as glycosylations, can impact that protein function. To complicate matters further, many proteins are inherently dynamic, so their structure-function relationship can give rise to dynamic functionality, with selection sometimes favoring very dynamic proteins. What is needed is a means to track protein conformation and its role in protein function, binding in particular, and this suggests we need a means to track the conformational state of a *single* protein. A method for using a nanopore transduction detector (NTD) is proposed for such an application. NTD transducers are molecules that serve to transduce the conformational or binding state of a molecule of interest into different channel current modulations, where the molecule of interest is tethered to a nanopore channel modulator. In previous work, using inexpensive biomolecular components, such as DNA hairpin channel-modulators, antibodies, and immuno-PCR linkages to antibodies, experiments were done to analyze individual antibodies and DNA molecules. Three complications were indicated before a general-use NTD platform could be established: (1) the convenient DNA-based modulators were often too short-lived for the binding study of interest; (2) the transducer's *bound* state often didn't modulate; and (3) the binding target often had a pI that didn't favor being drawn to the channel-tethered study molecule. In recent work NTD operation has been demonstrated for a wide range of pH, chaotrope concentration, and in the presence of interference agents, such that problem (3) is solved. While in the latest results shown here the other problems are solved as well: very long-lived channel modulators are demonstrated using locked nucleic acid (LNA) nucleosides; and induced modulations are demonstrated by engineering transducers to receive laser-tweezer impulses by means of a linked magnetic bead (another commoditized component). A general-use NTD platform is thereby possible using an alpha-hemolysin nanopore detector and performing the critical transducer engineering with readily available, inexpensive (commoditized), biomolecular components.

Proteins, such as enzymes, can have a high degree of variability. It has been demonstrated that enzyme turnover rate, for example, can differ at the single molecule level [254-256], with a single enzyme observed with one constant turnover rate, while another enzyme, differing only in conformation, or possibly by a difference in glycation, has a different, but still constant, substrate processing rate. And this is a simple example where there is only one interaction region and it is (mostly) unchanging in its conformation for the individual protein examined. Some allosteric proteins, on the other hand, with multiple binding sites for a particular target, change their binding affinity according to how many ligands they have bound. Antibodies are known to change conformation during binding to one (or two) antigens in such a significant manner that this is the basis for activation of the complement cascade of the adaptive immune response. The nanopore transduction detector (NTD) is presented here as a general method for informing our understanding of protein structure-function relationships at the single molecular variant/isoform level [1,9-11,13,14,16-39].

In previous work it has been observed that many antibodies directly exhibit modulatory channel blockades, and upon introduction of their antigen, their bound-state is directly transduced as a notably different channel modulation [29]. Determining the glycosylation profile of antibodies, and Fc glycosylation in particular, is critical to understanding antibody efficacy and blood circulation half-life, so the nanopore platform and the same signal processing methods for understanding NTD transducers can be directly applied to profiling antibody glycosylation blockade signals where the antibody is treated as an NTD transducer in and of itself. Direct antibody profiling would likely only work for part of the glycosylation (or glycation) profile, however, since the Fab N-terminus neutral glycosylation and glycations would probably still need to be assayed by use of antibody intermediates (as with the standard HbA1c test).

In this section we show results that support the hypothesis that the new laser-tweezer induced modulator motions are due to duplex DNA twist-dominated toggling (in addition to the previously observed conformation-dominated toggling). A proliferation in modulatory modes beyond shifting and twisting is not seen, however, thus the transducer blockade signal is found to be manageable (computationally) as is. The new understanding suggests that a straightforward, generally-applicable, method for transducer construction is possible with twist-mode dominated state tracking for large charge/mass biomolecules and their binding targets (with long-tether constructions), and configuration-switching dominated state tracking for small charge/mass biomolecules and their binding targets (using short-linkage constructions). General applications of the NTD method are described in what follows for trace-level biosensing, particularly for assaying on isomers.

In NTD binding results and testing results on NTD operation under extreme chaotrope conditions and over a wide range buffer pH [14], it was seen how various molecular strain conditions could lead to isomer splitting on the channel-modulatory molecule often used in transducer designs [14]. For experiments with high levels of chaotrope a clearly identifiable isomer splitting could be seen for the DNA hairpin molecules that were often used as distinctive nanopore blockade modulators. This result not only established further evidence of the ability to resolve isomers on the nanopore detector, but due to the special channel modulation role of the DNA hairpins examined, this result also clarified the nature of some of the complex channel blockade classes encountered under other strain conditions. The new, less-stable, channel modulations appear to be due to DNA hairpin conformations with variable loop/stem twist. The modulator's isomer 'twist' states typically have one isomer present under low-strain conditions and a second conformation when the molecule is under significant stress, whether due to a high applied potential, higher chaotrope concentration, higher pH, or large charge/mass torque when binding/transducing larger target molecules. The solution to the channel modulators having too short a melting time, and too much internal conformational freedom, turns out to be the same, to use locked nucleic acid (LNA) nucleosides. Chimeric LNA/DNA-based transducers and modulators are described in the results, along with further results on inducing a modulatory blockade by using a laser-tweezer. By establishing a general procedure for NTD transducer design a number of biosensing applications are made possible.

Constructions are thus indicated for using LNA/DNA chimeric three-way Y-transducers and four-way Holliday junction transducers, all locked with LNA's to the extent necessary, to evoke the desired twist-toggling or config-toggling modulations. The binding moiety of the transducer is typically antibody, aptamer, or annealing based. If working with aptamers or annealing, the entire transducer molecule could derive from nucleic acids. The more complex aptamer transducers are particularly relevant when considering therapeutic use of aptamer methods. Aptamer-based therapeutics have begun to get FDA approval in two settings: (1) dialysis therapy where aptamer-based filters are used to clean a patient's blood of accumulated kidney or liver toxins that are not being cleared due to damage to those organs; and (2) tissue or tumor directed treatments where the aptamer is linked to an antibody (encompassed by the aforementioned quadfunctional case) already known to target and localize to the tissue or tumor of interest. Long-term stable chimeric LNA/DNA transducers are shown for biotin-streptavidin binding detection.

Biotinylated 8bp LNA hairpin binding experiments with streptavidin
The results of the LNA/DNA chimera based NTD transducer/reporter redesign are shown in a series of screen captures of representative blockade events.

The biotinylated 8 base-pair DNA hairpin (DNA 8GC-Bt) has lifetime (until melting and channel translocation event) about 6s on average, with a wide range of observations from a fraction of a second to 15s that is dependent on buffer, and temperature, etc. (consistent with early work on DNA hairpin gauges in the nanopore). The biotinylated 8 base-pair LNA/DNA chimeric hairpin (LNA 8GC-Bt), on the other hand, has lifetime 12 minutes on average, ranging from about 3minutes to over 30 minutes for individual melting times. Similarly, 9 base-pair DNA hairpins have lifetimes going from about one minute with individual lifetimes from 2s to 120s. Compare this with LNA 9GC-Bt lifetimes that are typically greater than 60 minutes, even in 2M urea.

In Fig. 9.18 the nanopore detector software is set to only capture the first 5s of a blockade trace, then perform a polarity reversal to eject the captured analyte and proceed with a new capture.

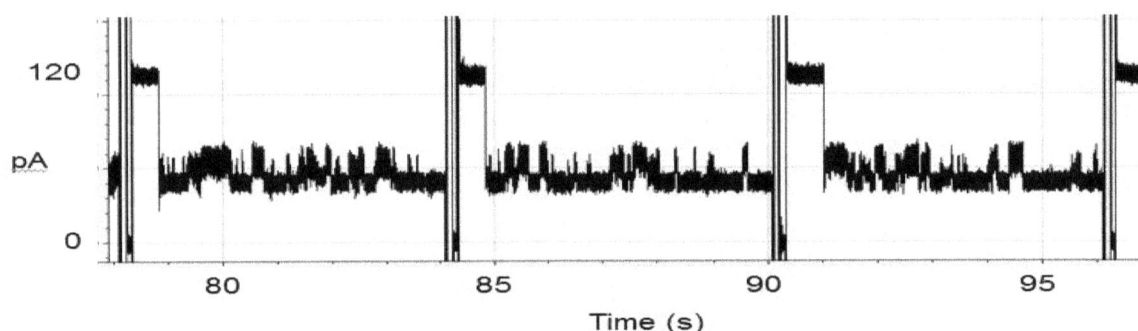

Fig. 9.18. LNA 8GC-Bt blockade signals, 5s blockade before auto-eject (shown as the vertical current reset pulses that occur during the polarity reversal). Very little open channel (less than 1s at 120pA) occurs before the next capture event. The concentration of LNA in the detector well is 2uM. Concentration of 12nM in the detector well produces similar blockades, but with significantly greater (~200s) open channel time between blockade events.

In Fig. 9.19 streptavidin is added in a 1:1 ratio to the LNA 8GC-Bt already present (e.g., the streptavidin concentration in the detector well is 2uM). The timescale is longer, but the hold time for the hairpin blockades is still held at 5s when comparing to Fig. 9.18. The result shown is typical for the first 10 minutes after introduction of streptavidin. The blockade signal structure is unaltered from that shown in Fig. 9.18, it is merely compressed by the larger timescale shown. Note the much longer intervals of open channel even though the LNA concentration hasn't changed. This is due to the streptavidin binding some of the LNA and sequestering it in solution, leaving effectively lower concentration of LNA free to report to the channel detector. The signals produced will continue to change as more LNA is sequestered, and eventually bound streptavidin is pulled to the nanopore detector (to 'report'). Unbound streptavidin is almost never seen to interact with the channel. Streptavidin has pI 7-8, so this was initially thought to be due to it having a possibly positive charge in the pH 8 of the standard experimental buffer setting, but in studies at pH 9 there is still no streptavidin blockade signal even in mM concentrations. Basically, non-glycosylated proteins, even if very negatively charged at pH 8, such as albumin with pI 4.7, will not interact with the channel. Certain proteins are found to strongly interact, however, such as some classes of antibodies (even with pI 8.5 in pH8 buffer), but this is not described further here.

188

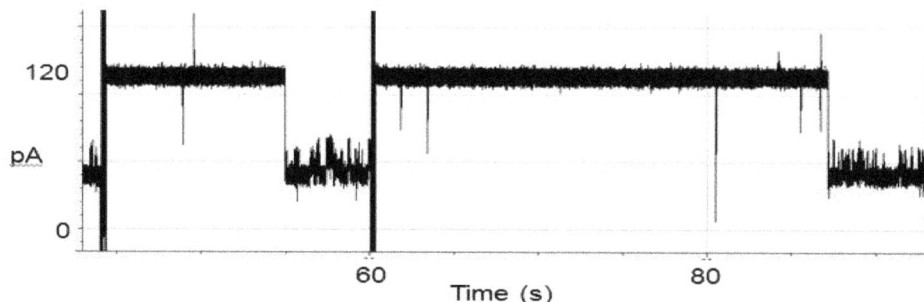

Figure 9.19. LNA 8GC-Bt blockade signals in the presence of streptavidin during the first 10 minutes after introduction of streptavidin. LNA and streptavidin are in a 1:1 ratio, with both at 2uM concentration in the detector well.

After another 10 minutes has passed since the introduction of streptavidin a new class of blockade begins to be seen (Fig. 9.20). The new class does not 'toggle' and is never seen (in runs with over 2000 LNA 8GC-Bt blockades) if streptavidin has not been added. After another 10 minutes has passed (30 minutes since the introduction of streptavidin) the free LNA sequestration is nearly complete (even though 1:1 streptavidin can bind up to 4 biotins). Fig. 9.21 shows one free LNA blockade (in middle), and two bound LNA blockades (one on either side).

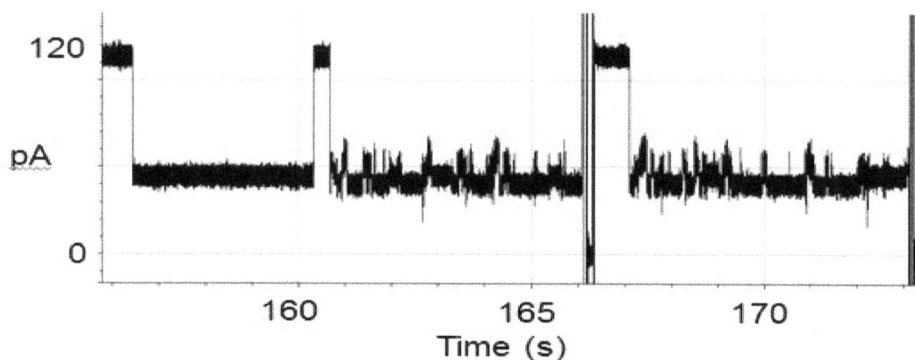

Fig. 9.20. LNA 8GC-Bt blockade signals in the presence of streptavidin during the second 10 minutes after introduction of streptavidin. A bound reporter signal is shown as the leftmost blockade event.

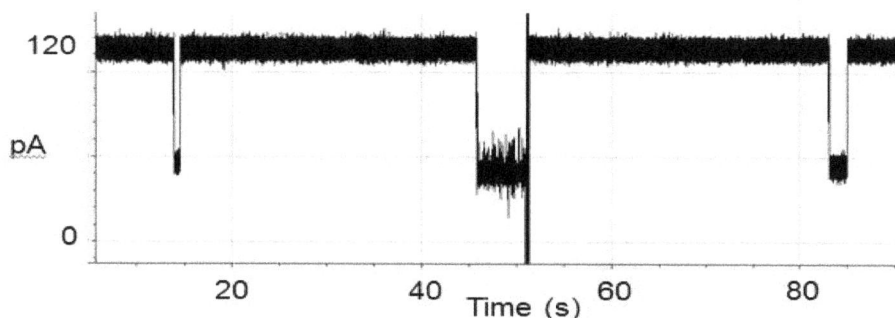

Fig. 9.21. LNA 8GC-Bt blockade signals in the presence of 1:1 streptavidin after about 30 minutes of reaction time. The central blockade is an

189

unbound reporter signal, the much shorter left and right blockades are bound reporter blockades.

After another 10 minutes has passed (roughly 40 minutes since the introduction of streptavidin) the free LNA sequestration is complete, free LNA will now be seen only rarely, with bound signal dominating (Fig. 9.22). Bound signal will now often be captured for sufficiently long that it reaches the 5s auto-eject time. This is likely because the captures will be dominated by streptavidin that is multiply-bound with biotinylated LNAs (providing an even greater pI shift than the singly bound streptavidin, thereby dominating the blockade events seen, and more strongly electrophoretically held at the channel). At later times and at the larger timescales (2.5 minutes shown in Fig. 9.22) 'melted' ssLNA translocation events are seen as short blockade 'spikes'.

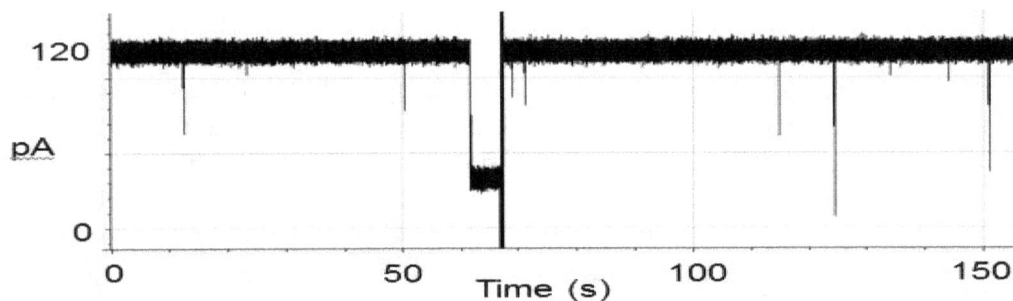

Fig. 9.22. Streptavidin bound LNA 8GC-Bt blockade signals after about 40 minutes of reaction time.

Chapter 10

The extra-element theorem and the single-cell 'bio-voltmeter'

10.1 System Biology at the single-molecule level

The system biologist is currently lacking a general-use method for a gene circuit 'voltmeter' or gene system algorithm 'print statement'. What is needed is a non-destructive, carrier non-modifying, means of testing 'live' biological systems at the single-molecule level. A method using the nanopore transduction detector (NTD) is described for single-molecule characterization in some situations, so may provide what is lacking. An important aspect of this approach is that use can be made of inexpensive antibody, protein, aptamer, duplex nucleic acid, or nucleic acid annealing molecules (for miRNA and viral monitoring) that have *specific* binding to the system component of interest. The NTD transducer's specific binding can also be designed to have low affinity binding as needed, such that there can be a 'catch and release' on low copy-number molecular components, such that there is not a disruption to the molecular system under study. NTD transducers are typically constructed by linking a binding moiety of interest to a nanopore current modulator, where the modulator is designed to be electrophoretically drawn to the channel and partly captured, with its captured end distinctively modulating the flow of ions through the channel. Using inexpensive (commoditized) biomolecular components, such as DNA hairpins, this allows for an easily constructed, versatile, platform for biosensing. High specificity high affinity binding also allows a very versatile platform for assaying at the single molecule level, even down to the single isoform level, including molecular substructure profiling, such as glycosylation profiling in antibodies. An inexpensive commoditized pathway for constructing nanopore transducers is thereby demonstrated. Nanopore transduction detector based reporter/event-transducer molecules may serve as a means to perform multicomponent mRNA-miRNA-protein and protein-protein systems analysis in general settings.

10.1.1 NTD application in programmable nanoblot

The NTD platform could be described as a programmable microarray. In essence, a programmable Southern Blot, Northern Blot, Western Blot, etc., is provided by the NTD given its direct computational coupling. Previous work with introducing PEG into the buffer also reveals strong size-exclusion chromatography fractionation effects, allowing species to be computationally grouped according to their PEG shift measurements [1] then presented as an ordered 'computational gel-separated' list of species (affording gel-separation and blot-identification entirely on the NTD apparatus, when the destructive aspect of adding a bunch of PEG is permissible). A method and system for using the nanopore transduction detector (NTD) is, thus, described for examining the binding and conformation changes of individual biomolecules in a non-destructive manner, and by (destructive) assay methods, involving urea and PEG for example, that provides a general tool for analysis of biomolecular systems.

10.1.2 The NTD sequential sampling constraint
The strengths of the device in terms of single molecule detection are also the weakness in the sense of event detection throughput. The previously mentioned PRI [20] informed sampling can eliminate blocking conditions at the (single) channel detector akin to having a Maxwell Demon for purposes of single-molecule classification and rejection; such that a nearly optimal use of the single-channel's sequential sampling operation can be accomplished, but this only goes so far. An array of nanopore detectors would significantly resolve this problem, and such has been done by other researchers and companies to some extent. It is unclear if the nanopore detectors in an array configuration have the necessary bandwidth for observing channel transduction enhancements, however, so this is still a largely unexplored area.

10.1.3 Stochastic carrier wave event encoding
The strengths of the NTD apparatus with stochastic carrier wave (SCW) event encoding are most evident when trying to have a discussion of noise problems. Transduced events have carrier-waves representations that are easily discerned under high noise conditions just as with any carrier-wave based communication scheme. It's as if an error-correcting encoding scheme is already built-in (that is realized using machine learning methods via an automated HMM feature extraction process and a SVM classification apparatus). Sensitivity and specificity for resolving highly similar control molecules is greater than 99.99% [1].

When the NTD transduction method makes use of transducer molecules that are DNA based and have annealing-based specific binding to a DNA target of interest things begin to sound like quantitative real-time PCR (qPCR). In qPCR the presence of a DNA molecule is revealed via a highly specific DNA probe annealing event where the DNA probe has a fluorophore attached that can be revealed under laser illumination at the appropriate frequency. Fluorophore excitation is a quantum statistical event at the single-fluorophore level, so this analysis is still typically a 'bulk', or aggregate, molecular analysis to some extent (although experiments to have truly single-molecule fluorescence have been performed, such as with FRET). The NTD probe comparison to qPCR probe would be much the same insofar as the annealing section of the probe nucleic acid. Instead of a fluorophore attachment for 'read-out', however, the NTD probe would have a portion that would be favored for channel-capture and for channel modulation. Just as different fluorophores offer multiplex capability, different channel modulators offer resolving capabilities for multiple applications. The NTD event transduction is inherently a single molecule detection event and has 'quantitative' at the single event level, so may offer more detailed evaluation of relative gene expression. The PCR part of qPCR can be co-opted on the NTD platform for enzymatically amplified detection events, for more discussion along these lines see the ELISA-like nanopore detector methods TERISA and TARISA in [1].

10.2 Non-destructive molecular biosystem analysis
The nanopore transduction detector (NTD) offers a means to examine the binding and conformational changes of individual biomolecules in a non-destructive manner that is well-suited to non-destructive analysis of biomolecular systems. The critical choice of transducer

in system biology NTD applications is for one with very high specificity but that is only weakly binding so as not to be disruptive to the biological system or gene circuit. It may also be possible to use the NTD method in live cell assays as well, via use of laser modulations, not for fluorophore excitation, however, but for noise state excitation for use by the NTD where the need to generate a steady channel *current* is avoided in detector operation (which would be destructive to the cell). The NTD method is typically based on a single protein-channel biosensor used with a patch clamp amplifier on a (synthetic cell membrane) lipid bilayer. In the live cell assay the patch clamp application would return to its origins, where it was developed for patch clamp measurements of currents and current gating through channels on live cells. In order for the NTD 'voltmeter' to operate on the biological system to work, however, the normal operational buffer of the NTD must also accommodate a change to the physiological or cellular buffer environment of the biological system of interest. In Ch. 8 we saw robust NTD operation with a variety of buffer pH and in the presence of high concentrations of interference agents [14].

In addition to the study of DNA, DNA-DNA interactions, and DNA-Protein interactions, the nanopore experimental setup has significant potential *vis-à-vis* the study of protein-protein interactions on the single molecule level. DNA-protein and protein-protein interactions are an integral component of gene-regulation and the cellular signaling apparatus. Cell signaling networks, gene regulation, and pathogen-induced genomic or transcriptome modifications, are areas of intense study since they are the basis for many disease states (ranging from metabolic disease, to cancer to autoimmunity). Fundamentally, the scientific benefits to molecular biology and a number of other fields (nanobiotechnology) are significantly impacted if nanopore detection methods can be utilized successfully in the system biology setting.

While cell biological, genetic, and structural biological approaches have contributed significantly to our understanding of signaling networks, we still do not have a clear understanding of the how these networks are regulated because of their inherent complexity. System wide approaches (yeast two-hybrid screens, bioinformatics approaches, for example) have emerged as powerful tools to map topologies of these signaling networks, but, unfortunately, are unable to tell us much about the nature of the links between individual nodes (activities). A complete understanding, therefore, requires that attention be paid to the single-molecule biochemistry and biophysics of the individual interacting species.

The NTD methods proposed are compatible with using the NTD method in live cell assay settings as well, with use of laser modulations for noise state excitation for use by the NTD. The NTD method is typically based on a single protein-channel biosensor implemented on a (synthetic cell membrane) lipid bilayer, but in the live cell assay it would be based on patch clamp measurements of current through a channel on a live cell. Measurement of single channels at the cellular level has been done for 30 years [257], since the development of the patch clamp amplifier (that was originally designed for use in channel studies on single cells, for which the Nobel was obtained in 1991 [258]). The biosensor conformation used in the typical nanopore detector, however, is based on channel current blockades at discernibly different *levels*, which implies that there is at least one current that isn't zero, which is incompatible with using the standard cell patch clamp for channel biosensor applications (the cell would rupture). In the nanopore transducer setting, however, a minimal charge current could be used that could be non-destructive to the cell if periodically reversed, where most of the critical signal information would now reside in the noise profile (where the noise state would be driven by a laser-tweezer tugging at a covalently attached magnetic bead). The key signal analysis method to use in reading the changing noise states has already been developed [1], and involves a collection of machine learning based signal processing methods comprising the stochastic carrier wave (SCW) platform.

10.3 System Level Analysis

A growing number of questions facing molecular and medicinal biology experts are systems biology questions, where the complex interaction of genes, mRNAs, proteins, miRNAs, and various metabolites is described at the 'system level'. System level problems are often described in terms of 'gene circuits' or 'metabolic algorithms'. These comparisons to system descriptions in electrical engineering and computer science offer some insights due to actual parallels, and some misleading comparisons due to oversimplification in comparison to actual biological systems.

A reductionist analysis of a biological system, not surprisingly, reveals that the sum is greater than its parts. But this is actually found to be the case in electrical circuits as well, where emergent properties, especially emergent noise and communications properties, are often found in circuits with feedback. Even simple physical systems involving just three bodies in classical orbital dynamics gives rise to chaotic behavior, which was not expected in early physics, where the sum was originally NOT thought to be greater than its parts. Iterative dynamical systems in general are found to exhibit chaotic behavior and emergent constructs such as strange attractors and limit cycles. Systems with feedback, thus, can do surprising things, and biological systems definitely have done some surprising things ranging from living systems in their amazing variety to complex phenomena such as intelligence, language, and consciousness.

The nanopore transduction detector (NTD) method is typically based on a single protein-channel biosensor implemented on a lipid bilayer (synthetic cell membrane), but it could also be implemented as a live cell assay by using the original patch clamp protocol for measuring current through a channel on a live cell [257]. In order for the NTD 'voltmeter' operating on the biological system to work, cell-based or not, the normal operational buffer of the NTD must accommodate a change to the physiological or cellular buffer environment of the biological system of interest, and, if cell-based, the 'carrier signal' that is the basis of the analysis can no longer be channel-current based, but channel-noise based with use of laser modulations for noise state excitation. Work with robust NTD operation with a variety of buffer pH and in the presence of high concentrations of interference agents reveals that operational stability with a wide range of buffers has been achieved [14]. Laser modulations have also been introduced to improve the NTD mechanism to have more general applicability [36], and for purposes of establishing an improved 'stochastic carrier wave' molecular state tracking capability [1], so many of the complications with returning to the single-cell application are mostly solved. What remains to be resolved for general applicability of the NTD system analyzer method, for both *in vitro* and an possible *in vivo* studies, is a standardized method for NTD transducer construction and operation [9], and progress along these lines will be shown in the Results. An inexpensive method for a NTD-based biological system 'voltmeter' is thus possible for both *in vivo* and *in vitro* applications.

10.4 Electrical/Biological circuits: the biosystem extra element theorem (BEET)

A reductionist analysis of electrical circuits involves a reduction to circuit elements that have linear responses. In this regard biology only compares weakly, as the components of a biological circuit are generally non-linear over much of their operational range. Even so, for some biological system settings sufficiently small perturbations in the biological components can often be made such that they provide a linear system response. Given the complexity of the biological feedback systems, however, this might seem to be small progress. It is very significant, however, given the existence of a sophisticated method from advanced circuit design and analysis that is applicable for linear response systems known as the 'extra element theorem' [259]. It interesting to note that this important circuit method from electrical system theory has not been imported into biological system discussions given its likely significant role in molecular evolutionary theory. The extra element theorem from electrical circuit theory allows simpler circuits, that are more easily understood, to have new components added (the 'extra' element), and if the new component happens to create a feedback loop, then the complexity of the feedback loop analysis can be much more easily evaluated and understood directly by way of the extra element theorem. In practice, very complex electrical amplifier circuits can be built-up and analyzed in this way, by repeated use of the extra element theorem. This offers the means to have a reductionist analysis while capturing the growing complexity of holistic irreducible systems. For a biological variant of the extra element theorem a patchwork of linear response regimes could be used in understanding a particular biological system.

The 'messengers' in biological and electrical systems differ greatly in many respects, which can make some gene circuit intuition entirely misguided. The carriers in an electrical circuit, for example, are remarkably simple by comparison with biological system signal carriers. Electrical charge moves through wires like a fluid. Granted, the electrical charge moves at a sizable fraction of the speed of light, but it is so like a fluid flow that some current flow discussions are basically plumbing discussions, where the description of the current flow is often compared to flow of water through pipes where pipe narrowness is akin to resistance, etc. The flow/interaction topology of electrical current is also self-evident in the connectivity that can be seen in the wiring of the circuit diagram. If the biological system is too interconnected in this comparison this is often where the analogy is shifted to discussions of a gene system *algorithm*. The electrical messengers, or charge carriers, are also vastly simpler than the biological system messengers. Electrical current carriers are of only one type (electrons), and don't have attractive self-interaction molecular carriers (as with dimerization … unless you are talking superconductivity), and don't have internal state (in the sense of the circuit model) like with biological secondary messengers. Biological system messengers, on the other hand, come in a huge variety, operate at the single molecule level, and depending on perspective, everything in the biological system might be considered a system messenger in a massive, living, autocatalytic cascade. The biological system carriers or messengers are also much fewer in number compared to their electrical counterparts. This actually makes things more complicated. In electronics having small currents is modelled as a noise source, where once the discreteness of the charge carriers begins to be discernible this puts one in the realm of stochastic 'shot' noise. In the biological comparison this stochastic underpinning, if significant, again favors a shift to the 'algorithm' analogy instead of the circuit analogy. To further complicate matters, the biological carriers of the system interactions interact with each other, and typically have internal states (e.g., proteins and riboswitches often have conformational states), so the picture of the carriers for biology introduces vastly greater complexity and interaction interconnectivity.

In electrical circuit analysis a good voltmeter is something that will not significantly 'load' or alter the circuit while measuring a particular component's voltage drop. Likewise, in analyzing a computer program, or resolving a runtime error (the closest analogy to analyzing a 'live' biological algorithm), one of the best tools available is to simply introduce a 'print statement' to track any internal state behavior of interest in the program. This is where the weakness of the circuit or algorithm analogy in biological systems is most profound. The system biologist doesn't have a gene circuit voltmeter or gene system algorithm print statement. Some of the closest biochemistry methods to offer such capabilities are fluorescence based, and in certain specialized applications remarkable results have been obtained along these lines, but they typically involve the introduction of constructs with a great deal of effort that won't scale well to the vast number of biological systems that need to be studied in the post-genomic era. What is needed is a non-destructive, carrier non-modifying, means of testing 'live' biological systems, possibly in their native cellular environment.

In the electrical engineering setting the extra element theorem (EET) allows circuits without feedback to be understood in the presence of feedback by choosing the extra element to be the feedback element. In electrical engineering this gives rise to an updated, quantitative, solution. In the stochastic Biosystem Extra-Element Theorem (BEET) setting, feedback complexity can be handled similarly. The BEET method allows a balance to be struck between reductionist and holistic approaches. In this setting it is possible to work with the 'black box' giving rise to the emergent behavior and consider perturbations to that system. BEET also shows how to evolve to gene circuits with more components via a series of small (evolutionary) changes.

Using the NTD-method to perform analysis of "gene circuits" it is, thus, possible to have a 'voltmeter for the circuit' in a circuit analogy. The NTD-quantified gene-circuit analysis can then be enhanced with use of (BEET) method for analysis. In the NTD BEET setting, a collection of NTD reporter molecules with specific binding to different molecules can be used to perform multiplex analysis of the system molecular profile by differentiating the reporter molecules according to their different channel modulation signals. The NTD BEET system could also employ multiple component modulation, and molecular knock-outs (by having strong binding) to effect double null injection to the equivalent gene circuit for a variety of extra element theorem testing procedures akin to their electrical engineering counterparts.

Preliminary work examining TBP binding to TATA binding site sequences placed in one arm of the Y-transducer construct [29] suggest a similar construct could be employed for purposes of miRNA binding site validation. The Y-transducer for miRNA binding site profiling on mRNAs would take the hypothesized sequence of the miRNA binding region, typically from the mRNA's 3'UTR region, and incorporate it either into one arm of a Y-transducer, or incorporate it such that it crosses the Y-nexus, the latter case potentially offering the greatest sensitivity to binding events, as was seen in the Y-SNP construct described previously. The latter case may not allow sufficient steric freedom for miRNA binding, however, when complexed with argonaute protein, so the arm variant may still be necessary for analysis of some miRNAs. This approach to miRNA target validation also benefits from validation at the actual annealing step of the interaction, thereby accounting for possible modification to the miRNA such as may occur with adenosine deaminases, where adenosine deaminases that act on RNA catalyze the conversion of adenosine to inosine residues in some double-stranded RNA substrates. A subset of miRNAs have been found to have modulated processing efficiency when deaminated at particular residues [260], and this is now thought to impact a significant fraction of miRNAs.

The RNAi probe examination could also be reversed, where the miRNA is sought that is associated with a suspected miRNA binding site (such as when the 3'UTR motif has an anomalous rate of occurrence and is shared across homologous genes in multiple organisms). Software to perform the aforementioned motif analysis has been developed for when only (pre-genomic) EST data is available, but this discussion is outside the scope of this paper so won't be discussed further.

From a system biology perspective, the port-oriented Extra Element theorem (EET) circuit analysis method [259] is interesting because it describes how a circuit can be built one element at a time (the "extra element"), where the element can be anywhere in the circuit system. This is done, via EET approach, in a way where the circuit is fully analyzed at each step insofar as gain and input impedance, etc., and thereby easily optimized…. before then adding another element. What EET allows is highly complex systems to be understood in terms of simpler versions of themselves that are subsequently built out to the fully complex system via repeated use of EET to add elements. This might seem conceivable in situations not changing the topology or connectivity of the circuit, but you might not believe that such a method could work for an extra element that, say, completed a feedback pathway on an amplifier circuit…. but it does, as will be shown in App. B.

Earlier versions of EET methodology can trace back to port-oriented analysis in [265] but it wasn't until Middlebrook's description of EET [259] that we see the general extension of the method to any *linear* system (so linear operational regimes of biosystems as well), whether with passive elements (resistors, etc.) or active elements (with gain factors, like transistors). Operating in the linear regime for an amplifier circuit is often a huge part of the problem. To even get into the non-saturated, non-oscillatory, gain regime of the amplifier can be difficult (akin to "lock range" from PLL). Once operating in the linear regime, however, the full power of the EET methodology can be brought to bear to understand the circuit (and, thus, be able to optimize its operation), or add to it (via further iterations of EET) for further functionalization.

The utility (or novelty) in biosystem analysis is that EET not only allows an understanding to be built from simpler versions of the biosystem, but also might indicate how an evolutionary process could have arrived at a more complex system by selection optimization of an extra added biological element. Further details and an example of how to use the EET method in an amplifier is shown in App. B.

10.5 Validation of miRNA's and miRNA binding sites via NTD

The discovery of the RNA interference (RNAi) immune response and translational regulation mechanism has led to an explosion in the number of identified microRNAs (miRNAs) and their mRNA binding sites. An understanding of miRNAs and their binding sites, typically in the 3' untranslated region (3' UTRs) of mRNAs, is helping to explain a wide range of complex phenomena, ranging from latency control by viruses during infection (such as with HIV) [261], to complex regulation in system syndromes such as in diabetes and in the effects of aging [262], to the general trans-regulation of mRNAs at the translational level (complementing transcription factor and promoter cis-regulation at the transcriptional level) [263]. The examination of miRNAs, and especially miRNA binding sites, is confounded by the small size of the miRNAs, however: 21-25 nucleotides in length for typical mature miRNAs, and only 7-8 base ssRNA seed regions in the guide-strand RNA incorporated into the RNAi's RISC complex for actual binding/repression to complementary 7-8 base sequence in the 3'UTR region of the target mRNA [264]. For the latter case of verification for miRNA/RISC derived sequence binding with a 7 base sequence in a mRNA's 3'UTR there is further complication given possible posttranscriptional modifications, such as via inosine substitution for adenosine due to adenosine deaminases with inosine recognition as guanine in terms of base-pairing that can alter the actual target sequence of the miRNA/RISC binding [260]. This is in addition to the obvious complication of identifying the presence of RNA annealing when the annealing only involves 7 bases of RNA.

Preliminary work with NTD-based detection on short DNA annealing suggests a possible means to examine the miRNA/RISC binding to target 3'UTR region with or without the RISC complexes argonaute proteins intact, where results are expected to improve even more upon refinement using locked nucleic acid transducer/reporter probes. NTD based detection of DNA annealing has been demonstrated on DNA sequences as short as 5 bases [1], and in the presence of a variety of interference agents and chaotropes [14]. NTD based detection has also been demonstrated in a variety of buffer conditions so could be established in a buffer conducive to the RISC complex remaining intact and where the annealing to 3'UTR complement sequence occurs with the binding strength found *in vivo*. NTD detection can also operate on small volumes since it makes use of a *single* protein channel interaction, thereby inherently operating at the single-molecule interaction level. NTD detection can, thus, identify single-molecule binding events in a non-destructive manner that may be conducive to the 'live' characterization of many critical, transient, interactions.

For biosensing or bioassays applications in general, not all miRNA or miRNA binding site analyses need be in cellular or physiological buffer either. In a 'destructive setting' more forceful miRNA validation assays, and analysis of annealing-based events, can be pursued by use of chaotropes such as urea. Clearer identification of collective binding events, such as for highly complementary annealing interactions, is found to occur upon introduction of chaotropes that eliminate non-specific DNA interactions, and many 'simple' binding interactions, not involving collective interactions of many components as with annealing [27].

10.6 Transducers designed for biosensing or non-destructive bEET analysis

LNA/DNA chimeras are shown to allow a much more robust long-lived NTD reporter molecule. The engineered NTD transducer/reporter molecule, minimally, has two functions, specific-binding and channel-modulation, and in the general setting, a third function to receive excitations such that channel modulation can be induced for all states of the transducer whether bound or not. Results shown here introduce excitations using a magnetic bead attachment in the presence of a laser-tweezer pulsing. A simple NTD transducer design via LNA/DNA chimeras or via mAb selection is also described. Operability of the NTD platform over a wide range of chaotrope concentration and inexpensive LNA/DNA transducer design allows simple nucleic acid testing via the NTD platform for purposes of miRNA profiling or viral RNA monitoring. For protein or nondestructive system type monitoring, the LNA/DNA transducer chimeras offer the necessary long-lived reporting capability needed to have a biological gene circuit 'voltmeter'.

Appendix A

The Nanoscope Kit

The NTD Kit could provide the means for a powerful biosensing platform for DNA (or RNA) detection, DNA aptamer-based detection, or detection via DNA-linked binding moieties (where the DNA part is designed to be drawn into the detector and establish a modulatory channel blockade signal). The NTD Kit also enables a variety of assaying protocols to be performed.

By periodically reversing the applied potential, hundreds of molecules can be sampled every second (Fig. A.1, Left). Standard Data acquisition hardware is then used to take the analog signal from the patch clamp amplifier and deliver it to a computer (Fig. A.1, Right).

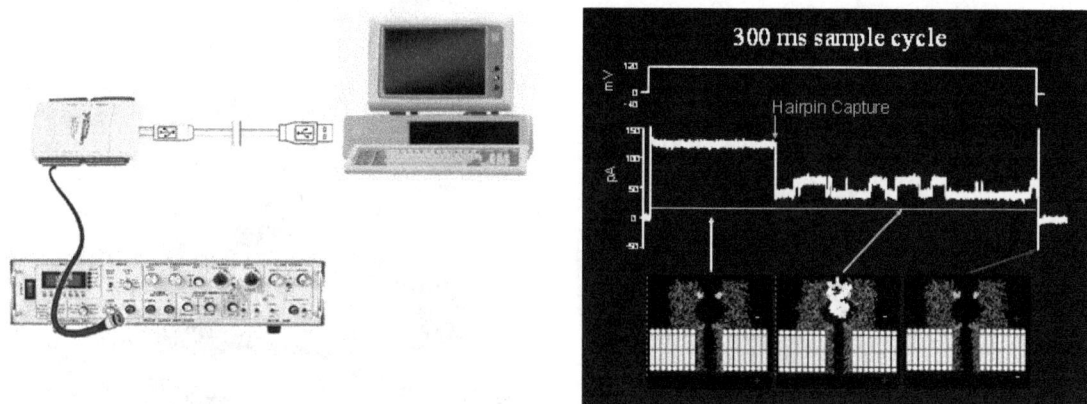

Fig. A.1 Sampling Protocol (left) and DAQ hardware (right).

NTD KIT Components and Software
Kit Package:
(1) NTD Kit Device Components and chemicals
(2) NTD Set-up Manual
(3) configured NTD Kit Computer with local SSA software
(4) buffer controls
(5) model systems/controls
(6) Data analysis and data repository services
(7) Setup, calibration, and troubleshooting services

A.1 Kit Components
In Fig. 5 is shown the NTD kit core Assembly with Peltier-based temperature controller model 350B manufactured by Newport Temperature. The A-M systems patch clamp amplifier model 2400 is used to impose a constant voltage across electrodes while monitoring picoampere fluctuations. A micro manipulator is used to deliver profusion tubing to *cis* chamber during channel acquisition. A TMC bench top vibration isolation table is used to support the core assembly and reduce vibration shock. With the NTD Kit we, thus, achieve a nanoscale single-molecule lab construction. This is obviously a low-cost method to perform nanoscale experiments. There is also a remarkable versatility in the incorporation of transducer, as will become more apparent in what follows.

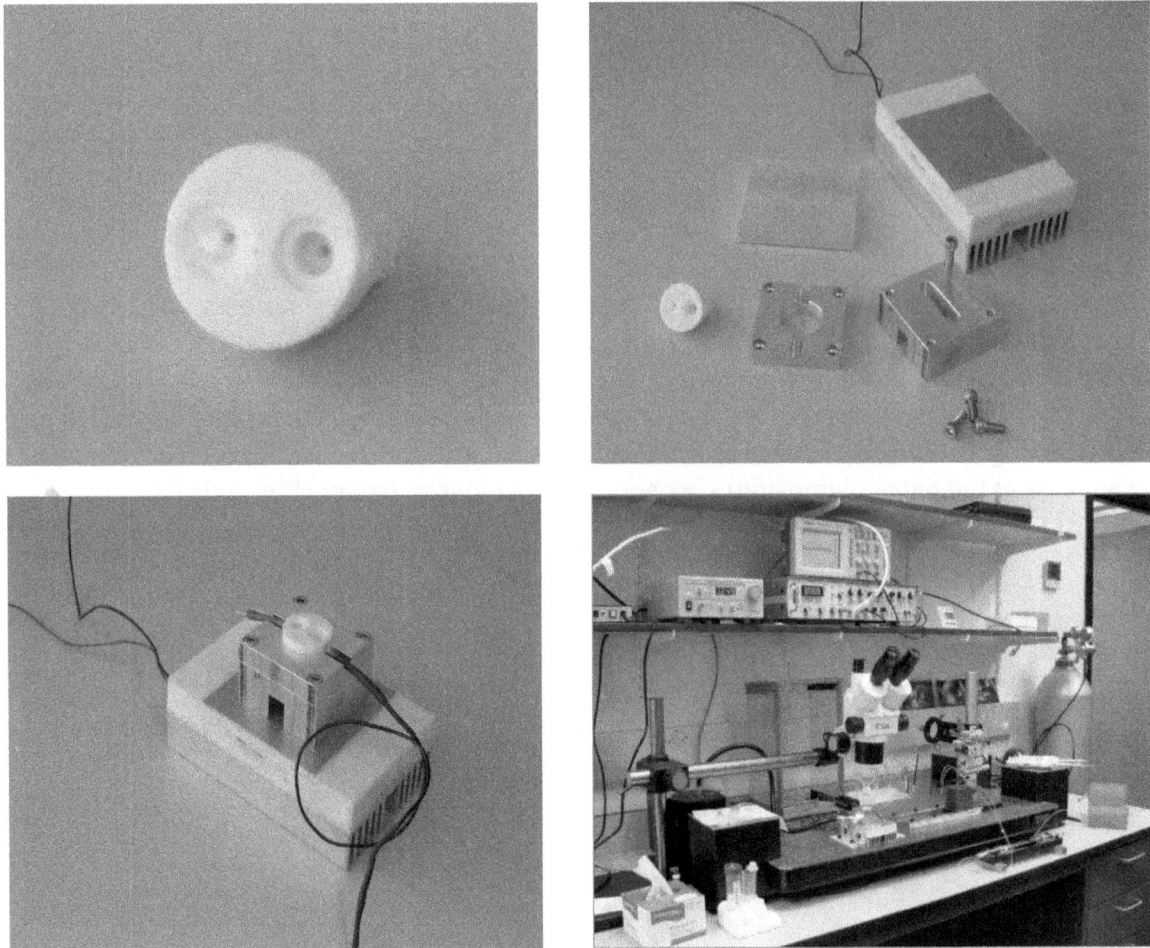

Fig. A.2. The NTD Kit Components. (Upper Left) One inch diameter Teflon core machined to contain two chambers that are connected by a segment of Teflon tubing. This tubing has one end open while the other is capped to restrict it's opening to a tiny pin hole (Aperture) approximately 25 microns in diameter (not shown). (Upper Right) Aluminum square block is machined to support the core and provide access for illumination and electrode ports, and have easy assembly and mounting on the Peltier device. (Lower Left) The Teflon and aluminum components are shown seated on the Peltier device for temperature control of the wells. Silver chloride (Ag/AgCl) pellet electrodes are shown inserted into the cis and trans chambers port wells. (Lower Right) A NTD Nanoscope kit arrangement that uses a table-top vibration isolation table, has perfusion apparatus, Faraday cage, and linkage to a computer (to left of workstation, not shown).

A.2 The SSA Toolset

Although the nanopore transduction detector can be a self-contained 'device' in a lab, external information can be used, for example, to update and broaden the operational information on control molecules ('carrier references'). For the general NTD 'kit' user, carrier reference signals and other systemically-engineered constructs can be used, for example, for a wide range of thin-client arrangements (where they typically have minimal local computational resource). The paradigm for both nanopore device and NTD kit implementations involve system-oriented interactions, where the kit implementation may operate on more of a data service/data repository level and thus need 'real-time' (high bandwidth) system processing of data-service requests or data-analysis requests. Although not as system-dependent on database-server linkages, the more self-contained 'device' implementation will still typically have, for example, local networked (parallelized) data-warehousing, and fast-access, for distributed processing speedup on real-time experimental operations.

For the general-user and the NTD-Kit user, the server interfaces for the SSA Server facilities are being designed to provide services ranging from web-interface based services to an active, streaming, experimental calibration/troubleshooting effort. Nanopore Device and Nanopore Kit users will have the signal acquisition module, the tFSA, bundled with their hardware (on a preconfigured computer, see Table A.1). This allows for bare-bones functionality with the bundling as a stand-alone device. When there is need for more refined signal processing, such as when using the weakness recovery protocol to acquire signal when the FSA fails, the SSA Server will have the full panoply of methods available to 'get a lock' on the signal, and examine it further. The transfer of data at this juncture will also be efficiently pre-processed (e.g., the FSA acquisition will already done) and will thereby allow for much lower bandwidth and latency in the networked signal processing. Biology has invented such structure repeatedly, such as with the retinal pre-processing in the human eye.

KIT-user on-site Software Toolset	General-user Web-Portal Toolset
FSA	FSA, FSA-tuning
HMM*	HMM, HMMD, etc.
SVM*	generalized SVM
	Tuning Metaheuristics
	NTD Calibration & Troubleshooting Services
	CCC Shared Reference Signal Library
(small data for HMM and SVM, no distributed processing)	(large datasets for HMM and SVM training, with use of distributed processing)

Table A.1. Software Toolsets. *specialized enhancement packages to be made available as needed.

The nanopore transduction detector is a powerful new tool for single molecule detection. Automation of the detector, with pattern recognition informed feedback, promises to greatly enhance the potential of this device technology. In prototype experiments a kit appears to be possible, with certain components bought separately (vibration isolation table; patch clamp amplifier, etc) for as little as $10,000 in total. Two elements are missing in a practical deployment of this very new technology: (1) model test systems – in response to this test buffers and test molecules will be used with the kit as test controls along the lines of the successful experiments described in [1]; and (2) software with both local acquisition tools (see [1]) and off-site server or web-interfaced tool-sets (see [1]), the latter with more elaborate data analysis methods and services can be made available via server-linked signal processing.

Appendix B

Middlebrook's extra element theorem

Most Electrical Engineering (EE) programs, even at graduate level, describe how to perform circuit analysis and design where the circuits are analyzed according to the standard reduction to a set of coupled equations that must be solved. This approach can become overwhelming in complex situations (amplifiers with feedback, for example) and often group the variables in a way that is not informative in design decisions. In the background given in App. B.1 and the example application to analysis of an amplifier with feedback in App. B.2, the advantages of port-oriented methods of circuit analysis become apparent. There are the input and output ports of an amplifier, for example, or a port created anywhere internally in the circuit. The culmination of the port-oriented circuit analysis methodology is Middlebrook's Extra Element theorem (EET) [259]. (In the Preface this is the same Middlebrook that was my EE Advisor. I was also his TA for precisely this course, on port-oriented style circuit analysis, for two years while an undergraduate).

B.1 General Background for the Extra Element Theorem

Port-oriented circuit analysis was in use by advanced circuit designers, for certain applications, as far back as the 1960's. What Middlebrook realized is that the circuit could be viewed in terms of a complex-valued transfer function between any ports of interest, and that any modification on the linear system would (by complex mapping arguments) have as added transfer function factor a simple bilinear transform term. The bilinear transfer term has two key parameters, and they relate to the circuit properties at the position where the extra element is being placed. What Middlebrook then figured out was a derivation of these two terms for the bilinear transform. A form of this derivation follows from a simple Thevenin-equivalent type analysis of the simple circuit diagram shown in Fig. B.1.

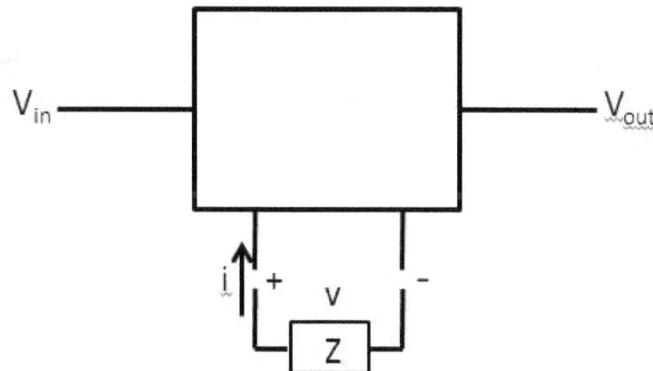

Fig. B.1. Port-oriented circuit analysis at level of Thevenin equivalent.

The port where we are adding an 'extra element' is shown with extra element Z and with the voltage v and current I as labeled. The port v-I relationship is simply the classic: $v = -Zi$, except care is taken to have sign convention negative since the standard port convention is also employed (with current going in).

Without Z and the port modification, the input voltage V_{in} is related to the output voltage V_{out} with gain given by:

$$V_{out} = G_{oc}*V_{in}+T_i*i$$
$$V=T_v*V_{in}+Z_D*i$$

Where,

$G_{oc} = V_{out}/V_{in}$ (gain for open circuit, e.g., i=0)

$Z_D = v/I$ (V_{in}=0, Z-port driving point impedance)

$T_i = V_{out}/I$ (when V_{in}=0)

$T_v = v/V_{in}$ (when i=0)

Grouping the port equations we get:

$V_{out}/V_{in} = G_{oc} -T_i *T_v /(Z+Z_D)$ (gain for circuit in general, e.g.,' i' need not =0)

At this point it seems we've gotten nowhere if we can't eliminate T_i and T_v, but this is where the port analysis has more to offer to complete the methodology. Consider that with the port methodology we can also subject the port to various conditions (conditions very useful and accessible to analysis).

Consider the condition that the output voltage is 'nulled' (adjusted to zero) by some choice of impedance at the Z-port (i.e., nulling doesn't mean zeroing out by shorting or opening the circuit, but by adjusting Z-port parameters such that a 'nullification' cancellation to zero occurs on V_{out}):

$0 = G_{oc}*V_{in}+T_i*i$

Using the other port equation in this circumstance then allows 'v/I' to be computed under nulling conditions, a value we will refer to as Z_N (N for nulling impedance):

$Z_N = v/i=Z_D-Ti*Tv/G_{oc}$ when V_{out}=0 via nulling.

Thus,

$Ti*Tv=G_{oc} *(Z_D-Z_N)$.

So, now we can eliminate T_i and T_v, since the multiplicative grouping, $T_i *T_v$, is precisely what is needed, and we get:

$V_{out}/V_{in} = G_{oc} [(1+Z_N /Z)/(1+Z_D /Z)]$.

B.2 Example Application of the Extra Element Theorem

This is a problem assigned by Prof. Middlebrook in his EE114 class in 1985: Consider a single-stage bias-stabilized transistor amplifier with unbypassed emitter resistance 'R' shown in Fig. B.2 below:

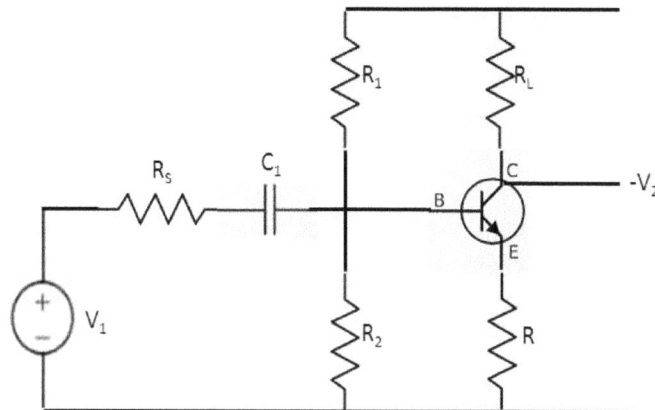

Fig. B.2. A single-stage bias-stabilized transistor amplifier.

The reduced small-signal midband model is shown in Fig. B.3:

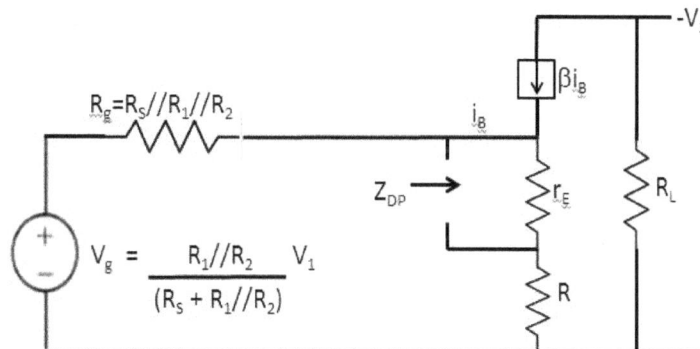

Fig. B.3 The reduced small-signal midband model.

In the analysis of the circuit we first get the midband gain given by V_2/V_1 from the mid-band model in Fig. B.3.

$A_m = V_2/V_1 = (\alpha\ i_E\ R_L)/V_1$;

$i_E[(r_E+R)+R_g/(1+\beta)\ ] = V_g = V_1(R_1//R_2)/(R_S+R_1//R_2)$;

note: $(R_1//R_2)=(R_1*R_2)/(R_1+R_2)$; $R_g=R_S//(R_1//R_2)$.

Solving:

$A_m = (R_1//R_2)/(R_S+R_1//R_2) * (\alpha\ R_L)/ [(r_E+R)+[R_S//(R_1//R_2)]/(1+\beta)\ ]$

The problem with the circuit is that at high frequencies, the gain declines due to the reactive effects of the diffusion capacitance (C_D) inside the transistor. In order to account for this we redo the standard analysis but with the capacitor as the extra element to go at the port indicated in Fig. B.3.

So now consider applying the EET:

$V_{out}/V_{in} = A_m [(1+Z_N/Z)/(1+Z_D/Z)]$, where $Z=1/sC_D$

First compute the driving-point impedance when output is nulled ($V_2=0$):

$Z_n = Z_{DP}|_{V2=0}$.

The null propagates: $V_2=0$ means no current thru R_L thus $\beta i_B = i_C = 0$, thus $i_B = 0$. So, we are driving the system at the point where the extra element will be placed so as to get a null at V_2, which means that

$V_{rE} = (1+\beta) i_B r_E = 0$, thus there is a virtual short at the driving-point and $Z_n = 0$.

Now let's compute $Z_d = Z_{DP}|_{V1=0}$, for this refer to Fig. B.4:

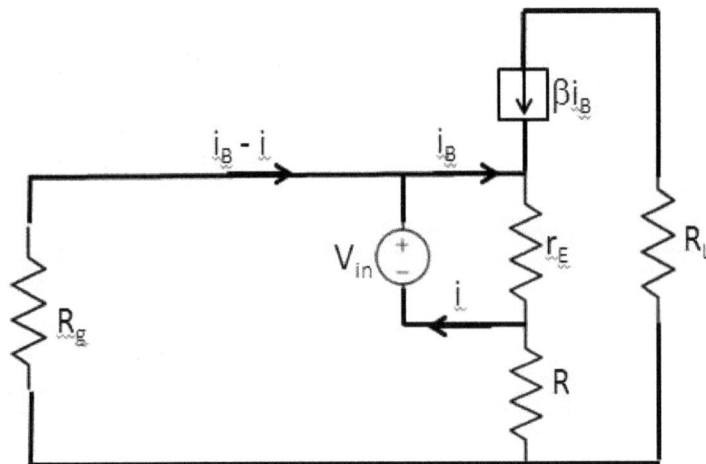

Fig. B.4. Circuit to compute $Z_d = Z_{DP}|_{V1=0}$.

$V_{in} = (1+\beta) i_B r_E = -[(1+\beta) i_B - i_{in}]R - (i_B - i_{in})R_g = -(1+\beta)R i_B - i_{in}R_g$, where $R_g = R_S//R_1//R_2$.

Thus,

$Z_d = (1+\beta) r_E (R+R_g)/[(1+\beta)(R+r_E)+R_g]$

Using the values for Z_d and Z_n :

$A=V_{out}/V_{in} = A_m [(1+Z_N/Z)/(1+Z_D/Z)]$, where $Z=1/sC_D$

$A = A_m / [1 + sC_D \, r_E \, (R+R_g)/(\, (r_E+R)+R_g/(1+\beta) \,)] = A_m / [1 + (s/\omega_\beta)(R+R_g)/((1+\beta)(r_E+R)+R_g)]$

Where we've introduced the variable $\omega_\beta = 1/(\, C_D \, r_E \, (1+\beta))$, known as the "$\beta$ cutoff frequency".

At this point, the design objective is to manage the corner frequencies introduced by the reactive extra element, and then see if there would be some benefit to bypassing R given the results (there will be) and it's a design process such as this that is employed by a circuit analyst. In the port-based methods this is much clearer than in the traditional approaches, as will soon be clear.

There is only one corner frequency, a pole:

$\omega_1 = \omega_\beta \, [((1+\beta)(r_E+R)+R_g) / (R+R_g)]$.

This can be simplified by approximating $(1+\beta)(r_E+R)$ as $(1+\beta)R$:

$\omega_1 = \omega_\beta \, [1 + \beta R / (R+R_g)]$.

So, the overall influence of increasing R_g on the corner frequency, ω_1, is to lower the corner frequency. For $R_g \gg \beta R$, have $\omega_1 \cong \omega_\beta$. The overall influence of decreasing R_g on the corner frequency, on the other hand, is to increase ω_1. For $R_g \to 0$, $\omega_1 \cong 1/(\, C_D \, r_E \,)$, which is independent of β, thus having negligible variance with temperature (highly desirable).

Now consider bypassing R (effectively, setting R=0):

$\omega_1|_{\text{bypassed}} \cong \omega_\beta + 1/(\, R_g \, C_D \,)$, which offers great advantage in that we can control ω_1 to fall between such extremes as ω_β when R_g is large and $1/(\, R_g \, C_D \,)$ when R_g is small. Again, notice that R_g chosen small means $\omega_1|_{\text{bypassed}} \cong 1/(\, R_g \, C_D \,)$ and there is no ω_β dependence…so the corner is pushed out to large value and is temperature stable, all to the good.

References

1. Winters-Hilt, S. "Machine-Learning based sequence analysis, bioinformatics & nanopore transduction detection." ISBN: 978-1-257-64525-1. (2011).

2. Winters-Hilt, S. Unified propagator theory and a non-experimental derivation for the fine-structure constant. Advanced Studies in Theoretical Physics, Vol. 12, 2018, no. 5, 243-255. https://doi.org/10.12988/astp.2018.8626.

3. Winters-Hilt, S. The 22 letters of reality: chiral bisedenion properties for maximal information propagation. Advanced Studies in Theoretical Physics, Vol. 12, 2018, no. 7, 301-318. https://doi.org/10.12988/astp.2018.8832.

4. Winters-Hilt, S. RNA-dependent RNA polymerase encoding artifacts in eukaryotic transcriptomes. Int. J. Mol. Genet Gene Ther 2(1), 2017: doi http://dx.doi.org/10.16966/2471-4968.108.

5. Winters-Hilt S, Evanilla J (2017) Characterization of Fish Stock Diversity via EST Based miRNA Trans-Regulation Profiling. Int J Mol Genet Gene Ther 3(1): doi http://dx.doi.org/10.16966/2471-4968.110.

6. Winters-Hilt, S. Distributed SVM Learning and Support Vector Reduction. International Journal of Computing and Optimization, Vol. 4, 2017, no. 1, 91 – 114.

7. Winters-Hilt, S. Clustering via Support Vector Machine boosting with simulated annealing. International Journal of Computing and Optimization, Vol. 4, 2017, no. 1, 53 – 89.

8. Winters-Hilt, S. Finite State Automaton based signal acquisition with Bootstrap Learning. International Journal of Computing and Optimization, Vol. 4, 2017, no. 1, 159 – 186.

9. Winters-Hilt, S. Nanopore Transducer Engineering and Design. Int J MolBiol Med 2(1): doi http://dx.doi.org/10.16966/ijmbm.108 (2017).

10. Winters-Hilt, S. Biological System Analysis Using a Nanopore Transduction Detector: from miRNA Validation, to Viral Monitoring, to Gene Circuit Feedback Studies. Advanced Studies in Medical Sciences, 5(1), 13 – 53. doi.org/10.12988/asms.2017.722 (2017).

11. Winters-Hilt, S. Isomer-Specific Trace-Level Biosensing Using a Nanopore Transduction Detector. Clinical and Experimental Medical Sciences, 5(1), 35-66. doi.org/10.12988 cems.2017.722 (2017).

12. Winters-Hilt, S. and A. Lewis. Alt-Splice Gene Predictor Using Multitrack-Clique Analysis: Verification of Statistical Support for Modelling in Genomes of Multicellular Eukaryotes. Informatics 2017, 4, 3; doi:10.3390/informatics4010003 (2017).

13. Winters-Hilt, S. Exploring protein conformation-binding relationships and antibody glyco-profiles using a nanopore transduction detector. Molecules & Medicinal Chemistry 2016; 2: e1378. doi: 10.14800/mmc.1378 (2016).

14. Winters-Hilt, S. and A. Stoyanov. Nanopore Event-Transduction Signal Stabilization for Wide pH Range under Extreme Chaotrope Conditions. Molecules 2016, 21(3), 346 (2016).

15. Winters-Hilt, S. Feynman-Cayley Path Integrals select Chiral Bi-Sedenions with 10-dimensional space-time propagation, Advanced Studies in Theoretical Physics, Vol. 9, 2015, no. 14, 667-683 (2015).

16. Winters-Hilt, S. Channel current cheminformatics and stochastic carrier-wave signal processing. International Journal of Computing and Optimization, Vol. 4, 2017, no. 1, 115 – 157.

17. Winters-Hilt, S., E. Horton-Chao, and E. Morales. The NTD Nanoscope: potential applications and implementations. BMC Bioinformatics; 12 (Suppl 10): S21 (2011).

18. Winters-Hilt, S. and C. Baribault. A Meta-state HMM with application to gene structure identification in eukaryotes. EURASIP Journal of Advances in Signal Processing, Special Issue on Genomic Signal Processing (2010).

19. Winters-Hilt, S., Jiang, Z., and C. Baribault. Hidden Markov model with duration side-information for novel HMMD derivation, with application to eukaryotic gene finding EURASIP Journal of Advances in Signal Processing, Special Issue on Genomic Signal Processing, 2010.

20. Eren AM, Amin I, Alba A, Morales E, Stoyanov A, and Winters-Hilt S. Pattern Recognition Informed Feedback for Nanopore Detector Cheminformatics. Accepted paper in book "Advances in Computational Biology", to be published by Springer in Advances in Experimental Medicine and Biology, AEMB 2010 book series.

21. Winters-Hilt S and Jiang Z. A hidden Markov model with binned duration algorithm. IEEE Trans. on Sig. Proc., Vol. 58 (2), Feb. 2010.

22. Winters-Hilt S. Nanopore Cheminformatics based Studies of Individual Molecular Interactions. Ch. 19 in Yanqing Zhang and Jagath C. Rajapakse, editors, Machine Learning in Bioinformatics, John Wiley & Sons, 2009.

23. Alexander Churbanov, Stephen Winters-Hilt, Eugene V Koonin and Igor B Rogozin. Accumulation of GC donor splice signals in mammals. Biology Direct 2008, 3:30.

24. Churbanov, Alexander and S. Winters-Hilt. Implementing EM and Viterbi algorithms for Hidden Markov Model in linear memory. BMC Bioinformatics 2008, 9:228.

25. Churbanov A and Winters-Hilt S. Clustering ionic flow blockade toggles with a Mixture of HMMs. BMC Bioinf. 9 S9, S13 (2008).

26. Roux B and Winters-Hilt S. Hybrid SVM/MM Structural Sensors for Stochastic Sequential Data. BMC Bioinf. 9 S9, S12 (2008).

27. Winters-Hilt S. The alpha-Hemolysin Nanopore Transduction Detector -- single-molecule binding studies and immunological screening of antibodies and aptamers. BMC Bioinf. 8 S7, S12 (2007).

28. Thomson K, Amin I, Morales E, and Winters-Hilt S. Preliminary Nanopore Cheminformatics Analysis of Aptamer-Target Binding Strength. BMC Bioinf. 8 S7, S14 (2007).

29. Winters-Hilt S., Davis, A, Amin, I, and Morales E. The Nanopore Cheminformatics of Individual Transcription Factor Binding Site Interactions. BMC Bioinf. 8 S7, S10 (2007).

30. Winters-Hilt S, Morales E, Amin, I., and Stoyanov, A. Nanopore Cheminformatics Analysis of Single Antibody-Channel Interactions and Antibody-Antigen Binding. BMC Bioinf. 8 S7, S20 (2007).

31. Winters-Hilt S and Merat S. SVM Clustering. BMC Bioinf. 8 S7, S18 (2007).

32. Churbanov A, Baribault C, Winters-Hilt S. Duration learning for nanopore ionic flow blockade analysis. BMC Bioinf. 8 S7, S14 (2007).

33. Landry M, Winters-Hilt S. Analysis of nanopore detector measurements using machine learning methods, with application to single-molecule kinetic analysis. BMC Bioinf. 8 S7, S12 (2007).

34. Winters-Hilt, S and C Baribault. A novel, fast, HMM-with-Duration implementation – for application with a new, pattern recognition informed, nanopore detector. BMC Bioinf. 8 S7, S19 (2007).

35. Winters-Hilt, S., Landry M, Akeson M, Tanase M, Amin I, Coombs A, Morales E, Millet J, Baribault C, and Sendamangalam S. Cheminformatics Methods for Novel Nanopore analysis of HIV DNA termini. BMC Bioinformatics 2006, Sept. 26, 7 S2: S22.

36. Winters-Hilt, S: Nanopore Detector based analysis of single-molecule conformational kinetics and binding interactions. BMC Bioinformatics 2006, Sept. 26, 7 S2: S21.

37. Winters-Hilt S, Yelundur A, McChesney C, Landry M: Support Vector Machine Implementations for Classification & Clustering. BMC Bioinformatics 2006, Sept. 26, 7 S2: S4.

38. Iqbal R, Landry M, Winters-Hilt S: DNA Molecule Classification Using Feature Primitives. BMC Bioinformatics 2006, Sept. 26, 7 S2: S15.

39. Winters-Hilt S: Hidden Markov Model Variants and their Application. BMC Bioinformatics 2006, 7 S2: S14.

40. Deamer, David W. and S. Winters-Hilt, "Nanopore analysis of DNA." Encyclopedia of Nanoscience and Nanotechnology, Ed. H. S. Nalwa. 2005. Vol. 7, pgs 229-235.

41. Winters-Hilt, S., "Single-molecule Biochemical Analysis Using Channel Current Cheminformatics," *Fourth International Conference on Unsolved Problems of Noise and Fluctuations, June 6–10, 2005.*

42. Winters-Hilt, S. and M. Akeson, "Nanopore cheminformatics," *DNA and Cell Biology, Vol. 23 (10), Oct. 2004.*

43. Winters-Hilt, S., "Nanopore detection using channel current cheminformatics," *SPIE Second International Symposium on Fluctuations and Noise, 25-28 May, 2004.*

44. Winters-Hilt S, "Highly Accurate Real-Time Classification of Channel-Captured DNA Termini," *Third International Conference on Unsolved Problems of Noise and Fluctuations, 2003.*

45. Winters-Hilt, S., W. Vercoutere, V. S. DeGuzman, D. Deamer, M. Akeson, and D. Haussler, "Highly Accurate Classification of Watson-Crick Base-Pairs on Termini of Single DNA Molecules," *Biophys. J.* Vol. 84, pg 967, 2003.

46. DeGuzman V, Winters-Hilt S, Solbrig A, Sughrue W, Deamer D, Haussler D, Akeson M. Sequence-dependent fraying of single DNA molecules measured in real time at 5 angstrom resolution using an ion channel. *Biophys J.* 2003 84(2):490A-490A Part 2 Suppl.

47. W. Vercoutere, S. Winters-Hilt, V. S. DeGuzman, D. Deamer, S. Ridino, J. T. Rogers, H. E. Olsen, A. Marziali, and M. Akeson, "Discrimination Among Individual Watson-Crick Base-Pairs at the Termini of Single DNA Hairpin Molecules," *Nucl. Acids Res. Vol.31, 1311-1318, 2003.*

48. W. Vercoutere, S. Winters-Hilt, H. Olsen, D. Deamer, D. Haussler, and M. Akeson, "Rapid Discrimination Among Individual DNA Molecules at Single Nucleotide Resolution Using an Ion Channel," *Nature Biotechnology,* Vol. 19, pg 248, 2001.

49. Winters-Hilt S, I. H. Redmount, and L. Parker, "Physical distinction among alternative vacuum states in flat spacetime geometries," *Phys. Rev. D* 60, 124017 (1999).

50. Friedman J. L., J. Louko, and S. Winters-Hilt, "Reduced Phase space formalism for spherically symmetric geometry with a massive dust shell," *Phys. Rev. D* 56, 7674-7691 (1997).

51. Louko J, J. Z. Simon, and S. Winters-Hilt, "Hamiltonian thermodynamics of a Lovelock black hole," *Phys. Rev. D* 55, 3525-3535 (1997).

52. Louko J and S. Winters-Hilt, "Hamiltonian thermodynamics of the Reissner-Nordstrom-anti de Sitter black hole," *Phys. Rev. D* 54, 2647-2663 (1996).

53. Winters-Hilt, S. "Informatics and Machine Learning, from Martingales to Metaheuristics." (2019)

54. Winters-Hilt, S. "Data Analytics, Bioinformatics, and Machine Learning." (2019)

55. Winters-Hilt, S. "Lagrangian Physics and Unified Propagator Theory." (2020).

56. Winters-Hilt, S. "Immune Repertoire Profiling Using Nanopore Transduction," Patent Pending 2019.

57. Winters-Hilt, S. "Environmental profiling Using Nanopore Transduction," Patent Pending 2019.

58. Winters-Hilt, S. "Biomolecule conformation, cofactor, and binding Profiling Using Nanopore Transduction," Patent Pending 2019.

59. Winters-Hilt, S. "Molecular system profiling using multiple Nanopore Transduction Detectors and active probing of a cell or bioreactor," Patent Pending 2019.

60. Winters-Hilt, S. and Morales, E. "Aperture construction method for Nanopore Transduction Detectors," Patent Pending 2019.

61. Winters-Hilt, S. "Cannamimetic Profiling Using Nanopore Transduction," Patent Pending 2019.

62. Winters-Hilt, S. "The nanodrop DNA microarray: molecular system profiling using a Nanopore Transduction Detector and an annealing-based aptamer probe-set," Patent Pending 2019.

63. Winters-Hilt, S. "The nanodrop Protein microarray: molecular system profiling using a Nanopore Transduction Detector and a non-annealing-based aptamer probe-set," Patent Pending 2019.

64. Winters-Hilt, S. "The mAb nanodrop microarray: molecular system profiling using a Nanopore Transduction Detector and a mAb based probe-set," Patent Pending 2019.

65. Adelman, R. and Winters-Hilt, S. "Cannabinoid Profiling Using Nanopore Transduction," Non-provisional patent Dec, 2018.

66. Winters-Hilt, S. "Method and System for miRNA binding site profiling using a nanopore transduction detector," Patent Pending 2017.

67. Winters-Hilt, S. and Adelman, R. "Method and System for high-specificity trace-level molecular testing, biosensing, diagnostics, and therapeutic testing using a nanopore transduction detector," Patent Pending 2017.

68. Winters-Hilt, S and Adelman, R. "Method and System for isomer assaying by use of a nanopore transduction detector," Patent Pending 2017.

69. Winters-Hilt, S. "Method and system for profiling protein conformation-binding relationships using a nanopore transduction detector," Patent Pending 2017.

70. Winters-Hilt, S. "Method for microRNA binding site transcriptome analysis for phenotype-diversity evaluation." Patent Pending 2017.

71. Winters-Hilt, S. "Method for microRNA binding site transcriptome analysis for individual transcript fingerprinting and system-level analysis." Patent Pending 2017.

72. Winters-Hilt, S. and Adelman, R. "Method and system for assaying mixtures of cannabinoids, terpenes, terpenoids, pyrethrins, and similar molecular weight biomolecules by use of a nanopore transduction detector," Patent Pending 2017.

73. Winters-Hilt, S. "Method and system for cooling positively pressurized spaces, rooms, and tents, using driven evaporation, and capillary-effect, heat-wicking and using a device controller with machine-learning based system monitoring by tracking the noise state of the system," Patent Pending 2017.

74. Winters-Hilt, S and Adelman, R. "Method and System for pyrethrin and insecticide assaying, chelator targeting & design, and chelation-column filtration by use of a nanopore transduction detector," Patent Pending 2017.

75. Winters-Hilt, S. "Method and System for single molecule analysis using Nanopore Transduction and Stochastic Carrier Wave Signal Processing," Non-provisional patent June, 2016.

76. Winters-Hilt, S. "Channel current cheminformatics and bioengineering methods for immunological screening, single-molecule analysis, and single molecular-interaction analysis."Patent Awarded and Published by the European Patent Office on March 24, 2015.

77. Winters-Hilt, S. "Method and System for Stochastic Carrier Wave Communications, Radio-Noise Embedded Steganography, and Robust Self-Tuning Signal Discovery and Data-Mining." Patented June, 2015.

78. Winters-Hilt, S. "Method and System for miRNA profiling and haplotyping, protein post-translational modification assaying, aptamer design and optimization, protein-protein interaction analysis, and analysis of gene circuits and complex biosystems." Patented June, 2015.

79. Winters-Hilt, S. "Method and System for Sequencing Nucleic Acids using Nanopore Transduction, Laser Excitations, and tracking on learned Nanopore Noise States." Patented June, 2015.

80. Winters-Hilt, S. "Method and System for Noise-state Transduction Detection, Ion channel State Tracking without translocation current, and Nanopore-based Non-Destructive Live-Cell Cytosol Assaying." Patented June, 2015.

81. Winters-Hilt, S., & Adelman, R. Methods and systems for sequential analysis and nanopore detector signal analysis using stochastic sequential analysis (SSA) methods such as hidden Markov models (HMMs).2011.

82. Winters-Hilt, S., & Adelman, R. Methods and systems for nanopore biosensing. 2011.

83. Winters-Hilt, S., & Adelman, R. Methods and systems for classification, clustering, pattern recognition, and nanopore detector cheminformatics, using Support Vector Machines (SVMs). 2011.

84. Winters-Hilt, S., & Adelman, R. Method and System for Characterizing or Identifying Molecules and Molecular Mixtures. Patented, August 2010.

85. Winters-Hilt, S., Nanopore transduction of DNA sequence information using enzymes covalently bound to channel modulators. Patented, Feb. 2010.

86. Winters-Hilt, S., Nanopore-based single-molecule DNA sequencing via simultaneous, single-molecule, discrimination of dsDNA terminus identification and dsDNA strand length, via resonance modulation with appropriate choice of buffer and/or dsDNA modifications. Patented, February 2010.

87. Winters-Hilt, S., Hidden Markov model based structure identification using (i) HMM-with-Duration with positionally dependent emissions and incorporation of side-information into an HMMD via the ratio of cumulants method; and/or (ii) meta-HMMs and higher-order HMMs with gap and sequence-specific (hash) interpolated Markov models and Support Vector Machine signal boosting; and/or (iii) topological structure identification; and/or (iv) multi-track, parallel, or holographic HMMs; and/or (v) distributed HMM methods via Viterbi-path based reconstruction and verification; and/or (vi) adaptive null-state binning for $O(TN)$ computation. Patented, February 2010.

88. Winters-Hilt, S., Biosensing processes with substrates, both immobilized (Immuno-absorbant matrices) and free (enzyme substrate): Transducer Enzyme-Release with Immuno-absorbent Assay (TERISA); Transducer Accumulation and Release with Immuno-absorbent Assay (TARISA); Electrophoretic contrast substrate. PATENT filing August 2009.

89. Winters-Hilt, S., Post-translational protein modification assaying and transient complex characterization via nanopore detection and nanopore transduction detection. PATENT filing August 2009.

90. Winters-Hilt S and Zhang J. An efficient implementation for HMM with duration. PATENT filing August 2009.

91. Winters-Hilt, S., NTD-based methods for: (I) electrophoresis-separation based on nanopore acquisition rate and nanopore-based classification; (II) multi-channel sensitivity gain and affinity gain, and related architectural refinements; and (III) multicomponent and nano-manipulation refinements. PATENT filing August 2009.

92. Stoyanov A and Winters-Hilt S. Method of electrophoresis for biopolymer separation in gel media with immobilized charges according to molecular size or asymptotic electrophoretic mobility and its multi-dimensional applications. PATENT filing August 2008.

93. Winters-Hilt, S., Pattern Recognition Informed (PRI) Nanopore Detection for Sample Boosting, Nanomanipulation, and Device Stabilization; and PRI Device Stabilization Methods in General. PATENT filing August 2009.

94. Winters-Hilt, S., Channel current cheminformatics and bioengineering methods for immunological screening, single-molecule analysis, and single molecular-interaction analysis. PATENT, UNO filing, 2005.

95. Winters-Hilt, S., and Pincus, S. Nanopore-based biosensing. PATENT, UNO filing, 2004.

96. Winters-Hilt, S., and Pincus, S. Nanopore-based antibody characterization and antibody-antigen efficacy screening. PATENT, UNO filing, 2004.

97. Winters-Hilt, S. Channel current cheminformatics based immunological screening of pore inhibiting agents. PATENT, UNO filing, 2004.

98. Winters-Hilt, S. Channel current cheminformatics based assayer of cytosolic antigen delivery. PATENT, UNO, 2004.

99. Akeson, M. and Winters-Hilt, S. Methods and devices for manipulating single biomolecules. PATENT, UCSC, 2003.

100. Akeson, M., Winters-Hilt, S., Vercoutere, W., Deamer, D., and Haussler, D. Methods and devices for characterizing duplex DNA molecules. PATENT, UCSC filing, 2000.

101. Goodstein, D.L. 1975. States of Matter. Prentice-Hall, New Jersey.

102. Morse, P.M. 1969. Thermal physics (2nd Ed.). Benjamin/Cummings, Reading, MA.

103. Katchalsky, A. and P. F. Curran. 1965. Nonequilibrium Thermodynamics in Biophysics. Harvard University Press, Cambridge, MA.

104. Onsager, L. 1931. Phys. Rev. 37, 405.

105. Onsager, L. 1931. Phys. Rev. 38, 2265.

106. Prigogine, I. 1970. Dynamical Foundations of Thermodynamics and Statistical Mechanics. *In* A Critical Review of Thermodynamics, E. Stuart, B. Gal-Or, and A. Brainard, eds. Mono Books, Baltimore.

107. Prigogine, I. 1973. A United Foundation of Dynamics and Thermodynamics. Chemica Scripta 4: 5-32.

108. Progigine, I. 1984. Order out of Chaos. Bantam Books, New York.

109. Landau, L.D. and E.M. Lifshitz. 1987. Fluid Mechanics (2nd Ed.). Pergamon Press, New York.

110. Glaser, R. 2001. Biophysics. Springer-Verlag, New York

111. Doi, M. and S.F. Edwards. 1986. The theory of polymer dynamics. Oxford University Press, New York.

112. Rief, M., H. Clausen-Schaumann, and H.E. Gaub. 1999. Sequence-dependent mechanics of single DNA molecules. *Nat Struct Biol* 6, 346-349.

113. Smith, S.B., Y. Cui, and C. Bustamante. 1996. Overstretching B-DNA: the elastic response of individual double-stranded and single-stranded DNA molecules. *Science* 271, 795-799.

114. H R Drew, H R, R M Wing, T Takano, C Broka, S Tanaka, K Itakura, and R E Dickerson. Structure of a B-DNA dodecamer: conformation and dynamics. PNAS April 1, 1981 vol. 78 no. 4 2179-2183

115. El Hassan, M.A. and C.R. Calladine, "The assessment of the geometry of dinucleotide steps in double-helical DNA; a new local calculation scheme". Journal of Molecular Biology , 251, 648-664 (1995).

116. Hille, B. 1992. Ionic Channels of Excitable Membranes (2nd Ed.). Sinauer Associates, Sunderland, MA.

117. Howorka, S., S. Cheley, and H. Bayley. 2001. Sequence-specific detection of individual DNA strands using engineered nanopores. Nat Biotechnol 19:636-9

118. Song L., M.R. Hobaugh, C. Shustak, S. Cheley, H. Bayley, and J.E. Gouaux, 1996. Structure of Staphylococcal Alpha-Hemolysin, a Heptameric Transmembrane Pore. Science 274 (5294):1859-1866.

119. Coulter, W. H. 1953. U.S. Patent No. 2.656.508, issued 20 Oct. 1953.

120. DeBlois, R.W. and Bean, C.P. 1970. Counting and sizing of submicron particles by the resistive pulse technique. Rev. Sci. Instr. 41: 909-916.

121. Bean, C.P. 1972. The physics of porous membranes – neutral pores. *In:* Membranes. G. Eisenman, ed. pp. 1-54. Marcel Dekker, New York.

122. DeBlois, R.W., Bean, C.P. and Wesley, R.K.A. 1977. Electrokinetic measurements with submicron particles and pores by the resistive pulse technique. J. Coll. Interface Sci. 61: 323-335.

123. Hladky, S.B. and D.A. Haydon. 1972. Ion transfer across lipid membranes in the presence of gramicidin A. Biochim. Biophys. Acta 274: 294-312.

124. Bezrukov, S.M., I. Vodyanoy, V.A. Parsegian. 1994. Counting polymers moving through a single ion channel. Nature 370 (6457), pgs 279-281.

125. Braha, O., B. Walker, S. Cheley, J. Kasianowicz, L. Song, J.E. Gouaux, and H. Bayley. 1997. Designed protein pores as components for biosensors. In: Chemistry & Biology (London). 4: 497-505.

126. Bayley, H. 2000. Pore planning: Functional membrane proteins by design. J. Gen. Physiol. 116. 1a.

127. Bayley, H., Braha, O. & Gu, L.Q. 2000. Stochastic sensing with protein pores. *Advan Mater* 12, 139-142.

128. Kasianowicz, J.J., E. Brandin, D. Branton, and D.W. Deamer. 1996. Characterization of Individual Polynucleotide Molecules Using a Membrane Channel. Proc. Natl. Acad. Sci. USA 93(24), 13770-73.

129. Li, J., C., D. McMullan, D. Stein, D. Branton, and J. Golovchenko. 2001. Solid state nanopores for single molecule detection. Biophys. J. 80. 339a.

130. Li, J., D. Stein, C. McMullan, D. Branton, M. J. Aziz, and J. A. Golovchenko, 2001. Ion Beam Sculpting on the Nanoscale. *Nature* .

131. Song L., M.R. Hobaugh, C. Shustak, S. Cheley, H. Bayley, and J.E. Gouaux, 1996. Structure of Staphylococcal Alpha-Hemolysin, a Heptameric Transmembrane Pore. Science 274 (5294):1859-1866.

132. Füssle R, Bhakdi S, Sziegoleit A, Tranum-Jensen J, Kranz T, Wellensiek HJ. On the mechanism of membrane damage by *S. aureus* alpha-toxin. *J Cell Biol.* 1981;**91**:83–94

133. Ballard JD, Collier RJ, Starnbach MN. Anthrax toxin mediated delivery of a cytotoxic T-cell epitope in vivo. *Proc Natl Acad Sci USA.* 1996;93:12531–12534.

134. Doling AM, Ballard JD, Shen H, Krishna KM, Ahmed R, Collier RJ, Starnbach MN. Cytotoxic T-lymphocyte epitopes fused to anthrax toxin induce protective antiviral immunity. *Infect Immun.* 1999;67:3290–3296.

135. Lu Y, Friedman R, Kushner N, Doling A, Thomas L, Touzjian N, Starnbach NM, Lieberman J. Genetically modified anthrax lethal toxin safely delivers whole HIV protein antigens into cytosol to induce T cell immunity. *Proc Natl Acad Sci USA.* 2000;97:8027–8032.

136. Abrami L, Fivaz M, van der Goot FG. Surface dynamics of aerolysin on the plasma membrane of living cells. *Int J Med Microbiol.* 2000;290:363–367.

137. Brown DA, London E. Functions of lipid rafts in biological membranes. *Annu Rev Cell Dev Biol.* 1998;14:111–136.

138. Valeva A, Walev I, Pinkernell M, Walker B, Bayley H, Palmer M, Bhakdi S. Trans-membrane β-barrel of staphylococcal α-toxin forms in sensitive but not in resistant cells. *Proc Natl Acad Sci USA.* 1997;94:11607–11611.

139. Sakmann, B. and E. Neher, Eds. 1995. Single-Channel Recording (2nd Ed.). Plenum Press, New York.

140. Akeson M, D. Branton, J.J. Kasianowicz, E. Brandin, D.W. Deamer. 1999. Microsecond Time-Scale Discrimination Among Polycytidylic Acid, Polyadenylic Acid, and Polyuridylic Acid as Homopolymers or as Segments Within Single RNA Molecules. Biophys. J. 77(6):3227-3233.

141. Johnson, J.B. 1927. Thermal agitation of electricity in conductors. Phys. Rev. 29, 367-368.

142. Johnson, J.B. 1928. Thermal agitation of electricity in conductors. Phys. Rev. 32, 97-109.

143. Nyquist, H. 1927. Thermal agitation in conductors. Phys. Rev. 29, 614.

144. Nyquist, H. 1928. Thermal agitation of electrical charge in conductors. Phys. Rev. 32, 110-113.

145. van der Ziel, A. 1954. Noise. Prentice-Hall, New Jersey.

146. van der Ziel, A. 1970. Noise: Sources, Characterization, Measurement. Prentice-Hall, New Jersey.

147. van der Ziel, A. 1976. Noise in Measurements. Wiley-Interscience, New York.

148. Horowitz, P. and W. Hill. 1980. The Art of Electronics. Cambridge University Press, New York.

149. Schottky, W. 1918. Uber spontane Stromschwankungen in verschiedenen Elektrizitatsleitern. Ann. Phys. 57, 541-567.

150. Sinai, Y. 1976. Introduction to Ergodic Theory. Princeton University Press.

151. Sklar, L. 1993. Physics and Chance. Cambridge Univerisity Press.

152. Kittel, C. and H. Kroemer. 1980. Thermal Physics. W.H. Freeman and Company, New York.

153. Kogan, Sh. 1996. Electronic Noise and Fluctuations in Solids. Cambridge University Press, Cambridge, UK.

154. Johnson, J.B. 1925. The Schottky effect in lowfrequency circuits. Phys. Rev. 26, 71-85.

155. Schottky, W. 1926. Small-shot effect and flicker effect. Phys. Rev. 28, 74-103.

156. van der Ziel, A. 1959. Fluctuation Phenomena in Semiconductors. Academic Press, New York.

157. DeFelice, L. J. 1981. Introduction to membrane noise. Plenum Press, New York.

158. Kullman, L., M. Winterhalter, and S. M. Bezrukov. 2002. Transport of Maltodextrins through Maltoporin: A Single-Channel Study. Biophys. J. 82, 803-812.

159. Bezrukov, S.M., I. Vodyanoy, R.A. Brutyan and J.J. Kasianowicz. 1996. Dynamics and free energy of polymers partitioning into a nanoscale pore. Macromolecules 29: 8517-8522.

160. Henrickson, S.E., M. Misakian, B. Robertson, and J.J. Kasianowicz. 2000. Driven DNA Transport into an Asymmetric Nanometer-Scale Pore. Phys. Rev. Lett. 85, 3057.

161. Meller A, L. Nivon, and D. Branton, 2001. Voltage-driven DNA translocations through a nanopore. Phys. Rev. Lett. 86(15):3435-3438

162. Nakane J.J., Akeson,M. and Marziali,A. (2003) Nanopore sensors for nucleic acid analysis. J. Phys. Condens. Matter, 15, R1365–1393.

163. Sanger, F., *et al.* 1977. DNA sequencing with chain-terminating inhibitors. *Proc. Natl. Acad. Sci. U.S.A.* 74, 5463 – 5467.

164. Cantor, C.R., and P.R. Schimmel. 1980. Biophysical Chemistry, Part III: The Behavior of Biological Macromolecules. W.H. Freeman, San Francisco.

165. Muthukumar, M. 1999. Polymer translocation through a hole. *J. Chem. Physics* 111:10371-10374.

166. Muthukumar, M. 2002. Theory of sequence effects on DNA translocation through proteins and nanopores. Electrophoresis 23, 1417-1420.

167. Bezrukov, S.M. 2000. Ion Channels as Molecular Coulter Counters to Probe Metabolite Transport. J. Membrane Biol. 174, 1-13.

168. Lubensky, D.K. and D.R. Nelson. 1999. Driven polymer translocation through a narrow pore. Biophys. J. 77: 1824-1838.

169. Lubensky, D.K. and D.R. Nelson. 2002. Single molecule statistics and the polynucleotide unzipping transition. Phys. Rev. E 65, 031917.

170. Erie, D. A., A.K. Suri, K.J. Breslauer, R.A. Jones, and W.K. Olson. 1993. Theoretical predictions of DNA hairpin loop conformations: Correlations with thermodynamics and spectroscopic data. Biochem. 32: 436-454.

171. Antao, V P and I. Tinoco Jr. 1992. Thermodynamic parameters for loop formation in RNA and DNA hairpin tetraloops. Nuc. Acids Res. 20: 819-824

172. Senior, M.M., R.A. Jones, and K.J. Breslauer. 1988. Influence of loop residues on the relative stabilities of DNA hairpin structures. *Proc Natl Acad Sci U S A* 85, 6242-6246.

173.Ralph, D.C., C.T. Black,and M. Tinkham. 1995 . Spectroscopic measurements of discrete electronic states in single metal particles. *Phys. Rev. Lett.* 74, 3241-3244.

174. Meller A, L. Nivon, E. Brandin, J. Golovchenko, and D. Branton, 2000. Rapid nanopore discrimination between single polynucleotide molecules. Proc. Natl. Acad. Sci. USA 97(3):1079-1084.

175. Baker, D.R. 1995. Capillary Electrophoresis. Wiley-Interscience, New York.

176. Clausen-Schaumann H, Rief M, Tolksdorf C, Gaub HE. Mechanical stability of single DNA molecules. *Biophys J.* 2000;78:1997–2007.

177. EssevazRoulet B, Bockelmann U, Heslot F. Mechanical separation of complementary strands of DNA. *Proc Nat Acad Sci USA.* 1997;94:11935–11940

178. Fisher TE, Marszalek PE, Fernandez JM. Stretching single molecules into novel conformations using the atomic force microscope. *Nature Struct Biology.* 2000;7:719–724.

179. Wang MD, Schnitzer MJ, Yin H, Landick R, Gelles J, Block SM. Force and velocity measured for single molecules of RNA polymerase. *Science.* 1998;282:902–907.

180. Chung, S-H., and P. W. Gage. 1998. Signal processing techniques for channel current analysis based on hidden Markov models. *In* Methods in Enzymology; Ion channels, Part B. P. M. Conn editior. Academic Press, Inc., San Diego. 420-437.

181. Krogh, A., I. S. Mian, and D. Haussler. 1994. A hidden Markov model that finds genes in E. coli DNA. Nucl. Acids Res. 22. 4768-4778.

182. Stormo, G. D. 2000. Gene-finding approaches for eukaryotes. Genome Res. 10. 394-397.

183. Durbin, R., S. Eddy, A. Krogh, and G. Mitchison. 1998. Biological sequence analysis: probalistic models of proteins and nucleic acids. Cambridge, UK New York: Cambridge University Press.

184. Winters-Hilt, S. Machine Learning Methods for Channel Current Cheminformatics, Biophysical analysis, and Bioinformatics. (2003). PhD Dissertation, UCSC.

185. Oppenheim, A. V., A. S. Willsky, and I. T. Young. 1983. Signals and Systems. Prentice-Hall, New Jersey.

186. Ziemer, R. E. and W. H. Tranter. 1985. Principles of Communications; Systems, Modulation, and Noise (2nd Ed.). Houghton Mifflin Company, Boston.

187. L.R. Rabiner. A tutorial on hidden markov models and selected application in speech recognition. Proceedings of the IEEE, 77:257-286, 1989.

188. S. Mian A. Krogh and D. Haussler. A hidden markov model that finds genes in e. coli dna. Nucleic Acids Research, 22:68-78, 1994.

189. Perry A. Stoll and Jun Ohya. Applications of hmm modeling to recognizing human gestures in image sequences for a man-machine interface. IEEE International Workshop on Robot and Human Communication, pages 129-134, 1995.

190. Jorg Appenrodt Mahmoud Elmezain, Ayoub Al-Hamadi and Bernd Michaelis. A hidden markov model-based continuous gesture recognition system for hand motion trajectory. IEEE Conference Proceeding, 2008.

191. J. Appenrodt M. Elmezain, A. Al-Hamadi and B. Michaelis. A hidden markov model-based isolated and meaningful hand gesture recognition. International Journal of Electrical, Computer, and Systems Engineering, 3:156-163, 2009.

192. E. Augustin S. Knerr and D. Price. Hidden markov model based word recognition and its application to legal amount reading on french checks. Computer Vision and Image Understanding, page 404, 1998.

193. M. Schenkel and M. Jabri. Low resolution, degraded document recognition using neural networks and hidden markov models. Pattern Recognition Letters, 3:365{371, 1998.

194. J. Vlontzos and S. Kung. Hidden markov models for character recognition. IEEE Transactions on Image Processing, 1992.

195. A. Najmi J. Li and R.M. Gray. Image classification by a two-dimensional hidden Markov model. IEEE Transactions on Signal Processing, 48, 2000.

196. R. A. Olshen J. Li, R. M. Gray. Multiresolution image classification by hierarchical modeling with two-dimensional hidden Markov models. IEEE Transactions on Information Theory, 2000.

197. M.S. Wu C.L. Huang and S.H. Jeng. Gesture recognition using the multi-pdm method and hidden markov model. Image and Vision Computing, 18:865, 2000.

198. J. Garcia-Frias and P. M. Crespo. Hidden markov models for burst error characterization in indoor radio channels. IEEE Transactions on Vehicular Technology, 1997.

199. J.P. Hughes E. Bellone and P. Guttorp. A hidden markov model for downscaling synoptic atmospheric patterns to precipitation amounts. Climate Research, 2000.

200. C. Raphael. Automatic segmentation of acoustic musical signals using hidden markov models. IEEE Transactions on Pattern Analysis and Machine Intelligence, 21:1998, 360.

201. Joseph A. Kogan and Daniel Margoliash. Automated recognition of bird song elements from continuous recordings using dynamic time warping and hidden markov models: A comparative study. Journal of the Acoustical Society of America, 1998.

202. Andrey Andreyevich Markov. Theory of Algorithms. Academy of Sciences of the USSR, 1954.

203. Moises Burset and Roderic Guigo. Evaluation of gene structure prediction programs. Genomics, 34:353-367, 1996.

204. Mathé C., M.-F. Sagot, T. Schiex and P. Rouzé. Current methods of gene prediction, their strengths and weaknesses. Nucleic Acids Research, 2002, Vol. 30, No. 19 4103-4117

205. Stanke, M., R. Steinkamp, S. Waack, and B. Morgenstern. AUGUSTUS: a web server for gene finding in eukaryotes. Nucleic Acids Research, 2004, Vol. 32, W309-W312.

206. Stanke, M. and Waack, S. Gene prediction with a hidden Markov model and new intron submodel. (2003) Bioinformatics, 19(Suppl. 2) ii215-ii225.

207. Guigo, R., Agarwal, P., Abril, J., Burset, M. and Fickett, J.W. An assessment of gene prediction accuracy in large DNA sequences. (2000) Genome Res., 10, 1631-1642.

208. Bousso, Raphael. "The Holographic Principle". http://arxiv.org/pdf/hep-th/0203101.Rev. Mod. Phys. 74, 825–874 (2002)

209. Donoho D. Compressed sensing. IEEE Trans. On Information Theory, 52(4), pp. 1289 - 1306, April 2006.

210. J.D. Ferguson. Variable duration models for speech. Proceedings of Symposium on the Application of Hidden Markov models to Text and Speech, pages 143-179, 1980.

211. P. Ramesh and J.G. Wilpon. Modeling state durations in hidden markov models for automatic speech recognition. Proceedings of IEEE International Conference on Acoustics, Speech and Signal Processing, 1:381-384, 1992.

212. SZ Yu and H. Kobayashi. An efficient forward-backward algorithm for an explicit-duration hidden markov model. IEEE Signal Processing Letters, 10:11-14, 2003.

213. M.T. Johnson. Capacity and complexity of hmm duration modeling techniques. IEEE Signal Processing Letters, 12:407-410, 2005.

214. Ellington AD, Szostak J: In vitro selection of RNA molecules that bind specific ligands. *Nature* 1990 , 346:818-22.

215. Tuerk C, Gold L: Systematic evolution of ligands by exponential enrichment: RNA ligands to bacteriophage T4 DNA polymerase. *Science* 1990 , 249:505-10.

216. Jayasena SD: Aptamers: an emerging class of molecules that rival antibodies in diagnostics. *Clin Chem* 1999 , 45:1628-1650.

217. Proske D, Blank M, Buhmann R, Resch A: Aptamers – basic research, drug development, and clinical applications. *Appl Microbiol Biotechnol* 2005 , 69:367-374.

218. Hamaguchi N, Ellington A, Stanton M: Aptamer beacons for the direct detection of proteins. *Anal Biochem* 2001 , 294:126-131.

219. Ulrich H, Martins AH, Pesquero JB: RNA and DNA aptamers in cytomics analysis. *Cytometry Part A* 2004 , 59A:220-231.

220. Brody EN, Gold L: Aptamers as therapeutic and diagnostic reagents. *J Biotechnol* 2000 , 74:5-13.

221. Yamamoto R, Baba T, Kumar PK: Molecular beacon aptamer fluoresces in the presence of Tat protein of HIV-1. *Genes Cells* 2000 , 5:389-396.

222. Yamamoto R, Katahira M, Nishikawa S, Baba T, Taira K, Kumar PK: A novel RNA motif that binds efficiently and specifically to the Tat protein of HIV and inhibits the transactivation by Tat of transcription in vitro and in vivo. *Genes Cells* 2000 , 5(5):371-388.

223. Ding, S., C. Gao, Li-Qun Gu. Capturing single molecules of immunoglobulin and Ricin with an aptamer-encoded glass nanopore. Ana. Chem 2009 ac9006705.

224. Ikebukuro, K., Y. Okumura, K. Sumikura, I. Karube. A novel method of screening thrombin-inhibiting DNA aptamers using an evolution-mimicking algorithm. Nucl. Acids Res. 2005, Vol. 33 (12).

225. Bettina Wagner, Donald C. Miller, Teri L. Lear,and Douglas F. Antczak. The Complete Map of the Ig Heavy Chain Constant Gene Region Reveals Evidence for Seven IgG Isotypes and for IgD in the Horse. J Immunol 2004; 173:3230-3242.

226. Daryl Fernandes. Demonstrating Comparability of Antibody Glycosylation during Biomanufacturing. European Biopharmaceutical Review. Summer 2005. pp 106 -110.

227. Kinchen, K. S., Sadler, J., Fink, N., Brookmeyer, R., Klag, M.J., Levey, A.S. and Powe, N.R. (2002), The Timing of Specialist Evaluation in Chronic Kidney Disease and Mortality. Ann Intern Med. 2002;137:479-486

228. Brenner B. M., Cooper, M.E., De Zeeuw, D., Keane, W.G., Mitch, W.E., Parving, H.H., Remuzzi, G., Snapinn, S.M., Zhang, Z., and Shahinfar, S. (2001). Effects of Losartan on Renal and Cardiovascular Outcomes in Patients with Type 2 Diabetes and Nephropathy. N Engl J Med, Vol. 345, No. 12

229. MacGregor, M. S., Boag, D. E., and Innes, A. (2006). Chronic kidney disease: evolving strategies for detection and management of impaired renal function. Q J Med 2006; 99:365–375

230. Selinger, S.L., Zhan, M., Hsu, V.D., Walker, L.D. and Fink.J.C. (2008). Chronic Kidney Disease Adversely Influences Patient Safety. J Am Soc Nephrol 19: 2414-2419

231. Race, R.E., A. Raines, T.G. M. Baron, M.W. Miller, A. Jenny, and E.S.Williams. Comparison of Abnormal Prion Protein Glycoforms Paterns from Transmissable Spongiform Encepphalography Agent-Infected Deer, Elk, Sheep, and Cattle. Journal of Virology, Dec. 2002, p. 12365-12368.

232. Righetti, P.G.; Stoyanov, A.V.; Zhukov, M. *The Proteome Revisited*; Elsevier: Amsterdam, The Netherlands, 2001.

233. Cann, J.R.; Stimpson, D.I.; Cox, D.J. Isoelectric focusing of interacting systems: Carrier ampholyte-induced macromolecular association or dissociation into subunits. *Anal. Biochem.* 1978, *86*, 34–49.

234. Kabytaev, K.; Durairaj, A.; Shin, D.; Rohlfing, C.L.; Connolly, S.; Little, R.R.; Stoyanov, A.V. Two-step ion-exchange chromatographic purification combined with reversed-phase chromatography to isolate C-peptide for mass spectrometric analysis. *J. Sep. Sci.* 2016, *39*, 676–681.

235. Stoyanov, A.V.; Rohlfing, C.L.; Connolly, S.; Roberts, M.L.; Nauser, C.L.; Little, R.R. Use of cation exchange chromatography for human C-peptide isotope dilution–mass spectrometric assay. *J. Chromatogr. A* 2011, *1218*, 9244–9249.

236. Stoyanov, A.V.; Righetti, P.G. Ampholyte dissociation theory and properties of ampholyte aqueous solutions. *Electrophoresis* 1987, *18*, 1944–1950.

237. Andreev, V.P.; Makarova, E.; Pliss, N.S. New capability of electroinjection analysis: Investigation of chemical reaction kinetics. *Anal. Chem.* 2001, *73*, 1316–1323.

238. Patterson, D.H.; Harmon, B.J.; Regnier, F.E. Dynamic modeling of electrophoretically mediated microanalysis. *J. Chromatogr. A* 1996, *732*, 119–132.

239. Shalongo, W.; Jagannadham, M.; Stellwagen, E. Kinetic analysis of the hydrodynamic transition accompanying protein folding using size exclusion chromatography. Comparison of spectral and chromatographic kinetic analyses. *Biopolymers* 1993, *33*, 135–145.

240. Stoyanov, A.V.; Righetti, P.G. Steady-state electrolysis of a solution of nonamphotheric compounds. *Electrophoresis* 1999, *20*, 718–722.

241. Stoyanov, A.V. Water electrolysis in mono-and hetero-phase low conductive systems and a secondary pH gradient establishment. In *Electrolysis: Theory, Types and Applications*; Kuai, S., Meng, J., Eds.; Nova Science Publishers Inc.: Hauppauge, NY, USA, 2010; pp. 511–516.

242. Stoyanov, A.V.; Pawliszyn, J. Buffer composition changes in background electrolyte during electrophoretic run in capillary zone electrophoresis. *Analyst* 2004, *129*, 979–982.

243. Stoyanov, A. IEF-based multidimensional applications in proteomics: Toward higher resolution. *Electrophoresis* 2012, *33*, 3281–3290.

244. Stoyanov, A.V.; Rogatsky, E.; Stein, D.; Connolly, S.; Rohlfing, C.L.; Little, R.R. Isotope dilution assay in peptide quantification: The challenge of microheterogeneity of internal standard. *Proteom. Clin. Appl.* 2013, *7*, 825–828.

245. Shade K-TC and Anthony RM. Antibody Glycosylation and Inflammation. Antibodies 2013, 2, 392-414.

246. Radaev S and Sun PD. The role of Fc glycosylation and the binding of peptide inhibitors. J. of Biological Chemistry Vol. 276, No. 19, May 11, pp. 16478–16483, 2001.

247. Ha S, Ou Y, Vlasak J, Li Y, Wang S, Vo K, et al. Isolation and Characterization of IgG1 with Asymmetrical Fc Glycosylation. Glycobiology. 2011 Aug;21(8):1087-96.

248. Zauner G, Selman MHJ, Bondt A, Rombouts Y, Blank D, Deelder AM, et al. Glycoproteomic Analysis of Antibodies. Mol Cell Proteomics. 2013 Apr;12(4):856-65.

249. Jerrard M. Hayes, Eoin F. J. Cosgrave, Weston B. Struwe, Mark Wormald, Gavin P. Davey, Roy Jefferis and Pauline M. Rudd. Glycosylation and Fc Receptors. Curr Top Microbiol Immunol. 2014;382:165-99.

250. Karlsen, K.K., and J. Wengal. Locked Nucleic Acid and Aptamers. Nucl. Acid Therapeutics 22(6), 2012.

251. Xiang, D, Shigdar S, Qiao G, Wang T, Kouzani AZ, Zhou S-F, et al. Nucleic Acid Aptamer-Guided Cancer Therapeutics and Diagnostics: the Next Generation of Cancer Medicine. Theranostics 5(1): 23-42 (2015).

252. Germer K, Leonard M, Zhang X. RNA aptamers and their therapeutic and diagnostic applications. Int J Biochem Mol Biol 2013;4(1):27-40.

253. Ming X, Laing B. Bioconjugates for targeted delivery of therapeutic oligonucleotides. Adv Drug Deliv Rev. 2015 Jun 29; 87: 81-89.

254. Craig DB, Arriaga E, Wong JCY, Lu H, and Dovichi NJ. The life and death of a single enzyme molecule. *Anal Chem* 1998, 70, 39A-43A.

255. Craig DB, Arriaga E, Wong JCY, Lu H, and Dovichi NJ. Studies on single alkaline phosphatase molecules: reaction rate and activation energy of a reaction catalyzed by a single molecule and the effect of thermal denaturation-The death of an enzyme. *J Am Chem Soci* 1996, 118, 5245-5253.

256. Chen DY. and Dovichi NJ. Single-molecule detection in capillary electrophoresis: molecular shot noise as a fundamental limit to chemical analysis. *Anal Chem 1996,* 68, 690-696.

257. Sakmann B, Neher E. Patch clamp techniques for studying ionic channels in excitable membranes. Annu Rev Physiol. 1984;46:455-72.

258. "The Nobel Prize in Physiology or Medicine 1991". nobelprize.org. Nobel Media AB.

259. Middlebrook, RD. Null Double Injection and the Extra Element Theorem. IEEE Trans Educ 32, No. 3 I67 (1989).

260. Yang W, Chendrimada TP, Wang Q, Higuchi M, Seeburg PH, Shiekhattar R, Nishikura K. Modulation of microRNA processing and expression through RNA editing by ADAR deaminases. Nat Struct Mol Biol (2006) 13: 13-21.

261. Zhang, H. Reversal of HIV-1 Latency with Anti-microRNA Inhibitors. Int J Biochem Cell Biol. 2009 Mar; 41(3): 451–454.

262. Liang, Ruqiang, David J Bates, and Eugenia Wang. Epigenetic Control of MicroRNA Expression and Aging. Curr Genomics. 2009 May; 10(3): 184–193.

263. Bantounas, I., L A Phylactou1 and J B Uney. RNA interference and the use of small interfering RNA to study gene function in mammalian systems. J Mol Endocrinol. 2004 Dec;33(3):545-57.

264. Agarwal, Vikram, George W Bell, Jin-Wu Nam, David P Bartel. Predicting effective microRNA target sites in mammalian mRNAs. eLife 2015;4:e05005.

265. Chow SA, Vincent KA, Ellison V, Brown PO. Reversal of integration and DNA splicing mediated by integrase of human immunodeficiency virus. *Science.* 1992;**255**:723–726.

266. Scottoline BP, Chow S, Ellison V, Brown PO. Disruption of the terminal base pairs of retroviral DNA during integration. *Genes Dev.* 1997;**11**:371–382.

cell	13,14,27,38,64,98,125,126, 127,144,145,163,164,165,1 72,191,193,194		compression	61,89
cells	4,13,14,16,27,28,38,39,98, 102,133,134,140,141,145,1 63,193		Compressive	88
			compressive	88,89
			copolymer	130
cellular	13,141,145,193,194,100		copolymers	24
chaotrope	2,117,151,152,157,150		cosmological	13
chaotropes	27,29,113,151,152,153,156 ,158,178,198		Coulomb	11
			Coulter	4,14,18,27,40,94,161
chaotropic	114,119,152,153,156		coulter	93
chelation	172,181,182		Crick	6,12,25,31,37,40,44,120,12 2,169,171,172,182
Chelator	167,181		
Cheminfor- matics	43,46,80		crystal	12,13,35
			crystalline	11,14,27
cheminformat- ics	28,44,46,65,80,83,90,99,15 3,185		crystallizable	127
			crystallization	42
chimera	176,187		crystallized	12,13,125,126
chimeras	171,176,179,185,198		crystallograph- ic	38,99
Chimeric	174,187		
chimeric	167,169,171,172,176		crystallog- raphy	1,9,12
Cholesterol	16,164,165		
cholesterol	16,165		crystals	12
chopper	176		cyclase	17
chromato- graphic	154,156		cytometry	98
			cytosol	144,163
chromatog- raphy	151,165,191		cytosolic	133,144
			cytotoxic	144
chromophore	42		DAQ	199
chromosomal	12		DAQs	56
chromosomes	12		Database	85
chunking	64		database	82,85,86,201
ciphertext	56		deaminases	196,198
Circuit	206		deaminated	196
circuit	17,18,19,37,40,191,193,19 4,195,196,197,198,203,204 ,205,207		Debye	11,15
			decane	16
			Decoding	68
circuits	191,194,195,196,203		decoding	63,68,80,84,88,89,90,91,92 ,108
clique	63,66		
cluster	11,74,76,144,153		decodings	91
clustered	85		decomposed	82
Clustering	11,76,81		decomposition	60
clustering	38,43,56,66,72,73,75,76,77 ,81,84,85,166		decompressed	89
			decompres- sion	89
clusters	11,73,76,82,143,153,166		
cofactor	102,133,146		deconvolution	40
cofactors	8,96,146		decoupled	40,63
cognate	99		deglycosylated	140,141
cognates	80		Deletion	134
			delimit	18
			denatured	156

PCR	2,28,39,98,99,101,113,119, 124,142,143,149,151,169,1 70,178,186,192
PEG	14,21,24,28,107,143,144,1 51,162,165,166,191,192
peptide	128,144
peptides	98,144
perceptrons	72
phage	12
phagocyte	127
phagocytes	127
phagocytic	98
phenomeno-logical	10,12,23,24
phenomenol-ogies	88
phenomenol-ogy	9,10,23,41
phenomolo-gies	89
phosphatidyl-choline	28,160
phosphoam-idite	171
phos-phodiester	15,35
phos-phodiesterase	17
phospho-glyceride	16
phospholipas-es	17
phosphory-lated	3
phosphoryla-tion	17,36,98
piezoelectric	48
planetary	13
pleiotropic	17
PLL	11,50,92,197
pMM	66
polyA	22,23
polyC	22,23
Polyethylene	151,165
polyethylene	14,21,28
Polymer	9,24
polymer	5,9,10,11,12,14,21,22,23,2 4,28,37,38,39,41,93,94,95
polymerase	137,144,146

polymerases	12,42
polymerization	147
Polymers	9,11,21,22,23
polymers	9,21,23,24,28,102
polymorphism	28,39,99,181
polypeptide	130
polypeptides	162
polypurine	22
polypyrimidine	22
polysaccharide	21
polystyrene	14,27
Pore	9,24,144
pore	4,5,14,15,16,21,22,23,24,2 7,28,32,33,35,36,37,38,40, 41,100,111,129,144,160
pores	13,14,15,16,23,24,27,28,29 ,38,39,40
posttranscrip-tional	198
Potassium	162
prion	140
proline	126,127
PSA	103
psa	156
PTMs	125
public	8,103,140
purine	23
pyrethrin	3
pyrimidine	23
qPCR	192
quadfunction-al	172,187
quadrature	20
quadruplex	118
quadruplexes	173,182
racemic	113,114
racemization	152
recursive	68,69,70
recursively	70
regulate	144
regulated	139,193
regulating	159
regulation	102,133,193,197
regulatory	102,133,140,144
reject	103
rejected	32,33,34
rejecting	34,50
Rejection	77,78,79
rejection	32,33,45,53,55,56,77,78,19

threshold	32,33,83	transverse	102
thresholding	49	Triglyceride	164
thrombin	123	troponin	4
thymidine	169,175	TSH	4,102,133,139,140
Thymine	128	Urea	152,155,156,157,159
thymine	124,138,148	urea	29,113,114,117,118,100
Thyroid	140	UTR	196,197,198
thyroid	139	UTRs	197
Toxin	113	UV	47,96,97,100,105,107,141, 142,161
toxin	5,13,16,30,100,103,111,14 4	virulence	144
toxins	16,98,103,144,187	virus	119
Transcription	133,144	Viruses	14
transcription	98,116,121,137,138,100	viruses	144,197
transcriptional	144,197	viscoelasticity	11
Transcriptome	133,144	Viscosity	23
transcriptome	193	viscosity	10,11,23,156,159
Transform	61	Viscous	10
transform	20,44,52,60,61,62,82,151,1 67,203	viscous	10,11
transient	1,4,36,41,61,80,97,98,101, 131,133,134,139,140,163,1 98	Viterbi	43,46,63,65,66,67,68,69,70 ,71,83,84,88,89,90,131,132 ,138
transposition	12	Zika	119
transposons	12	zinc	144

www.ingramcontent.com/pod-product-compliance
Lightning Source LLC
Chambersburg PA
CBHW061404210326

41598CB00035B/6094